VOLUME THREE HUNDRED AND FORTY FOUR

INTERNATIONAL REVIEW OF CELL AND MOLECULAR BIOLOGY

Nucleic Acid Sensing and Immunity, Part A

INTERNATIONAL REVIEW OF CELL AND MOLECULAR BIOLOGY

VOLUME THREE HUNDRED AND FORTY FOUR

INTERNATIONAL REVIEW OF CELL AND MOLECULAR BIOLOGY

Nucleic Acid Sensing and Immunity, Part A

Edited by

CLAIRE VANPOUILLE-BOX
Weill Cornell Medical College, New York, NY, United States

LORENZO GALLUZZI
Weill Cornell Medical College, New York, NY, United States

ELSEVIER

ACADEMIC PRESS
An imprint of Elsevier

Academic Press is an imprint of Elsevier
50 Hampshire Street, 5th Floor, Cambridge, MA 02139, United States
525 B Street, Suite 1650, San Diego, CA 92101, United States
The Boulevard, Langford Lane, Kidlington, Oxford OX5 1GB, United Kingdom
125 London Wall, London, EC2Y 5AS, United Kingdom

First edition 2019

ISBN: 978-0-12-815979-8
ISSN: 1937-6448

For information on all Academic Press publications
visit our website at https://www.elsevier.com/books-and-journals

Publisher: Zoe Kruze
Acquisition Editor: Ashlie M. Jackman
Editorial Project Manager: Teresa Pons-Ferrer
Production Project Manager: James Selvam
Cover Designer: Alan Studholme

Typeset by SPi Global, India

CONTENTS

CONTRIBUTORS

Takayuki Abe
Department of Systems Biology; Department of Microbiology and Immunology, Columbia University, New York, NY, United States

Elva Abril
Clinical Pathology and Microbiology, San Gallicano Dermatological Institute IRCCS, Rome, Italy

Alejandro Alice
Earle A. Chiles Research Institute, Robert W. Franz Cancer Center, Providence Portland Medical Center, Portland, OR, United States

Paras K. Anand
Infectious Diseases and Immunity, Imperial College London, London, United Kingdom

Jason R. Baird
Earle A. Chiles Research Institute, Robert W. Franz Cancer Center, Providence Portland Medical Center, Portland, OR, United States

Valentina Bordignon
Clinical Pathology and Microbiology, San Gallicano Dermatological Institute IRCCS, Rome, Italy

Ilaria Cavallo
Clinical Pathology and Microbiology, San Gallicano Dermatological Institute IRCCS, Rome, Italy

Giovanna D'Agosto
Clinical Pathology and Microbiology, San Gallicano Dermatological Institute IRCCS, Rome, Italy

Enea Gino Di Domenico
Clinical Pathology and Microbiology, San Gallicano Dermatological Institute IRCCS, Rome, Italy

Fabrizio Ensoli
Clinical Pathology and Microbiology, San Gallicano Dermatological Institute IRCCS, Rome, Italy

Victoria Godfrey
Department of Pathology, UT Southwestern Medical Center, Dallas, TX, United States

Michael J. Gough
Earle A. Chiles Research Institute, Robert W. Franz Cancer Center, Providence Portland Medical Center, Portland, OR, United States

Hong-Ming Hu
Earle A. Chiles Research Institute, Robert W. Franz Cancer Center, Providence Portland Medical Center, Portland, OR, United States

Taro Kawai
Laboratory of Molecular Immunobiology, Division of Biological Science, Graduate School of Science and Technology, Nara Institute of Science and Technology, Ikoma, Japan

Takumi Kawasaki
Laboratory of Molecular Immunobiology, Division of Biological Science, Graduate School of Science and Technology, Nara Institute of Science and Technology, Ikoma, Japan

Shahanshah Khan
Department of Pathology, UT Southwestern Medical Center, Dallas, TX, United States

Nikolai N. Khodarev
Department of Radiation and Cellular Oncology and Ludwig Center for Metastasis Research, The University of Chicago, Chicago, IL, United States

Christopher R. Lupfer
Department of Biology, Missouri State University, Springfield, MO, United States

Terry Medler
Earle A. Chiles Research Institute, Robert W. Franz Cancer Center, Providence Portland Medical Center, Portland, OR, United States

Jaina M. Patel
Earle A. Chiles Research Institute, Robert W. Franz Cancer Center, Providence Portland Medical Center, Portland, OR, United States

Martina Pontone
Clinical Pathology and Microbiology, San Gallicano Dermatological Institute IRCCS, Rome, Italy

Meagan D. Rippee-Brooks
Department of Biology, Missouri State University, Springfield, MO, United States

Sagi D. Shapira
Department of Systems Biology; Department of Microbiology and Immunology, Columbia University, New York, NY, United States

Elisabetta Trento
Clinical Pathology and Microbiology, San Gallicano Dermatological Institute IRCCS, Rome, Italy

Md. Hasan Zaki
Department of Pathology, UT Southwestern Medical Center, Dallas, TX, United States

NUCLEIC ACID SENSING AT THE INTERFACE BETWEEN INNATE AND ADAPTIVE IMMUNITY

Innate immunity largely relies on the recognition of evolutionarily conserved structures on pathogens that are commonly known as microbe-associated molecular patterns (MAMPs) or pathogen-associated molecular patterns (PAMPs), through a limited number of pattern recognition receptors (PRRs) (Kawai and Akira, 2011). MAMPs are invariant among entire classes of microbes, are often (but not always) essential for their survival, and can be easily distinguished from "self" structures (Schlee and Hartmann, 2016). Importantly, PRRs can also sense endogenous factors, especially when these are present at ectopic locations (e.g., cytoplasmic nucleic acids and/or extracellular ATP) or in excessive amounts as a consequence of infection, inflammation, or other types of threats to cellular and organismal homeostasis (Galluzzi et al., 2018b; Vanpouille-Box et al., 2018). Endogenous factors that can activate PRRs are cumulatively known as damage-associated molecular patterns (DAMPs) (Krysko et al., 2012; Liston and Masters, 2017). Upon MAMP or DAMP recognition, PRRs deliver a danger signal to the host, generally resulting in the initiation of proinflammatory responses aiming at the preservation of organismal homeostasis (Cao, 2016). Ultimately, PRR engagement orchestrates the early response of the host to a threat (be it endogenous or exogenous in nature) and sets the stage for the activation of adaptive immunity (Iwasaki and Medzhitov, 2015).

The immunostimulatory properties of nucleic acids have been described as early as in 1963 by Warren and Isaac, who discovered that DNA and RNA molecules derived from microbes or host cells can induce interferon (IFN) production by fibroblasts (Jensen et al., 1963; Rotem et al., 1963). However, the existence of dedicated nucleic acid sensors was ignored at the time. Our current understanding of innate immunity began indeed to take shape only two decades later, following a hypothesis from Janeway (1989) that stimulated the discovery that Toll (a *Drosophila melanogaster* protein that was known for its role in embryonic development) can operate as a PRR capable to drive the secretion of antimicrobial peptides in response to fungal infection (Lemaitre et al., 1996). Shortly thereafter, the first human Toll-like receptor (TLR) was reported to stimulate the secretion of several proinflammatory cytokines on activation (Medzhitov et al., 1997), but it was only in the early 2000s that TLR3 and TLR9 were discovered to respond to viral double-stranded RNA and bacterial CpG-rich DNA, respectively (Alexopoulou et al., 2001; Hemmi et al., 2000). Since then, more than a dozen different human sensors for DNA and RNA have been discovered (Roers et al., 2016; Vanpouille-Box et al., 2018).

Given the central role of nucleic acid sensors in the control of invading pathogens and endogenous threats, it is not surprising that both microbes and malignant cells have evolved a large number of strategies to circumvent the activation of these PRRs (Chan and Gack, 2016; Chen et al., 2016; Galluzzi et al., 2017). As a stand-alone example, some cancer cells can evade the activation of cyclic GMP-AMP synthase (a sensor for cytosolic DNA) by upregulating three-prime repair exonuclease 1 (TREX1), resulting in limited type I IFN production in response to stress (be it endogenous or imposed by treatment) (Galluzzi et al., 2018a; Xia et al., 2016; Zitvogel et al., 2015), and hence a poorly inflammatory profile that supports immune evasion (Vanpouille-Box et al., 2017). Moreover, reflecting the potent proinflammatory responses driven by nucleic acid sensors, deregulated signaling via these PRRs has been associated with a plethora of pathologies, including rheumatic and autoimmune disorders (Crow et al., 2018; Muskardin and Niewold, 2018), neurological conditions (McGlasson et al., 2015), cardiac degeneration (King et al., 2017), and others (McNab et al., 2015). Thus, nucleic acid sensors and their downstream effectors have become attractive targets for the development of therapeutic agents with various applications (Junt and Barchet, 2015; Parker et al., 2016; Weiss et al., 2017).

In *Nucleic Acid Sensing and Immunity* (a special issue of the *International Review of Cell and Molecular Biology* series) world-class experts in the field offer an overview of (1) the major molecular and cellular aspects of nucleic acid sensing across the eukaryotic kingdom; (2) the complexity of innate and adaptive immune responses and their key role in the maintenance of cellular and organismal homeostasis; and (3) the intimate connections between deregulated nucleic acid sensing and human pathology. Each of the thirteen chapters of *Nucleic Acid Sensing and Immunity* focuses on one specific aspect of nucleic acid sensing, encompassing signaling, regulation, and pathophysiological relevance. The book (which consists of two parts, Part A and Part B) is equally addressed to expert investigators who may wish to extend their knowledge on nucleic acid sensing at the interface between innate and adaptive immunity, and to newcomers to this exciting and quickly progressing field of investigation.

ACKNOWLEDGMENTS

LG is supported by a Breakthrough Level 2 grant from the US Department of Defense (DoD), Breast Cancer Research Program (BRCP) (#BC180476P1), by a startup grant from the Dept. of Radiation Oncology at Weill Cornell Medicine (New York, US), by industrial collaborations with Lytix (Oslo, Norway) and Phosplatin (New York, US), and by donations from Phosplatin (New York, US), the Luke Heller TECPR2 Foundation (Boston, US) and Sotio a.s. (Prague, Czech Republic).

DISCLOSURES

LG provides remunerated consulting to OmniSEQ (Buffalo, NY, USA), Astra Zeneca (Gaithersburg, MD, USA), VL47 (New York, NY, USA) and the Luke Heller TECPR2 Foundation (Boston, MA, USA), and he is member of the Scientific Advisory Committee of OmniSEQ (Buffalo, NY, USA).

Claire Vanpouille-Box
Department of Radiation Oncology, Weill Cornell Medical College, New York, NY, United States
claire.vanpouille.box@gmail.com
Lorenzo Galluzzi
Department of Radiation Oncology, Weill Cornell Medical College, New York, NY, United States
Sandra and Edward Meyer Cancer Center, New York, NY, United States
Université Paris Descartes/Paris V, Paris, France
deadoc80@gmail.com

REFERENCES

Alexopoulou, L., Holt, A.C., Medzhitov, R., Flavell, R.A., 2001. Recognition of double-stranded RNA and activation of NF-kappaB by Toll-like receptor 3. Nature 413, 732–738.

Cao, X., 2016. Self-regulation and cross-regulation of pattern-recognition receptor signalling in health and disease. Nat. Rev. Immunol. 16, 35–50.

Chan, Y.K., Gack, M.U., 2016. Viral evasion of intracellular DNA and RNA sensing. Nat. Rev. Microbiol. 14, 360–373.

Chen, Q., Sun, L., Chen, Z.J., 2016. Regulation and function of the cGAS-STING pathway of cytosolic DNA sensing. Nat. Immunol. 17, 1142–1149.

Crow, M.K., Olferiev, M., Kirou, K.A., 2018. Type I interferons in autoimmune disease. Annu. Rev. Pathol. https://doi.org/10.1146/annurev-pathol-020117-043952.

Galluzzi, L., Buque, A., Kepp, O., Zitvogel, L., Kroemer, G., 2017. Immunogenic cell death in cancer and infectious disease. Nat. Rev. Immunol. 17, 97–111.

Galluzzi, L., Chan, T.A., Kroemer, G., Wolchok, J.D., Lopez-Soto, A., 2018a. The hallmarks of successful anticancer immunotherapy. Sci. Transl. Med. 10 (459), pii: eaat7807. https://doi.org/10.1126/scitranslmed.aat7807.

Galluzzi, L., Yamazaki, T., Kroemer, G., 2018b. Linking cellular stress responses to systemic homeostasis. Nat. Rev. Mol. Cell Biol. 19, 731–745.

Hemmi, H., Takeuchi, O., Kawai, T., Kaisho, T., Sato, S., Sanjo, H., Matsumoto, M., Hoshino, K., Wagner, H., Takeda, K., Akira, S., 2000. A Toll-like receptor recognizes bacterial DNA. Nature 408, 740–745.

Iwasaki, A., Medzhitov, R., 2015. Control of adaptive immunity by the innate immune system. Nat. Immunol. 16, 343–353.

Janeway Jr., C.A., 1989. Approaching the asymptote? Evolution and revolution in immunology. Cold Spring Harb. Symp. Quant. Biol. 54 (Pt. 1), 1–13.

Jensen, K.E., Neal, A.L., Owens, R.E., Warren, J., 1963. Interferon responses of chick embryo fibroblasts to nucleic acids and related compounds. Nature 200, 433–434.

Junt, T., Barchet, W., 2015. Translating nucleic acid-sensing pathways into therapies. Nat. Rev. Immunol. 15, 529–544.

Kawai, T., Akira, S., 2011. Toll-like receptors and their crosstalk with other innate receptors in infection and immunity. Immunity 34, 637–650.

King, K.R., Aguirre, A.D., Ye, Y.X., Sun, Y., Roh, J.D., Ng Jr., R.P., Kohler, R.H., Arlauckas, S.P., Iwamoto, Y., Savol, A., Sadreyev, R.I., Kelly, M., Fitzgibbons, T.P., Fitzgerald, K.A., Mitchison, T., Libby, P., Nahrendorf, M., Weissleder, R., 2017. IRF3 and type I interferons fuel a fatal response to myocardial infarction. Nat. Med. 23, 1481–1487.

Krysko, D.V., Garg, A.D., Kaczmarek, A., Krysko, O., Agostinis, P., Vandenabeele, P., 2012. Immunogenic cell death and DAMPs in cancer therapy. Nat. Rev. Cancer 12, 860–875.

Lemaitre, B., Nicolas, E., Michaut, L., Reichhart, J.M., Hoffmann, J.A., 1996. The dorsoventral regulatory gene cassette spätzle/Toll/cactus controls the potent antifungal response in Drosophila adults. Cell 86, 973–983.

Liston, A., Masters, S.L., 2017. Homeostasis-altering molecular processes as mechanisms of inflammasome activation. Nat. Rev. Immunol. 17, 208–214.

McGlasson, S., Jury, A., Jackson, A., Hunt, D., 2015. Type I interferon dysregulation and neurological disease. Nat. Rev. Neurol. 11, 515–523.

McNab, F., Mayer-Barber, K., Sher, A., Wack, A., O'Garra, A., 2015. Type I interferons in infectious disease. Nat. Rev. Immunol. 15, 87–103.

Medzhitov, R., Preston-Hurlburt, P., Janeway Jr., C.A., 1997. A human homologue of the Drosophila Toll protein signals activation of adaptive immunity. Nature 388, 394–397.

Muskardin, T.L.W., Niewold, T.B., 2018. Type I interferon in rheumatic diseases. Nat. Rev. Rheumatol. 14, 214–228.

Parker, B.S., Rautela, J., Hertzog, P.J., 2016. Antitumour actions of interferons: implications for cancer therapy. Nat. Rev. Cancer 16, 131–144.

Roers, A., Hiller, B., Hornung, V., 2016. Recognition of endogenous nucleic acids by the innate immune system. Immunity 44, 739–754.

Rotem, Z., Cox, R.A., Isaacs, A., 1963. Inhibition of virus multiplication by foreign nucleic acid. Nature 197, 564–566.

Schlee, M., Hartmann, G., 2016. Discriminating self from non-self in nucleic acid sensing. Nat. Rev. Immunol. 16, 566–580.

Vanpouille-Box, C., Alard, A., Aryankalayil, M.J., Sarfraz, Y., Diamond, J.M., Schneider, R.J., Inghirami, G., Coleman, C.N., Formenti, S.C., Demaria, S., 2017. DNA exonuclease Trex1 regulates radiotherapy-induced tumour immunogenicity. Nat. Commun. 8, 15618.

Vanpouille-Box, C., Demaria, S., Formenti, S.C., Galluzzi, L., 2018. Cytosolic DNA sensing in organismal tumor control. Cancer Cell 34, 361–378.

Weiss, J.M., Guerin, M.V., Regnier, F., Renault, G., Galy-Fauroux, I., Vimeux, L., Feuillet, V., Peranzoni, E., Thoreau, M., Trautmann, A., Bercovici, N., 2017. The STING agonist DMXAA triggers a cooperation between T lymphocytes and myeloid cells that leads to tumor regression. Oncoimmunology 6, e1346765.

Xia, T., Konno, H., Ahn, J., Barber, G.N., 2016. Deregulation of STING signaling in colorectal carcinoma constrains DNA damage responses and correlates with tumorigenesis. Cell Rep. 14, 282–297.
Zitvogel, L., Galluzzi, L., Kepp, O., Smyth, M.J., Kroemer, G., 2015. Type I interferons in anticancer immunity. Nat. Rev. Immunol. 15, 405–414.

CHAPTER ONE

Discrimination Between Self and Non-Self-Nucleic Acids by the Innate Immune System

Takumi Kawasaki[1], Taro Kawai[1]

Laboratory of Molecular Immunobiology, Division of Biological Science, Graduate School of Science and Technology, Nara Institute of Science and Technology, Ikoma, Japan

[1]Corresponding authors: e-mail address: kawast01@bs.naist.jp; tarokawai@bs.naist.jp

Contents

International Review of Cell and Molecular Biology, Volume 344
ISSN 1937-6448
https://doi.org/10.1016/bs.ircmb.2018.08.004

Abstract

During viral and bacterial infections, the innate immune system recognizes various types of pathogen-associated molecular patterns (PAMPs), such as nucleic acids, via a series of membrane-bound or cytosolic pattern-recognition receptors. These include Toll-like receptors (TLRs), RIG-I-like receptors (RLRs), AIM2-like receptors (ALRs), and cytosolic DNA sensors. The binding of PAMPs to these receptors triggers the production of type I interferon (IFN) and inflammatory cytokines. Type I IFN induces the expression of interferon stimulated genes (ISGs), which protect surrounding cells from infection. Some ISGs are nucleic acids-binding proteins that bind viral nucleic acids and suppress their replication. As nucleic acids are essential components that store and transmit genetic information in every species, infectious pathogens have developed systems to escape from the host nucleic acid recognition system. Host cells also have their own nucleic acids that are frequently released to the extracellular milieu or the cytoplasm during cell death or stress responses, which, if able to bind pattern-recognition receptors, would induce autoimmunity and inflammation. Therefore, host cells have acquired mechanisms to protect themselves from contact with their own nucleic acids. In this review, we describe recent research progress into the nucleic acid recognition mechanism and the molecular bases of discrimination between self and non-self-nucleic acids.

1. INTRODUCTION

Invading microbes are initially recognized by the host cell's innate immune system via germ line-encoded pattern-recognition receptors (PRRs). PRRs recognize conserved pathogen-specific molecular signatures such as viral nucleic acids and fungal and bacterial cell-wall components, which are termed pathogen-associated molecular patterns (PAMPs). Because activation of the innate immune system triggered by PRR recognition is a critical host defense mechanism, many infectious pathogens have developed the ability to suppress PRR-mediated responses in order to escape detection. The binding of PAMPs to PRRs initiates a signal transduction cascade culminating in the induction of innate immune responses through the production of inflammatory cytokines, type I interferon (IFN), and other mediators. Cytokines transmit a signal to neighboring cells that induce the expression of an array of antiviral and antibacterial genes. These factors prevent pathogen replication by binding viral or bacterial components or by impairing host cell protein synthesis. In addition to triggering immediate host defensive responses, cytokines also prime and orchestrate antigen-specific adaptive immune responses (Kawai and Akira, 2011; Rathinam and Fitzgerald, 2011; Wu and Chen, 2014).

Nucleic acids, such as RNA and DNA, are also PAMPs. Mammals have several distinct classes of nucleic acid-recognizing PRRs, including Toll-like receptors (TLRs), RIG-I (retinoic-acid inducible gene I)-like receptors (RLRs), AIM2 (absent in melanoma 2)-like receptors (ALRs), and intracellular DNA sensors such as cyclic GMP-AMP synthase (cGAS). These receptors are classified according to their cellular localizations, the types of nucleic acids they recognize, and the specific signaling pathways that they activate. For instance, TLRs localize on plasma membrane or endocytic membrane and recognize extracellular nucleic acids, whereas other nucleic acid-recognizing PRRs are cytosolic DNA and RNA sensors; TLR3, TLR7, TLR8, TLR13, and RLRs recognize RNA, and TLR9 and DNA sensors recognize DNA; TLRs, RLRs, and DNA sensors induce the production of type I IFN and inflammatory cytokines, whereas ALRs induce the activation of the inflammasome and produce IL-1β and IL-18. Other than PRRs, several interferon stimulatory genes (ISGs) also bind to pathogen-derived nucleic acids to restrict viral and bacterial replication and prevent the spread of infection to other cells.

Endogenous nucleic acids, such as mRNA, tRNA, mitochondrial DNA, and genomic DNA, that are released during stress responses or cell death can also engage with PRRs to trigger inflammatory cytokines and type I IFN, leading to the initiation and perpetuation of chronic inflammation. Inappropriate or prolonged detection of endogenous nucleic acids underlies many autoimmune diseases. Therefore, host cells have innate mechanisms to protect them from unwanted contact with endogenous (self) nucleic acids. These mechanisms can be largely subdivided into three categories: the cellular localization of PRRs, nucleic acid modifications, and the degradation of unnecessary nucleic acids.

In this review, we describe the most recent research progress into the innate immune system's nucleic acid recognition mechanism and the molecular bases of its discrimination between self and non-self (pathogen) nucleic acids (Fig. 1).

2. INTRACELLULAR NUCLEIC ACID SENSORS

2.1 RLRs

The RLR family is composed of the proteins RIG-I and MDA5, which have similar multi-domain architectures: tandem N-terminal caspase activation and recruitment domains (CARDs) that propagate downstream signaling, a DExD/H box RNA helicase domain, and a C-terminal domain

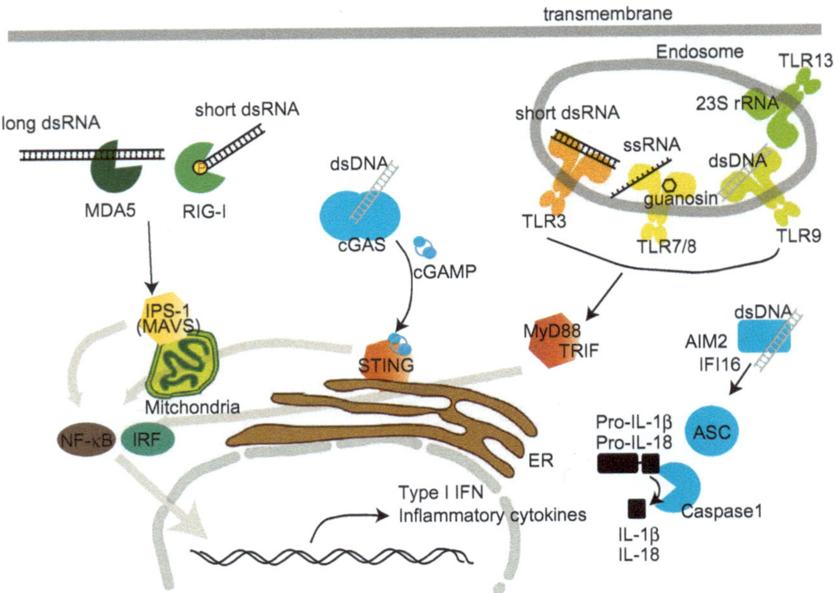

Fig. 1 Recognition of nucleic acids by PRRs. TLR3, 7/8, 9 and 13 are localized in endosomes. TLR3 recognizes dsRNAs, TLR7/8 recognizes ssRNAs and guanosine derivatives such as R837 and R848, TLR9 recognizes dsDNAs, and TLR13 recognizes 23S rRNAs. Ligand binding TLRs associate with MyD88 and/or TRIF in the cytosol. Intracellular RNA is recognized by RIG-I and MDA5, which bind to short dsRNAs and long dsRNAs, respectively. The binding of RIG-I and MDA5 to nucleic acids induces the formation of filaments that bind to mitochondria-localizing IPS-1 (MAVS). Intracellular dsDNAs bind to cGAS and produce cGAMP. cGAMP binds to STING which localizes in the endoplasmic reticulum (ER). Activation of the MyD88/TRIF, IPS-1 and STING adaptor proteins induces translocation of the transcription factors NF-κB and IRF to the nucleus following the production of type I IFN and inflammatory cytokines. AIM2 and IFI16 bind to dsDNAs and form an inflammasome with ASC and caspase-1. Activation of caspase-1 cleaves pro-IL-1β or pro-IL-18 to the mature forms of IL-1β and IL-18, respectively.

(CTD) (Yoneyama and Fujita, 2008). RIG-I binds to blunt ends of dsRNA displaying 5′-diphosphate and 5′-triphosphate moieties and to at least 10–19 bp of the adjacent dsRNA (Hornung et al., 2006; Schlee et al., 2009). RIG-I binds to the 5′ terminus of dsRNA as a monomer in the absence of ATP and forms filamentous oligomers near the dsRNA end in a ATP-dependent manner (Peisley et al., 2013; Wu et al., 2014). In contrast, individual molecules of MDA5 bind dsRNA internally in a highly cooperative manner and form long filaments around dsRNA in the absence of ATP (Berke et al., 2012; Wu et al., 2013). These data support the theory that RIG-I recognizes short dsRNAs whereas MDA5 preferentially binds longer dsRNAs. In the absence of dsRNAs, the CARDs of RIG-I and MDA5 interact with their helicase domain, forming an inhibitory complex (Kowalinski et al., 2011). Upon dsRNA recognition, RIG-I and MDA5 form filaments along dsRNA and their CARDs bind to the CARD of the adaptor protein interferon-β promoter stimulator 1 (IPS-1, also known as mitochondrial antiviral signaling (MAVS) protein). IPS-1 oligomerizes upon RIG-I and MDA5 binding and activates the downstream type I IFN pathway (Hou et al., 2011).

Genetic studies revealed that specific viruses were recognized by RIG-I and MDA5: Newcastle disease virus (NDV), Sendai virus (SeV), vesicular stomatitis virus (VSV), influenza virus, and Japanese encephalitis virus are specifically recognized by RIG-I. In contrast, picornaviruses, including encephalomyocarditis virus (EMCV), Mengo virus, and Theiler's virus, are recognized by MDA5. Dengue virus and West Nile viruses (WNV) are recognized by both RIG-I and MDA5 (Fredericksen et al., 2008; Gitlin et al., 2006; Loo et al., 2008; Rehwinkel et al., 2010). In addition to RNA viruses, RIG-I is involved in the recognition of DNA viruses including herpes simplex virus (HSV) and adenovirus (Chiu et al., 2009; Rasmussen et al., 2009). Furthermore, RIG-I binds short dsRNA intermediates containing a 5′-triphosphate group transcribed by DNA-dependent RNA polymerase III (Pol III) to induce IFNβ (Ablasser et al., 2009; Chiu et al., 2009). Recently, 5S ribosomal RNA pseudogene 141 (RNA5SP141) was found to bind to RIG-I during HSV-1 and influenza A virus (IAV) infections to enhance the expression of type I IFNs (Chiang et al., 2018).

2.2 ALRs

ALRs belong to the PYHIN protein family that typically consist of an N-terminal pyrin domain and one or two C-terminal hematopoietic IFN-inducible nuclear protein with 200 amino acids (HIN-200) domains. The HIN-200 domain contains a oligonucleotide/oligosaccharide-binding fold (OB fold), which is a common DNA-binding motif. Although only 4 *PYHIN* genes are present in humans, 13 are encoded in mice. AIM2 is the only ALR that is conserved between humans and mice. The AIM2 ALR acts as a cytosolic dsDNA sensor that oligomerizes on cytosolic foreign dsDNA and nucleates the polymerization of ASC (adaptor protein apoptosis-associated speck-like protein) to form a caspase-1 activating inflammasome, a filamentous supramolecular signaling platform that is required to instigate innate immune responses. Activated caspase-1 then cleaves pro-IL-1β or pro-IL-18 enabling the release of the mature active cytokines IL-1β and IL-18 (Fernandes-Alnemri et al., 2009;

Hornung et al., 2009; Roberts et al., 2009). AIM2 also has a crucial role in the development of the innate immune response against pathogen infections that include *Francisella tularensis* (Fernandes-Alnemri et al., 2010), HSV-1 (Maruzuru et al., 2018) and human papillomavirus (Reinholz et al., 2013). IFI16 is also a member of the ALR family. It is able to sense pathogenic DNA in both the cytoplasm and the nucleus, and recognizes DNA from viruses including human immunodeficiency virus 1 (HIV-1) (Monroe et al., 2014) and Kaposi sarcoma-associated herpesvirus (KSHV) (Kerur et al., 2011).

AIM2 is potentially able to bind to small DNA fragments below 20 bp in length; however, binding and oligomerization of AIM2 on long DNA fragments are required to initiate filament formation and mediate a robust innate immune response. In contrast, a transient association between the IFI16 HIN-200 domains and dsDNA larger than 60 bp permits at least four IFI16 molecules to cluster, initiating filament assembly (Morrone et al., 2014).

2.3 cGAS

cGAS is a cytosolic DNA sensor that recognizes DNA derived from pathogens or host cells (Gao et al., 2013; Sun et al., 2013; Zhang et al., 2013). It is involved in the immune recognition of HSV-1; retroviruses such as HIV-1, murine leukemia virus, and simian immunodeficiency virus; and reverse transcribed (+) ssDNA (Reinert et al., 2016; Sun et al., 2013). It is currently unclear how cGAS recognizes retroviruses, but cGAS-deficient mice exhibit reduced type I IFN expression following retrovirus infection (Gao et al., 2013). cGAS also plays an important role in eliciting anti-mycobacterial immunity by recognizing DNA released by *Mycobacterium tuberculosis* (Collins et al., 2015; Wassermann et al., 2015; Watson et al., 2015).

Biochemical and structural analysis of cGAS showed that it is able to bind to dsDNAs smaller than 20 bp in length (Civril et al., 2013; Li et al., 2013); however, 20 bp dsDNAs are unable to activate cGAS *in vivo*. Another study demonstrated that effective activation of cGAS occurs in the presence of dsDNA over 40 bp in length (Herzner et al., 2015). Recent work revealed that cGAS is activated in a length-dependent manner and that cGAS dimers form a ladder-like networks with DNA over 45 bp in length, which is sufficient for activation of innate immune responses *in vitro* and *in vivo* (Andreeva et al., 2017).

cGAS is a cytoplasmic nucleotidyltransferase, which, upon dsDNA binding, catalyzes the synthesis of cGAMP from GMP and AMP. In turn, cGAMP binds to and activates the STING adaptor protein, which initiates signaling to induce type I IFN production (Sun et al., 2013). cGAMP also functions as a second messenger for transducing alarm signals to surrounding cells: it is delivered from producing cells to neighboring cells through gap junctions, where it promotes STING activation and antiviral immunity independent of type I IFN signaling (Ablasser et al., 2013). During virus infection and expansion, cGAS-synthesized cGAMP can be packaged into viral particles and extracellular vesicles. The viral particles then deliver cGAMP to target cells such as macrophages and dendritic cells (DCs). This triggers innate immunity and antiviral defenses (Gentili et al., 2015; Xu et al., 2016).

2.4 ISGs

Type I IFNs produced upon pathogen infection protect host cells by inducing the expression of ISGs. Some ISGs mediate antibacterial responses, while others mediate antiviral responses. As already discussed, RIG-I and MDA5 are well-known RNA sensors; however, these factors are also key ISGs that stimulate antiviral responses against hepatitis C virus (HCV), WNV, and Venezuelan equine encephalitis virus (VEEV). It has been proposed that RIG-1 and MDA5 inhibit viral replication by binding to the viral RNA genome (Schoggins et al., 2011). In addition, the RNA-activated protein kinase R (PKR), the tetratricopeptide repeats (IFITs) family protein, oligoadenylate synthase (OAS), and the Moloney leukemia virus 10 (MOV10) protein are also known to bind to pathogenic or endogenous RNA. PKR is activated by RNA having bulges and internal loops. Upon activation, PKR phosphorylates eukaryotic initiation factor 2α (eIF2α), which induces the inhibition of translation and leads to apoptosis (Zhang et al., 2001).

OAS has been identified as an enzyme that senses viral dsRNA. OAS is activated upon viral dsRNA binding and catalyzes the generation of the second messenger, 2′-5′-oligoadenylate (2′-5′A). 2′-5′A activates endoribonuclease L (RNase L), which degrades host and viral RNA (Chebath et al., 1987). Another OAS family member is OASL (oligoadenylate synthetase-like), which promotes antiviral activity by enhancing the sensitivity of RLR activation (Lee et al., 2013; Zhu et al., 2014). MOV10 is a member of the RNA helicase superfamily-1, and is also associated with Argonaute 2 (AGO2), a component of the RNA-induced silencing complex. MOV10 was originally identified as a protein that prevents production of the endogenous Moloney murine leukemia virus infection in mice (Goodier et al., 2012; Schnieke et al., 1983). Later studies determined that MOV10 is able to reduce the infectivity of HIV-1, simian immunodeficiency virus, and murine leukemia virus (Wang et al., 2010). MOV10 is thought to multifunctional protein that MOV10 binds to polymerase basic protein 2 (PB2) in nucleoprotein (NP) of IAV and inhibits IAV replication which is unnecessary for RNA helicase activity (Zhang et al., 2016a).

3. EXTRACELLULAR NUCLEIC ACID RECOGNITION

3.1 TLRs

TLRs recognize many types of PAMPs, including lipids, lipoproteins, proteins, and nucleotides. The TLR family comprises 10 members (TLR1–TLR10) in humans and 12 (TLR1–TLR9, TLR11–TLR13) in mice. TLRs are composed of an extracellular domain with leucine-rich repeats (LRRs) that mediates the recognition of PAMPs, a transmembrane domain, and a cytoplasmic Toll/IL-1 receptor (TIR) domain that initiates downstream signaling. The TLR family can be divided into two subgroups: TLRs that localize to the cell surface and TLRs that reside in intracellular compartments such as endosomes, lysosomes, and endolysosomes. Each subgroup recognizes distinct PAMPs. For example, the nucleic acid-sensing TLRs exclusively localize intracellularly; TLR3, TLR7, TLR8 and TLR13 recognize RNA, whereas TLR9 recognizes DNA. Binding of TLR to pathogenic nucleic acids induces an association with a specific set of TIR domain-containing adaptors, such as MyD88 and TRIF, which activates the transcription factors NF-κB and the interferon regulatory factor family (IRFs). Transcription factor activation then induces the production of type I IFN and inflammatory cytokines. Although different TLRs bind to distinct synthetic ligands, their sensitivity to microbial components overlaps. TLRs are highly expressed on innate immune cells such as macrophages and DCs, which are professional antigen presenting cells that produce inflammatory cytokines and type I IFN. One of these TLRs, TLR13, has recently been shown to recognize a 23S ribosomal rRNA sequence that is the binding site of macrolide, lincosamide, and streptogramin group antibiotics (including erythromycin) in bacteria (Oldenburg et al., 2012). However, other intracellular nucleic acid sensors, such as cGAS and RLRs, are expressed in a variety of cell types including epithelial cells, macrophages and DCs. Therefore, the cooperative recognition of invaded pathogens by distinct receptors is important in triggering a robust innate immune response.

3.2 TLR9

TLR9 was first identified as a DNA sensing receptor expressed in professional innate immune cells such as DCs and macrophages. TLR9 preferentially binds to hexamers containing unmethylated CpG (cytosine-phosphate-guanine dideoxynucleotide) motifs (Hemmi et al., 2000). Studies on TLR9-deficient mice have shown that TLR9 acts as a sensor for DNA viruses such as cytomegalovirus (MCMV), HSV-1 (Hochrein et al., 2004), HSV-2 (Lund et al., 2003), and adenovirus (Zhu et al., 2007) in DCs. However, innate immune recognition of these viruses in non-DCs is dependent on cytosolic DNA sensing of viral DNA and is TLR-independent.

Structural analysis of the TLR9 extracellular domain bound to a CpG-DNA 12-mer indicated that TLR9 and CpG-DNA form a symmetric complex with 2:2 stoichiometry, and CpG-DNA was found to bind to both protomers in the TLR9 dimer (Ohto et al., 2015). Investigation of the optimal DNA sequence motifs for TLR9 binding and activation suggested that TLR9 has an additional DNA-binding site to enhance its activation (Pohar et al., 2015a,b). A recent structural study confirmed that TLR9 has a second DNA-binding site that binds to DNA containing cytosine at the second position from the 5′ end (5′-xCx DNA), which cooperatively promotes the dimerization and activation of TLR9 (Ohto et al., 2018).

3.3 TLR3

TLR3 detects double-stranded RNA (dsRNA) and its synthetic analogue, polyinosinic-polycytidylic acid (poly I:C) (Alexopoulou et al., 2001). TLR3-deficient mice are susceptible to ssRNA viruses, such as WNV (Daffis et al., 2008), Semliki Forest virus (Schulz et al., 2005), and EMCV (Hardarson et al., 2007) as well as DNA viruses, such as MCMV (Tabeta et al., 2004) and HSV-1 (Zhang et al., 2007). TLR3 deficiency in humans also causes an increase in HSV-1 infection rates (Zhang et al., 2007).

Structural analysis of the mouse TLR3 ectodomain in complex with dsRNA indicated that TLR3 forms a homodimer with a flat horseshoe-like shape, and positively charged residues in the ectodomains of the two TLR3 protomers each bind the sugar-phosphate backbone of one dsRNA ligand (Liu et al., 2008). The positively charged residues of the ligand binding sites in TLR3 are in two discrete regions near the N- and C-termini of the horseshoe. The ligand binding sites of TLR3 are separated by approximately 120 Å, which is equivalent to dsRNA of approximately 45 bp in length, the minimal length that has been determined for stable binding of TLR3 to dsRNA.

3.4 TLR7 and TLR8

TLR7 and TLR8 recognize single-stranded RNA molecules (ssRNAs) (Diebold et al., 2004; Heil et al., 2004; Lund et al., 2004). Human TLR7, TLR8, and mouse TLR7 also recognize imidazoquinoline derivatives such as imiquimod (R837), resiquimod (R848), and guanine analogues, whereas mouse TLR8 does not. Cells derived from TLR7-deficient mice are unable to induce cytokine production in response to ssRNA viruses such as IAV, VSV, and HIV-1 (Hemmi et al., 2002; Lund et al., 2004). In addition, TLR7 detects bacterial ssRNAs from strains such as Group B *Streptococci*. However, it is not able to sense ssRNAs from other bacteria such as *Listeria monocytogenes* and Group A *Streptococci* (Mancuso et al., 2009). TLR7 is predominantly expressed in plasmacytoid DCs (pDCs) and is involved in the robust expression of IFNα in both humans and mice.

Structural analysis of TLR7 indicated that it forms an m-shaped dimer with two binding sites, which act as a dual receptor for ssRNA and guanosine. Binding of ssRNA to TLR7 enhances the affinity of the guanosine binding site for its ligand. The specific binding of both ligands to TLR7 is crucial for its dimerization and activation (Zhang et al., 2016b). Conversely, synthetic agonists for TLR7 (R837 and R848) and TLR8 (CL097, CL075 and R848) can, in isolation, directly bind and induce dimer formation of their respective receptors (Tanji et al., 2013; Zhang et al., 2016b) (Fig. 1).

4. MODIFICATIONS ENABLING DNA DISCRIMINATION

DNA methylation is utilized by bacterial restriction and modification systems to allow bacteria to distinguish between self and non-self-DNA. The mammalian genome contains cytosine (C) and guanosine (G) enriched regions, where the C and G bases are separated by a single phosphate group. This structural feature is known as a CpG motif, which is shorthand for 5′-C-phosphate-G-3′. Cytosine nucleotides in CpGs are commonly methylated by methyltransferases, and 70–80% of CpG sites in mammals are methylated. In contrast, the methylation levels of CpG sites in bacterial and viral genomes are significantly lower. Unmethylated CpG nucleotides are recognized by TLR9; therefore, methylation-based DNA modification is a well-established system that allows distinction between self and non-self-nucleic acids (Hemmi et al., 2000).

The oxidized base 8-hydroxyguanosine (8-OHG), a marker of DNA oxidative damage, promotes cytosolic immune recognition by decreasing its susceptibility to 3′ repair exonuclease 1 (TREX1)-mediated degradation. Reactive oxygen species (ROS) or UV exposure are responsible for the generation of 8-OHG, which stimulates the production of UV-exposed skin lesions in lupus-prone mice and is abundant in lupus-induced skin lesions in humans (Gehrke et al., 2013).

In addition to DNA modifications, host cells have also evolved DNA-modulating proteins that directly or indirectly regulate their defense against viruses and suppress non-coding endogenous retrotransposons and retrovirus elements (Fig. 2).

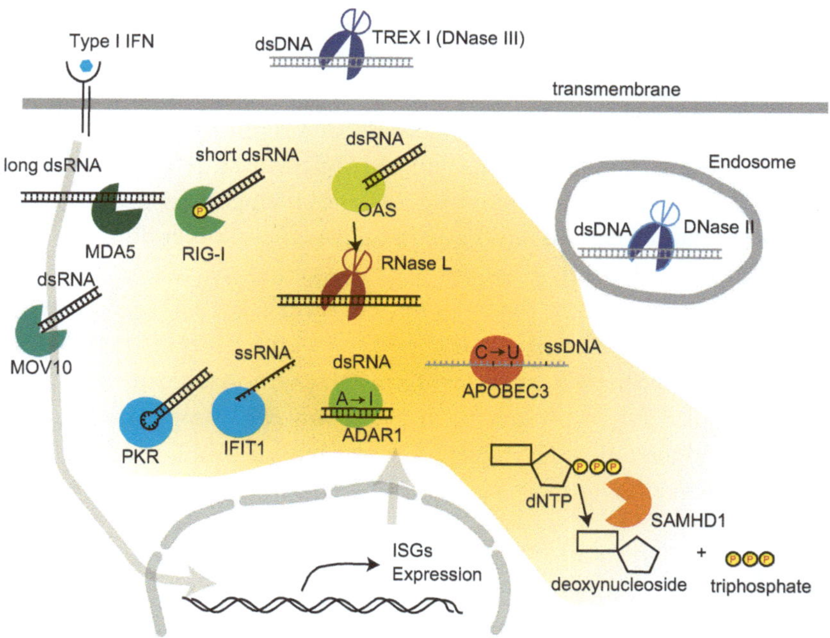

Fig. 2 Regulation of nucleic acid-related restriction by ISGs. Released type I IFNs induce various types of ISGs expression in surrounding cells. ISG expression protects cells from viral and bacterial infections. RIG-I and MDA5 work as a receptor and a viral restriction factor by binding to short dsRNAs and long dsRNAs, respectively. PKR binds to RNAs that have bulges and internal loops, MOV10 binds to dsRNAs, and IFIT1 binds to ssRNAs. OAS binds to dsRNAs and generates 2′-5′A to activate RNase L. APOBEC3 catalyzes the deamination of cytidine (C) to uridine (U) in ssDNA substrates. ADAR catalyzes the conversion of adenosine (A) to inosine (I) in dsRNA substrates. SAMHD1 is a triphosphohydrolase that converts deoxynucleoside triphosphates to deoxynucleoside and triphosphate, which reduces cellular dNTP levels. TREX1 (DNase III) degrades dsDNAs in extracellular spaces, and DNase II degrades endosomal dsDNAs derived from the macrophage engulfment of dead cells.

4.1 APOBEC3

Apolipoprotein B editing complex 3 (APOBEC3) family members are cytidine deaminases that catalyze the removal of an amino group from cytidine (C), forming uridine (U) in single-stranded DNA substrates. The human genome encodes seven APOBEC3 (A3) proteins. APOBEC3G (A3G) was initially identified as a suppressor of HIV-1 infectivity (Doyle et al., 2015). During synthesis of the first (minus) retroviral DNA strand, APOBEC-catalyzed deamination of cytosine (C) residues results in the accumulation of excessive levels of uracil (U). Subsequently, Uracil (U) is recognized as thymine (T) by the viral reverse transcriptase and adenine (A) is incorporated into the newly synthesized second (plus) DNA strand rather than guanine (G). Therefore, mutation levels in HIV-1 genomes are increased following replication; the consequential augmentation of mutant functional proteins results in a decrease in viral infectivity (Doyle et al., 2015).

Regions of non-coding endogenous retrotransposons and retrovirus elements in the human genome are remnants of ancestral retroviral germ line infections. The expansion and transposition of endogenous retroviruses are believed to induce cancers, autoimmune diseases, and other disorders in the host. Most retroelements have lost their ability to replicate because of the accumulation of inactivating mutations, but several, including some murine intracisternal A-particles (IAP) and MusD sequences, are still mobile. A3G plays a crucial role in restricting their retrotransposition (Esnault et al., 2005). The accumulation of endogenous viruses upon A3G suppression has been shown to induce type I IFN expression though a STING-mediated DNA sensing pathway (Stavrou et al., 2015).

4.2 SAMHD1

The sterile alpha-motif (SAM) and histidine-aspartate (HD) domain-containing protein 1 (SAMHD1) was identified as a factor that inhibits HIV-1 infection in myeloid cells (Laguette et al., 2011). SAMHD1 is a potent dNTP triphosphohydrolase that converts deoxynucleoside triphosphates (dNTPs) into their corresponding deoxynucleoside and inorganic tripolyphosphate moieties (Goldstone et al., 2011). SAMHD1 decreases the concentration of intracellular dNTPs in myeloid cells to a level that is unable to support reverse transcription, establishing a cellular state that is not permissive to HIV-1 infections (Lahouassa et al., 2012). Since DNA viruses also utilize cellular dNTPs, SAMHD1 additionally limits the replication of DNA viruses such as vaccinia virus and HSV-1 in myeloid target cells (Hollenbaugh et al., 2013).

As well as suppressing HIV-1 propagation, SAMHD1 was originally identified as a binding protein for the HIV-1 accessory protein, virion-associated protein (Vpr). HIV-2 genomes also express the accessory protein Vpx, which augments the ability of HIV-2 to infect myeloid cells and was initially proposed to counteract SAMHD1 (Sharova et al., 2008). Later studies showed that Vpx binds to SAMHD1 and induces its degradation (Hrecka et al., 2011; Laguette et al., 2011).

5. MODIFICATIONS ALLOWING RNA DISCRIMINATION

RNA classes include transfer RNA (tRNA), messenger RNA (mRNA), and non-coding RNA (ncRNA). Although over 100 different types of RNA modifications have been identified across all living organisms, little is known about their resultant functions. The more abundant host RNA modifications have been elucidated, which include pseudouridine (ψ) and 2′-*O*-methylation (2′OMe) in tRNA, and N^6-methyladenosine (m^6A) and 5-methyl-cytidine (m^5C) in mRNA. These modifications elicit various functional roles such as translation, localization, and stability. Among these, it is well known that discrimination of self and non-self-RNA is dependent upon naturally occurring modifications at the 5′ end of RNA. Mammalian cell mRNAs usually have a 5′ cap structure, called cap 0 (m7Gppp) that consists of a guanine nucleotide that is connected to mRNA via a reverse 5′–5′-triphosphate bridge. In addition, the guanosine is methylated at the N7 position. The methylated guanosine cap residues of mRNAs are recognized and bound by the eukaryotic translation initiation factor 4E (eIF4E) to recruit transcripts to the ribosome and initiate translation. In higher eukaryotes, the mRNA cap is further modified by ribose 2′-*O*-methylation on the first and sometimes second nucleotides adjacent to the cap, forming cap 1- or cap 2-mRNA (Werner et al., 2011).

5.1 2′-*O*-Methyltransferase

N7 methylation is important for various processes including transcriptional elongation, poly-adenylation, splicing, and nuclear export. Although the functional role of 2′-*O*-methylation in the cap 1 and cap 2 structures on mRNA is not generally understood, other studies have indicated that methylation serves as a molecular signature to distinguish self and non-self mRNA

(Daffis et al., 2010; Zust et al., 2011). Many viruses that replicate in the cytoplasm (for example, picornavirus, flavivirus, coronavirus, and poxvirus) encode their own RNA modification enzymes, including 2′-O-methyltransferase. The importance of viral RNA modifications in replication has been supported by studies that the infection of viruses lacking 2′-O-methyltransferase activity induces higher type I IFN expression, and unmodified genomic RNA was recognized by the host RNA sensor, MDA5 (Zust et al., 2011). The functional importance of 2′-O-methyltransferase activity for the infectivity of RNA viruses was also verified by studies on dengue virus (DENV) (Zust et al., 2018), Middle East respiratory syndrome coronavirus (MERS-CoV) (Aouadi et al., 2017), and ronivirus (Zeng et al., 2016). These viruses also encode 2′-O-methyltransferase and its disruption also causes type I IFN induction and virus elimination from the host following infection. Together, these results indicate that RNA viruses have evolved to encode their own methyltransferases that mimic host RNA methylation patterns to avoid recognition by the host's innate immune system.

5.2 ADAR

ADARs (adenosine deaminases acting on RNA) catalyze the conversion of adenosine (A) to inosine (I) in double-stranded RNA (dsRNA) substrates. Such "A-to-I editing" alters the coding of RNA, as inosine is read as guanosine (G) instead of adenosine by ribosomes during translation, leading to the production of non-functional proteins. In humans, two autoimmune disorders, Dyschromatosis Symmetrica Hereditaria (DSH) and Aicardi-Goutières Syndrome (AGS), are attributed to specific mutations that have been mapped to the *Adar1* gene (Rice et al., 2012). Furthermore, *Adar1* inducible knockout (KO) mice exhibit a high level of IFN production in neuronal tissues, a key characteristic of AGS. In healthy cells, ADAR1-mediated nucleotide conversions are believed to be important for the suppression of viral replication and, more importantly, endogenous RNA recognition by the RNA sensor, MDA5. Correspondingly, mutations in either MDA5 or MAVS rescue the embryonic lethal phenotype of *Adar1* KO mice (Liddicoat et al., 2015; Mannion et al., 2014; Pestal et al., 2015).

Approximately half of the mammalian genome is composed of non-coding retrotransposons, such as SINEs (short interspersed nuclear elements) and Alu elements, which typically form dsRNA duplexes (Lander et al., 2001; Waterston et al., 2002). Retrotransposons are subjected to extensive

A-to-I RNA editing by ADAR (Ramaswami et al., 2012). It is possible that the location of repetitive elements determines their immunogenicity. Retrotransposons located within introns do not persist in the cytosol and therefore are unable to activate MDA5. Repetitive elements in 3′ UTRs, although rare, can be retained and form duplexes to activates MDA5. Editing of self-dsRNA by ADAR1 generates multiple I-U mismatches to avoid MDA5 recognition (Berke et al., 2012; Wu et al., 2013). In the absence of ADAR1 activity, long dsRNA stem loops can activate MDA5. However, A-to-I editing also alters the RNA secondary structure, which may lead to recognition by innate immune receptors (Liddicoat et al., 2015; Pestal et al., 2015). Recent data have shown that ADAR1 predominantly edits Alu elements in RNA polymerase II (pol II)-transcribed mRNAs, but not those transcribed by pol III in human cells (Chung et al., 2018), and A-to-I conversion in Alus suppresses MDA5 activation *in vitro* (Ahmad et al., 2018).

6. ENDOGENOUS NUCLEIC ACIDS AS DAMPs

Apoptosis is a process of programmed cell death that occurs in multicellular organisms, where cellular components, such as RNA and DNA, are completely degraded. However, during non-apoptotic cell death events, including necrosis, intracellular contents are released into the surrounding tissues. Released endogenous RNA and DNA are potent inducers for the activation of innate immune signaling as they can bind to both endogenous and exogenous nucleic acid receptors. Endogenous nucleic acids are also released as damage-associated molecular patterns (DAMPs) during stress conditions such as chemotherapy, UV, and radiation that also induce non-apoptotic cell death.

6.1 Stress-Induced Release of Nucleotides

Radiation damage elicits the production of dsRNA, which activates TLR3 and causes inflammatory diseases in various tissues such as the skin and small intestine (Bernard et al., 2012; Takemura et al., 2014, 2018). Intracellular DNA is also released upon cellular stress. Chemicals such as mutagenic 7,12-dimethylbenz(a)anthracene (DMBA), cisplatin, and etoposide induce cytosolic leakage of nuclear DNA, which intrinsically induces type I IFN expression through the cGAS-STING pathway (Ahn et al., 2014). Mitochondrial (mt) DNA released into the cytosol upon mitochondrial stress is sensed by cGAS (West et al., 2015). In addition, DNA released by tumor chemotherapy and radiation therapy acts as an adjuvant that activates the cGAS-STING pathway by releasing type I IFN from DCs (Deng et al., 2014; Woo et al., 2014). Such inductions of the inflammatory response are further facilitated by nucleotide binding proteins that enhance DNA incorporation into immune cells and protect it from nucleases. Inflammatory diseases in mice and humans are induced when released intracellular DNA and RNA, including endogenous retroelements and non-coding RNA, cannot be adequately cleared.

6.2 The Role of Nucleotide Binding Proteins in Inflammation

Various nucleotide binding proteins have been identified as factors that exacerbate inflammation. Most nucleotide binding proteins are expected to protect DNA from degradation by DNase and RNase in the extracellular fluid and facilitate its incorporation into innate immune cells. High mobility group box 1 (HMGB1) is a nuclear protein that binds to DNA and participates in processes such as transcription and repair. However, when it is released in complex with DNA in response to cell damage or inflammation, it acts as a potent inflammatory cytokine. HMGB1/DNA complexes facilitate incorporation into innate immune cells by binding to RAGE (receptor for advanced glycosylation end) on the surface of B-cells and DCs, which leads to the recruitment of TLR9 and potentiates type I IFN expression (Yanai et al., 2009). Another report has demonstrated that HMGB1 and its related family members, HMGB2 and HMGB3, bind to DNA and these complexes activate TLRs. In addition, a recent finding indicated that HMGB1 is able to induce the formation of U-turns and bends in dsDNA that nucleate cGAS dimers, leading to the formation of cGAS ladder-like networks with DNA (Andreeva et al., 2017). Another factor that can convert self-DNA into a pathogenic ligand is LL37, which is the active form of cathelicidin, an endogenous anti-microbial peptide that is produced by keratinocytes and neutrophils. LL37 has been found to bind to self-DNA released from necrotic cells in psoriatic lesional skin. The LL37-DNA complex forms condensed structures that protect self-DNA from nuclease degradation. The LL37-DNA complex is then translocated into the endocytic compartments of pDCs, inducing TLR9 activation and resulting in a sustained type I IFN response (Lande et al., 2007, 2011). The production of LL37 is associated with the onset of inflammatory diseases such as Systemic Lupus Erythematosus (SLE).

Cells also secrete extracellular vesicles called exosomes. Exosomes are endosomal membrane vesicles with a diameter of ~40–150 nm and contain various molecular constituents, including proteins and nucleic acids, from their cells of origin. Nucleic acids in exosomes are also protected from nuclease degradation. Cancer cells release exosomes that contain DNA or RNA. These exosomes are captured by DCs which then induce type I IFN production (Kitai et al., 2017; Nabet et al., 2017).

7. ENDOGENOUS NUCLEOTIDES FACILITATE INFLAMMATION

Mutations in genes encoding SAMHD1, TREX1 (DNase III), DNase II, Ribonuclease (RNase) H2, and ADAR are associated with AGS and chilblain lupus, which are rare early-onset inflammatory diseases characterized by inappropriate immune activation and overproduction of type I IFN (Rice et al., 2009, 2012). In healthy cells, SAMHD1, TREX1, DNase II, RNase H2, and ADAR suppress the accumulation of endogenous extracellular DNA or intracellular RNA derived from non-coding retrotransposons such as SINEs, Alus, long-non coding RNAs (lncRNAs) and circular RNAs (circRNAs).

7.1 Self-DNA Accumulation and Inflammation

SAMHD1 deficient mice show reduced cellular dNTP concentrations and elevation of ISG expression in various cell types, which are similar characteristics exhibited by cells from AGS and SLE patients (Behrendt et al., 2013). However, one study unexpectedly found that SAMHD1 deficient mice show increases in ISG, but no SLE-like phenotype (Rehwinkel et al., 2013).

TREX1 is a 3′ repair exonuclease that digests cytosolic DNA. Mutations in TREX1 cause AGS and chilblain lupus (Crow et al., 2006, 2015; Stetson et al., 2008). The AGS-like symptoms in TREX1 deficient mice are rescued by genetic ablation of cGAS or STING (Gray et al., 2015).

Another major cellular nuclease is DNase II, which localizes in the lysosome and is largely responsible for the clearance of DNA from dead cells and expelled nuclei. *DNase II*$^{-/-}$ mice die in utero, and the embryos become severely anemic by embryonic day 17.5. However, these phenotypes were rescued by crossing with IRF3- or cGAS-deficient mice (Gao et al., 2015; Kawane et al., 2006). DNase II mutations have been found in human patients with inflammatory phenotypes (Rodero et al., 2017).

AGS is also caused by a mutation in RNase H2. Cells deficient in RNase H2 accumulate ribonucleotides within their genomic DNA, which results in single-stranded breaks. This DNA damage leads to activation of the cGAS-STING pathway (Mackenzie et al., 2016; Pokatayev et al., 2016).

7.2 Self-RNA Accumulation and Inflammation

Mutations in ADAR1 are also associated with AGS (Crow et al., 2015). ADAR1 deficiency in mice results in a lethal embryonic phenotype, which is accompanied by substantial expression of type I IFNs and hematopoietic failure (Wang et al., 2000). This embryonic lethality is partially rescued in *Adar1* and *Mavs* double mutants (Liddicoat et al., 2015; Mannion et al., 2014; Pestal et al., 2015). Cells deficient in ADAR1 induce type I IFN production, as self-dsRNA will not contain multiple I-U mismatches, enabling its recognition by MDA5 (Berke et al., 2012; Wu et al., 2013). A deficiency in IPS-1 (MAVS) prevents the recognition of long dsRNAs derived from 3′ UTRs or non-coding retrotransposons in ADAR1 deficient mice. In addition, treatment of cells with DNA methyltransferase inhibitors induces endogenous RNA virus transcription that enhances RLRs signaling; however, this response is decreased twofold in IPS-1 knockout cells (Chiappinelli et al., 2015; Roulois et al., 2015).

The microRNA let-7 is a highly abundant regulator of gene expression at the post-transcriptional level in the central nervous system. Extracellular let-7 activates the TLR7 RNA-sensor and induces neurodegeneration through neuronal TLR7 and is associated with progression of amyotrophic lateral sclerosis (ALS) (Lehmann et al., 2012). MicroRNA-21 is one of the most highly expressed members in various types of cells and has several functions such as cellular differentiation, cancer proliferation and cytokines production. MicroRNA-21 in macrophage reported to be induced by LPS and suppressed cytokines production in macrophages (Sheedy et al., 2010). Other microRNA such as microRNA-135-5p, microRNA-26b and microRNA-29a also reported to be contributed in inflammatory response after LPS stimulation (Li et al., 2016; Tang et al., 2017; Zhang et al., 2015), therefore increase or decrease of microRNA in macrophage and DC may causes the progression of inflammatory disease.

Over the past few years, thousands of lncRNA and circRNA have been identified in mammalian genome and shown to have various functions. Some of lncRNA and circRNA are proposed to regulate innate immune responses during TLR and RLR activation. Several lncRNAs, such as lncRNA-Cox2 and lncRNA-1992, are inducible expressed during innate immune activation and are suggested to regulate cytokines production by inhibiting regulatory proteins expression (Carpenter et al., 2013; Li et al., 2014). A recent report showed that self-recognition of lncRNA-Lsm3b with RIG-I inhibits RIG-I activation to prevent overproduction of type I IFN for maintenance of immune homeostasis (Jiang et al., 2018). CircRNAs are generated through back splicing and have been postulated to function as microRNA sponges. A recent report indicated that circRNA generated from foreign introns potently stimulates innate immune signaling and inhibits viral infection (Chen et al., 2017), and circRNA biogenesis is directly involved in viral restriction (Li et al., 2017).

8. CONCLUSION

In this review, we have summarized the innate immune sensing mechanisms for nucleic acids as well as antiviral and antibacterial restriction factors that modulate the replication of nucleic acids. Host cells have mechanisms to protect endogenous nucleic acids from unwanted contact with host PRRs. These mechanisms can be subdivided into three categories: the cellular localization of PRRs, nucleic acid modifications, and the degradation of unnecessary nucleic acids. Although retroelements, such as SINE, LINE, Alus, and retrotransposons, are derived from host cells and could potentially initiate an innate immune response via MDA5 and RIG-I recognition, the activation of ADAR1 and SAMHD1 suppresses the recognition and accumulation of these retroelements. Protection against endogenous nucleic acid recognition is also achieved by nucleic acid degradation. TREX1 degrades dsDNA in the extracellular milieu derived from host cells. In addition, DNase II localizes in the lysosome and is largely responsible for the clearance of DNA from dead cells and expelled nuclei. Self-nucleic acids may be released into the extracellular space upon cell necrosis. These are degraded by DNases and RNases before they come into contact with nucleic acid-sensing TLRs in endosomes. In addition, RNA 2′-*O* methylation and CpG-DNA methylation in the host are other important signatures for self-nucleic acid discrimination.

In recent years, our understanding of self and non-self-nucleic acid recognition mechanisms has greatly increased because of advances in structural analyses, *in vitro* experiments, and DNA/RNA sequencing. However, the detailed molecular mechanisms of self-nucleic acid recognition during development, inflammatory diseases or stress conditions are still unclear. Nevertheless, the current accumulation of information regarding nucleic acid discriminatory systems will be applicable for the treatment and management of immunological disorders in the future.

ACKNOWLEDGMENT

We thank Emma Andrew, PhD, from Edanz Group (www.edanzediting.com/ac) for editing a draft of this chapter.

REFERENCES

Ablasser, A., Bauernfeind, F., Hartmann, G., Latz, E., Fitzgerald, K.A., Hornung, V., 2009. RIG-I-dependent sensing of poly(dA:dT) through the induction of an RNA polymerase III-transcribed RNA intermediate. Nat. Immunol. 10, 1065–1072.

Ablasser, A., Schmid-Burgk, J.L., Hemmerling, I., Horvath, G.L., Schmidt, T., Latz, E., Hornung, V., 2013. Cell intrinsic immunity spreads to bystander cells via the intercellular transfer of cGAMP. Nature 503, 530–534.

Ahmad, S., Mu, X., Yang, F., Greenwald, E., Park, J.W., Jacob, E., Zhang, C.Z., Hur, S., 2018. Breaching self-tolerance to alu duplex RNA underlies MDA5-mediated inflammation. Cell 172, 797–810.e13.

Ahn, J., Xia, T., Konno, H., Konno, K., Ruiz, P., Barber, G.N., 2014. Inflammation-driven carcinogenesis is mediated through STING. Nat. Commun. 5, 5166.

Alexopoulou, L., Holt, A.C., Medzhitov, R., Flavell, R.A., 2001. Recognition of double-stranded RNA and activation of NF-kappaB by toll-like receptor 3. Nature 413, 732–738.

Andreeva, L., Hiller, B., Kostrewa, D., Lassig, C., De Oliveira Mann, C.C., Jan Drexler, D., Maiser, A., Gaidt, M., Leonhardt, H., Hornung, V., Hopfner, K.P., 2017. cGAS senses long and HMGB/TFAM-bound U-turn DNA by forming protein-DNA ladders. Nature 549, 394–398.

Aouadi, W., Blanjoie, A., Vasseur, J.J., Debart, F., Canard, B., Decroly, E., 2017. Binding of the methyl donor S-adenosyl-l-methionine to middle east respiratory syndrome coronavirus 2'-O-methyltransferase nsp16 promotes recruitment of the allosteric activator nsp10. J. Virol. 91, e02217-16.

Behrendt, R., Schumann, T., Gerbaulet, A., Nguyen, L.A., Schubert, N., Alexopoulou, D., Berka, U., Lienenklaus, S., Peschke, K., Gibbert, K., Wittmann, S., Lindemann, D., Weiss, S., Dahl, A., Naumann, R., Dittmer, U., Kim, B., Mueller, W., Gramberg, T., Roers, A., 2013. Mouse SAMHD1 has antiretroviral activity and suppresses a spontaneous cell-intrinsic antiviral response. Cell Rep. 4, 689–696.

Berke, I.C., Yu, X., Modis, Y., Egelman, E.H., 2012. MDA5 assembles into a polar helical filament on dsRNA. Proc. Natl. Acad. Sci. U.S.A. 109, 18437–18441.

Bernard, J.J., Cowing-Zitron, C., Nakatsuji, T., Muehleisen, B., Muto, J., Borkowski, A.W., Martinez, L., Greidinger, E.L., Yu, B.D., Gallo, R.L., 2012. Ultraviolet radiation damages self noncoding RNA and is detected by TLR3. Nat. Med. 18, 1286–1290.

Carpenter, S., Aiello, D., Atianand, M.K., Ricci, E.P., Gandhi, P., Hall, L.L., Byron, M., Monks, B., Henry-Bezy, M., Lawrence, J.B., O'neill, L.A., Moore, M.J., Caffrey, D.R., Fitzgerald, K.A., 2013. A long noncoding RNA mediates both activation and repression of immune response genes. Science 341, 789–792.

Chebath, J., Benech, P., Revel, M., Vigneron, M., 1987. Constitutive expression of (2′-5′) oligo A synthetase confers resistance to picornavirus infection. Nature 330, 587–588.

Chen, Y.G., Kim, M.V., Chen, X., Batista, P.J., Aoyama, S., Wilusz, J.E., Iwasaki, A., Chang, H.Y., 2017. Sensing self and foreign circular rnas by intron identity. Mol. Cell 67, 228–238.e5.

Chiang, J.J., Sparrer, K.M.J., Van Gent, M., Lassig, C., Huang, T., Osterrieder, N., Hopfner, K.P., Gack, M.U., 2018. Viral unmasking of cellular 5S rRNA pseudogene transcripts induces RIG-I-mediated immunity. Nat. Immunol. 19, 53–62.

Chiappinelli, K.B., Strissel, P.L., Desrichard, A., Li, H., Henke, C., Akman, B., Hein, A., Rote, N.S., Cope, L.M., Snyder, A., Makarov, V., Budhu, S., Slamon, D.J., Wolchok, J.D., Pardoll, D.M., Beckmann, M.W., Zahnow, C.A., Merghoub, T., Chan, T.A., Baylin, S.B., Strick, R., 2015. Inhibiting DNA methylation causes an interferon response in cancer via dsRNA including endogenous retroviruses. Cell 162, 974–986.

Chiu, Y.H., Macmillan, J.B., Chen, Z.J., 2009. RNA polymerase III detects cytosolic DNA and induces type I interferons through the RIG-I pathway. Cell 138, 576–591.

Chung, H., Calis, J.J.A., Wu, X., Sun, T., Yu, Y., Sarbanes, S.L., Dao Thi, V.L., Shilvock, A.R., Hoffmann, H.H., Rosenberg, B.R., Rice, C.M., 2018. Human ADAR1 prevents endogenous RNA from triggering translational shutdown. Cell 172, 811–824.e14.

Civril, F., Deimling, T., De Oliveira Mann, C.C., Ablasser, A., Moldt, M., Witte, G., Hornung, V., Hopfner, K.P., 2013. Structural mechanism of cytosolic DNA sensing by cGAS. Nature 498, 332–337.

Collins, A.C., Cai, H., Li, T., Franco, L.H., Li, X.D., Nair, V.R., Scharn, C.R., Stamm, C.E., Levine, B., Chen, Z.J., Shiloh, M.U., 2015. Cyclic GMP-AMP synthase is an innate immune DNA sensor for mycobacterium tuberculosis. Cell Host Microbe 17, 820–828.

Crow, Y.J., Hayward, B.E., Parmar, R., Robins, P., Leitch, A., Ali, M., Black, D.N., Van Bokhoven, H., Brunner, H.G., Hamel, B.C., Corry, P.C., Cowan, F.M., Frints, S.G., Klepper, J., Livingston, J.H., Lynch, S.A., Massey, R.F., Meritet, J.F., Michaud, J.L., Ponsot, G., Voit, T., Lebon, P., Bonthron, D.T., Jackson, A.P., Barnes, D.E., Lindahl, T., 2006. Mutations in the gene encoding the 3′-5′ DNA exonuclease TREX1 cause Aicardi-Goutieres syndrome at the AGS1 locus. Nat. Genet. 38, 917–920.

Crow, Y.J., Chase, D.S., Lowenstein Schmidt, J., Szynkiewicz, M., Forte, G.M., Gornall, H.L., Oojageer, A., Anderson, B., Pizzino, A., Helman, G., Abdel-Hamid, M.S., Abdel-Salam, G.M., Ackroyd, S., Aeby, A., Agosta, G., Albin, C., Allon-Shalev, S., Arellano, M., Ariaudo, G., Aswani, V., Babul-Hirji, R., Baildam, E.M., Bahi-Buisson, N., Bailey, K.M., Barnerias, C., Barth, M., Battini, R., Beresford, M.W., Bernard, G., Bianchi, M., Billette De Villemeur, T., Blair, E.M., Bloom, M., Burlina, A.B., Carpanelli, M.L., Carvalho, D.R., Castro-Gago, M., Cavallini, A., Cereda, C., Chandler, K.E., Chitayat, D.A., Collins, A.E., Sierra Corcoles, C., Cordeiro, N.J., Crichiutti, G., Dabydeen, L., Dale, R.C., D'arrigo, S., De Goede, C.G., De Laet, C., De Waele, L.M., Denzler, I., Desguerre, I., Devriendt, K., Di Rocco, M., Fahey, M.C., Fazzi, E., Ferrie, C.D., Figueiredo, A., Gener, B., Goizet, C., Gowrinathan, N.R., Gowrishankar, K., Hanrahan, D., Isidor, B., Kara, B., Khan, N., King, M.D., Kirk, E.P., Kumar, R., Lagae, L., Landrieu, P., Lauffer, H., Laugel, V., La Piana, R., Lim, M.J., Lin, J.P., Linnankivi, T., Mackay, M.T., Marom, D.R., Marques Lourenco, C., Mckee, S.A., Moroni, I., Morton, J.E., Moutard, M.L., Murray, K., Nabbout, R., Nampoothiri, S., Nunez-Enamorado, N., Oades, P.J., Olivieri, I., Ostergaard, J.R., Perez-Duenas, B., Prendiville, J.S., Ramesh, V., Rasmussen, M., Regal, L., Ricci, F., Rio, M., Rodriguez, D., et al., 2015. Characterization of human disease phenotypes associated with mutations in TREX1, RNASEH2A, RNASEH2B, RNASEH2C, SAMHD1, ADAR, and IFIH1. Am. J. Med. Genet. A 167A, 296–312.

Daffis, S., Samuel, M.A., Suthar, M.S., Gale Jr., M., Diamond, M.S., 2008. Toll-like receptor 3 has a protective role against west nile virus infection. J. Virol. 82, 10349–10358.

Daffis, S., Szretter, K.J., Schriewer, J., Li, J., Youn, S., Errett, J., Lin, T.Y., Schneller, S., Zust, R., Dong, H., Thiel, V., Sen, G.C., Fensterl, V., Klimstra, W.B., Pierson, T.C., Buller, R.M., Gale Jr., M., Shi, P.Y., Diamond, M.S., 2010. 2′-O methylation of the viral mRNA cap evades host restriction by IFIT family members. Nature 468, 452–456.

Deng, L., Liang, H., Xu, M., Yang, X., Burnette, B., Arina, A., Li, X.D., Mauceri, H., Beckett, M., Darga, T., Huang, X., Gajewski, T.F., Chen, Z.J., Fu, Y.X., Weichselbaum, R.R., 2014. STING-dependent cytosolic DNA sensing promotes radiation-induced type I interferon-dependent antitumor immunity in immunogenic tumors. Immunity 41, 843–852.

Diebold, S.S., Kaisho, T., Hemmi, H., Akira, S., Reis e Sousa, C., 2004. Innate antiviral responses by means of TLR7-mediated recognition of single-stranded RNA. Science 303, 1529–1531.

Doyle, T., Goujon, C., Malim, M.H., 2015. HIV-1 and interferons: who's interfering with whom? Nat. Rev. Microbiol. 13, 403–413.

Esnault, C., Heidmann, O., Delebecque, F., Dewannieux, M., Ribet, D., Hance, A.J., Heidmann, T., Schwartz, O., 2005. APOBEC3G cytidine deaminase inhibits retrotransposition of endogenous retroviruses. Nature 433, 430–433.

Fernandes-Alnemri, T., Yu, J.W., Datta, P., Wu, J., Alnemri, E.S., 2009. AIM2 activates the inflammasome and cell death in response to cytoplasmic DNA. Nature 458, 509–513.

Fernandes-Alnemri, T., Yu, J.W., Juliana, C., Solorzano, L., Kang, S., Wu, J., Datta, P., McCormick, M., Huang, L., McDermott, E., Eisenlohr, L., Landel, C.P., Alnemri, E.S., 2010. The AIM2 inflammasome is critical for innate immunity to Francisella tularensis. Nat. Immunol. 11, 385–393.

Fredericksen, B.L., Keller, B.C., Fornek, J., Katze, M.G., Gale Jr., M., 2008. Establishment and maintenance of the innate antiviral response to west nile virus involves both RIG-I and MDA5 signaling through IPS-1. J. Virol. 82, 609–616.

Gao, D., Wu, J., Wu, Y.T., Du, F., Aroh, C., Yan, N., Sun, L., Chen, Z.J., 2013. Cyclic GMP-AMP synthase is an innate immune sensor of HIV and other retroviruses. Science 341, 903–906.

Gao, D., Li, T., Li, X.D., Chen, X., Li, Q.Z., Wight-Carter, M., Chen, Z.J., 2015. Activation of cyclic GMP-AMP synthase by self-DNA causes autoimmune diseases. Proc. Natl. Acad. Sci. U.S.A. 112, E5699–E5705.

Gehrke, N., Mertens, C., Zillinger, T., Wenzel, J., Bald, T., Zahn, S., Tuting, T., Hartmann, G., Barchet, W., 2013. Oxidative damage of DNA confers resistance to cytosolic nuclease TREX1 degradation and potentiates STING-dependent immune sensing. Immunity 39, 482–495.

Gentili, M., Kowal, J., Tkach, M., Satoh, T., Lahaye, X., Conrad, C., Boyron, M., Lombard, B., Durand, S., Kroemer, G., Loew, D., Dalod, M., Thery, C., Manel, N., 2015. Transmission of innate immune signaling by packaging of cGAMP in viral particles. Science 349, 1232–1236.

Gitlin, L., Barchet, W., Gilfillan, S., Cella, M., Beutler, B., Flavell, R.A., Diamond, M.S., Colonna, M., 2006. Essential role of mda-5 in type I IFN responses to polyriboinosinic: polyribocytidylic acid and encephalomyocarditis picornavirus. Proc. Natl. Acad. Sci. U.S.A. 103, 8459–8464.

Goldstone, D.C., Ennis-Adeniran, V., Hedden, J.J., Groom, H.C., Rice, G.I., Christodoulou, E., Walker, P.A., Kelly, G., Haire, L.F., Yap, M.W., De Carvalho, L.P., Stoye, J.P., Crow, Y.J., Taylor, I.A., Webb, M., 2011. HIV-1 restriction factor SAMHD1 is a deoxynucleoside triphosphate triphosphohydrolase. Nature 480, 379–382.

Goodier, J.L., Cheung, L.E., Kazazian Jr., H.H., 2012. MOV10 RNA helicase is a potent inhibitor of retrotransposition in cells. PLoS Genet. 8, e1002941.

Gray, E.E., Treuting, P.M., Woodward, J.J., Stetson, D.B., 2015. Cutting edge: cGAS is required for lethal autoimmune disease in the Trex1-deficient mouse model of aicardi-goutieres syndrome. J. Immunol. 195, 1939–1943.

Hardarson, H.S., Baker, J.S., Yang, Z., Purevjav, E., Huang, C.H., Alexopoulou, L., Li, N., Flavell, R.A., Bowles, N.E., Vallejo, J.G., 2007. Toll-like receptor 3 is an essential component of the innate stress response in virus-induced cardiac injury. Am. J. Physiol. Heart Circ. Physiol. 292, H251–H258.

Heil, F., Hemmi, H., Hochrein, H., Ampenberger, F., Kirschning, C., Akira, S., Lipford, G., Wagner, H., Bauer, S., 2004. Species-specific recognition of single-stranded RNA via toll-like receptor 7 and 8. Science 303, 1526–1529.

Hemmi, H., Takeuchi, O., Kawai, T., Kaisho, T., Sato, S., Sanjo, H., Matsumoto, M., Hoshino, K., Wagner, H., Takeda, K., Akira, S., 2000. A Toll-like receptor recognizes bacterial DNA. Nature 408, 740–745.

Hemmi, H., Kaisho, T., Takeuchi, O., Sato, S., Sanjo, H., Hoshino, K., Horiuchi, T., Tomizawa, H., Takeda, K., Akira, S., 2002. Small anti-viral compounds activate immune cells via the TLR7 MyD88-dependent signaling pathway. Nat. Immunol. 3, 196–200.

Herzner, A.M., Hagmann, C.A., Goldeck, M., Wolter, S., Kubler, K., Wittmann, S., Gramberg, T., Andreeva, L., Hopfner, K.P., Mertens, C., Zillinger, T., Jin, T., Xiao, T.S., Bartok, E., Coch, C., Ackermann, D., Hornung, V., Ludwig, J., Barchet, W., Hartmann, G., Schlee, M., 2015. Sequence-specific activation of the DNA sensor cGAS by Y-form DNA structures as found in primary HIV-1 cDNA. Nat. Immunol. 16, 1025–1033.

Hochrein, H., Schlatter, B., O'keeffe, M., Wagner, C., Schmitz, F., Schiemann, M., Bauer, S., Suter, M., Wagner, H., 2004. Herpes simplex virus type-1 induces IFN-alpha production via Toll-like receptor 9-dependent and -independent pathways. Proc. Natl. Acad. Sci. U.S.A. 101, 11416–11421.

Hollenbaugh, J.A., Gee, P., Baker, J., Daly, M.B., Amie, S.M., Tate, J., Kasai, N., Kanemura, Y., Kim, D.H., Ward, B.M., Koyanagi, Y., Kim, B., 2013. Host factor SAMHD1 restricts DNA viruses in non-dividing myeloid cells. PLoS Pathog. 9, e1003481.

Hornung, V., Ellegast, J., Kim, S., Brzozka, K., Jung, A., Kato, H., Poeck, H., Akira, S., Conzelmann, K.K., Schlee, M., Endres, S., Hartmann, G., 2006. 5′-Triphosphate RNA is the ligand for RIG-I. Science 314, 994–997.

Hornung, V., Ablasser, A., Charrel-Dennis, M., Bauernfeind, F., Horvath, G., Caffrey, D.R., Latz, E., Fitzgerald, K.A., 2009. AIM2 recognizes cytosolic dsDNA and forms a caspase-1-activating inflammasome with ASC. Nature 458, 514–518.

Hou, F., Sun, L., Zheng, H., Skaug, B., Jiang, Q.X., Chen, Z.J., 2011. MAVS forms functional prion-like aggregates to activate and propagate antiviral innate immune response. Cell 146, 448–461.

Hrecka, K., Hao, C., Gierszewska, M., Swanson, S.K., Kesik-Brodacka, M., Srivastava, S., Florens, L., Washburn, M.P., Skowronski, J., 2011. Vpx relieves inhibition of HIV-1 infection of macrophages mediated by the SAMHD1 protein. Nature 474, 658–661.

Jiang, M., Zhang, S., Yang, Z., Lin, H., Zhu, J., Liu, L., Wang, W., Liu, S., Liu, W., Ma, Y., Zhang, L., Cao, X., 2018. Self-recognition of an inducible host lncRNA by RIG-I feedback restricts innate immune response. Cell 173, 906–919.e13.

Kawai, T., Akira, S., 2011. Toll-like receptors and their crosstalk with other innate receptors in infection and immunity. Immunity 34, 637–650.

Kawane, K., Ohtani, M., Miwa, K., Kizawa, T., Kanbara, Y., Yoshioka, Y., Yoshikawa, H., Nagata, S., 2006. Chronic polyarthritis caused by mammalian DNA that escapes from degradation in macrophages. Nature 443, 998–1002.

Kerur, N., Veettil, M.V., Sharma-Walia, N., Bottero, V., Sadagopan, S., Otageri, P., Chandran, B., 2011. IFI16 acts as a nuclear pathogen sensor to induce the inflammasome in response to Kaposi Sarcoma-associated herpesvirus infection. Cell Host Microbe 9, 363–375.

Kitai, Y., Kawasaki, T., Sueyoshi, T., Kobiyama, K., Ishii, K.J., Zou, J., Akira, S., Matsuda, T., Kawai, T., 2017. DNA-containing exosomes derived from cancer cells treated with topotecan activate a STING-dependent pathway and reinforce antitumor immunity. J. Immunol. 198, 1649–1659.

Kowalinski, E., Lunardi, T., Mccarthy, A.A., Louber, J., Brunel, J., Grigorov, B., Gerlier, D., Cusack, S., 2011. Structural basis for the activation of innate immune pattern-recognition receptor RIG-I by viral RNA. Cell 147, 423–435.

Laguette, N., Sobhian, B., Casartelli, N., Ringeard, M., Chable-Bessia, C., Segeral, E., Yatim, A., Emiliani, S., Schwartz, O., Benkirane, M., 2011. SAMHD1 is the dendritic- and myeloid-cell-specific HIV-1 restriction factor counteracted by Vpx. Nature 474, 654–657.

Lahouassa, H., Daddacha, W., Hofmann, H., Ayinde, D., Logue, E.C., Dragin, L., Bloch, N., Maudet, C., Bertrand, M., Gramberg, T., Pancino, G., Priet, S., Canard, B., Laguette, N., Benkirane, M., Transy, C., Landau, N.R., Kim, B., Margottin-Goguet, F., 2012. SAMHD1 restricts the replication of human immunodeficiency virus type 1 by depleting the intracellular pool of deoxynucleoside triphosphates. Nat. Immunol. 13, 223–228.

Lande, R., Gregorio, J., Facchinetti, V., Chatterjee, B., Wang, Y.H., Homey, B., Cao, W., Su, B., Nestle, F.O., Zal, T., Mellman, I., Schroder, J.M., Liu, Y.J., Gilliet, M., 2007. Plasmacytoid dendritic cells sense self-DNA coupled with antimicrobial peptide. Nature 449, 564–569.

Lande, R., Ganguly, D., Facchinetti, V., Frasca, L., Conrad, C., Gregorio, J., Meller, S., Chamilos, G., Sebasigari, R., Riccieri, V., Bassett, R., Amuro, H., Fukuhara, S., Ito, T., Liu, Y.J., Gilliet, M., 2011. Neutrophils activate plasmacytoid dendritic cells by releasing self-DNA-peptide complexes in systemic lupus erythematosus. Sci. Transl. Med. 3, 73ra19.

Lander, E.S., Linton, L.M., Birren, B., Nusbaum, C., Zody, M.C., Baldwin, J., Devon, K., Dewar, K., Doyle, M., Fitzhugh, W., Funke, R., Gage, D., Harris, K., Heaford, A., Howland, J., Kann, L., Lehoczky, J., Levine, R., Mcewan, P., Mckernan, K., Meldrim, J., Mesirov, J.P., Miranda, C., Morris, W., Naylor, J., Raymond, C., Rosetti, M., Santos, R., Sheridan, A., Sougnez, C., Stange-Thomann, Y., Stojanovic, N., Subramanian, A., Wyman, D., Rogers, J., Sulston, J., Ainscough, R., Beck, S., Bentley, D., Burton, J., Clee, C., Carter, N., Coulson, A., Deadman, R., Deloukas, P., Dunham, A., Dunham, I., Durbin, R., French, L., Grafham, D., Gregory, S., Hubbard, T., Humphray, S., Hunt, A., Jones, M., Lloyd, C., McMurray, A., Matthews, L., Mercer, S., Milne, S., Mullikin, J.C., Mungall, A., Plumb, R., Ross, M., Shownkeen, R., Sims, S., Waterston, R.H., Wilson, R.K., Hillier, L.W., McPherson, J.D., Marra, M.A., Mardis, E.R., Fulton, L.A., Chinwalla, A.T., Pepin, K.H., Gish, W.R., Chissoe, S.L., Wendl, M.C., Delehaunty, K.D., Miner, T.L., Delehaunty, A., Kramer, J.B., Cook, L.L., Fulton, R.S., Johnson, D.L., Minx, P.J., Clifton, S.W., Hawkins, T., Branscomb, E., Predki, P., Richardson, P., Wenning, S., Slezak, T., Doggett, N., Cheng, J.F., Olsen, A., Lucas, S., Elkin, C., Uberbacher, E., Frazier, M., et al., 2001. Initial sequencing and analysis of the human genome. Nature 409, 860–921.

Lee, M.S., Kim, B., Oh, G.T., Kim, Y.J., 2013. OASL1 inhibits translation of the type I interferon-regulating transcription factor IRF7. Nat. Immunol. 14, 346–355.

Lehmann, S.M., Kruger, C., Park, B., Derkow, K., Rosenberger, K., Baumgart, J., Trimbuch, T., Eom, G., Hinz, M., Kaul, D., Habbel, P., Kalin, R., Franzoni, E., Rybak, A., Nguyen, D., Veh, R., Ninnemann, O., Peters, O., Nitsch, R., Heppner, F.L., Golenbock, D., Schott, E., Ploegh, H.L., Wulczyn, F.G., Lehnardt, S., 2012. An unconventional role for miRNA: let-7 activates Toll-like receptor 7 and causes neurodegeneration. Nat. Neurosci. 15, 827–835.

Li, X., Shu, C., Yi, G., Chaton, C.T., Shelton, C.L., Diao, J., Zuo, X., Kao, C.C., Herr, A.B., Li, P., 2013. Cyclic GMP-AMP synthase is activated by double-stranded DNA-induced oligomerization. Immunity 39, 1019–1031.

Li, Z., Chao, T.C., Chang, K.Y., Lin, N., Patil, V.S., Shimizu, C., Head, S.R., Burns, J.C., Rana, T.M., 2014. The long noncoding RNA THRIL regulates TNFalpha expression through its interaction with hnRNPL. Proc. Natl. Acad. Sci. U.S.A. 111, 1002–1007.

Li, P., Fan, J.B., Gao, Y., Zhang, M., Zhang, L., Yang, N., Zhao, X., 2016. miR-135b-5p inhibits LPS-induced TNFalpha production via silencing AMPK phosphatase Ppm1e. Oncotarget 7, 77978–77986.

Li, X., Liu, C.X., Xue, W., Zhang, Y., Jiang, S., Yin, Q.F., Wei, J., Yao, R.W., Yang, L., Chen, L.L., 2017. Coordinated circRNA biogenesis and function with NF90/NF110 in viral infection. Mol. Cell 67, 214–227.e7.

Liddicoat, B.J., Piskol, R., Chalk, A.M., Ramaswami, G., Higuchi, M., Hartner, J.C., Li, J.B., Seeburg, P.H., Walkley, C.R., 2015. RNA editing by ADAR1 prevents MDA5 sensing of endogenous dsRNA as nonself. Science 349, 1115–1120.

Liu, L., Botos, I., Wang, Y., Leonard, J.N., Shiloach, J., Segal, D.M., Davies, D.R., 2008. Structural basis of toll-like receptor 3 signaling with double-stranded RNA. Science 320, 379–381.

Loo, Y.M., Fornek, J., Crochet, N., Bajwa, G., Perwitasari, O., Martinez-Sobrido, L., Akira, S., Gill, M.A., Garcia-Sastre, A., Katze, M.G., Gale Jr., M., 2008. Distinct RIG-I and MDA5 signaling by RNA viruses in innate immunity. J. Virol. 82, 335–345.

Lund, J., Sato, A., Akira, S., Medzhitov, R., Iwasaki, A., 2003. Toll-like receptor 9-mediated recognition of herpes simplex virus-2 by plasmacytoid dendritic cells. J. Exp. Med. 198, 513–520.

Lund, J.M., Alexopoulou, L., Sato, A., Karow, M., Adams, N.C., Gale, N.W., Iwasaki, A., Flavell, R.A., 2004. Recognition of single-stranded RNA viruses by toll-like receptor 7. Proc. Natl. Acad. Sci. U.S.A. 101, 5598–5603.

Mackenzie, K.J., Carroll, P., Lettice, L., Tarnauskaite, Z., Reddy, K., Dix, F., Revuelta, A., Abbondati, E., Rigby, R.E., Rabe, B., Kilanowski, F., Grimes, G., Fluteau, A., Devenney, P.S., Hill, R.E., Reijns, M.A., Jackson, A.P., 2016. Ribonuclease H2 mutations induce a cGAS/STING-dependent innate immune response. EMBO J. 35, 831–844.

Mancuso, G., Gambuzza, M., Midiri, A., Biondo, C., Papasergi, S., Akira, S., Teti, G., Beninati, C., 2009. Bacterial recognition by TLR7 in the lysosomes of conventional dendritic cells. Nat. Immunol. 10, 587–594.

Mannion, N.M., Greenwood, S.M., Young, R., Cox, S., Brindle, J., Read, D., Nellaker, C., Vesely, C., Ponting, C.P., McLaughlin, P.J., Jantsch, M.F., Dorin, J., Adams, I.R., Scadden, A.D., Ohman, M., Keegan, L.P., O'connell, M.A., 2014. The RNA-editing enzyme ADAR1 controls innate immune responses to RNA. Cell Rep. 9, 1482–1494.

Maruzuru, Y., Ichinohe, T., Sato, R., Miyake, K., Okano, T., Suzuki, T., Koshiba, T., Koyanagi, N., Tsuda, S., Watanabe, M., Arii, J., Kato, A., Kawaguchi, Y., 2018. Herpes simplex virus 1 VP22 inhibits AIM2-dependent inflammasome activation to enable efficient viral replication. Cell Host Microbe 23, 254–265.e7.

Monroe, K.M., Yang, Z., Johnson, J.R., Geng, X., Doitsh, G., Krogan, N.J., Greene, W.C., 2014. IFI16 DNA sensor is required for death of lymphoid CD4 T cells abortively infected with HIV. Science 343, 428–432.

Morrone, S.R., Wang, T., Constantoulakis, L.M., Hooy, R.M., Delannoy, M.J., Sohn, J., 2014. Cooperative assembly of IFI16 filaments on dsDNA provides insights into host defense strategy. Proc. Natl. Acad. Sci. U.S.A. 111, E62–E71.

Nabet, B.Y., Qiu, Y., Shabason, J.E., Wu, T.J., Yoon, T., Kim, B.C., Benci, J.L., Demichele, A.M., Tchou, J., Marcotrigiano, J., Minn, A.J., 2017. Exosome RNA unshielding couples stromal activation to pattern recognition receptor signaling in cancer. Cell 170, 352–366.e13.

Ohto, U., Shibata, T., Tanji, H., Ishida, H., Krayukhina, E., Uchiyama, S., Miyake, K., Shimizu, T., 2015. Structural basis of CpG and inhibitory DNA recognition by toll-like receptor 9. Nature 520, 702–705.

Ohto, U., Ishida, H., Shibata, T., Sato, R., Miyake, K., Shimizu, T., 2018. Toll-like receptor 9 contains two DNA binding sites that function cooperatively to promote receptor dimerization and activation. Immunity 48, 649–658.e4.

Oldenburg, M., Kruger, A., Ferstl, R., Kaufmann, A., Nees, G., Sigmund, A., Bathke, B., Lauterbach, H., Suter, M., Dreher, S., Koedel, U., Akira, S., Kawai, T., Buer, J., Wagner, H., Bauer, S., Hochrein, H., Kirschning, C.J., 2012. TLR13 recognizes bacterial 23S rRNA devoid of erythromycin resistance-forming modification. Science 337, 1111–1115.

Peisley, A., Wu, B., Yao, H., Walz, T., Hur, S., 2013. RIG-I forms signaling-competent filaments in an ATP-dependent, ubiquitin-independent manner. Mol. Cell 51, 573–583.

Pestal, K., Funk, C.C., Snyder, J.M., Price, N.D., Treuting, P.M., Stetson, D.B., 2015. Isoforms of RNA-editing enzyme ADAR1 independently control nucleic acid sensor MDA5-driven autoimmunity and multi-organ development. Immunity 43, 933–944.

Pohar, J., Kuznik Krajnik, A., Jerala, R., Bencina, M., 2015a. Minimal sequence requirements for oligodeoxyribonucleotides activating human TLR9. J. Immunol. 194, 3901–3908.

Pohar, J., Lainscek, D., Fukui, R., Yamamoto, C., Miyake, K., Jerala, R., Bencina, M., 2015b. Species-specific minimal sequence motif for oligodeoxyribonucleotides activating mouse TLR9. J. Immunol. 195, 4396–4405.

Pokatayev, V., Hasin, N., Chon, H., Cerritelli, S.M., Sakhuja, K., Ward, J.M., Morris, H.D., Yan, N., Crouch, R.J., 2016. RNase H2 catalytic core aicardi-goutieres syndrome-related mutant invokes cGAS-STING innate immune-sensing pathway in mice. J. Exp. Med. 213, 329–336.

Ramaswami, G., Lin, W., Piskol, R., Tan, M.H., Davis, C., Li, J.B., 2012. Accurate identification of human Alu and non-Alu RNA editing sites. Nat. Methods 9, 579–581.

Rasmussen, S.B., Jensen, S.B., Nielsen, C., Quartin, E., Kato, H., Chen, Z.J., Silverman, R.H., Akira, S., Paludan, S.R., 2009. Herpes simplex virus infection is sensed by both toll-like receptors and retinoic acid-inducible gene-like receptors, which synergize to induce type I interferon production. J. Gen. Virol. 90, 74–78.

Rathinam, V.A., Fitzgerald, K.A., 2011. Cytosolic surveillance and antiviral immunity. Curr. Opin. Virol. 1, 455–462.

Rehwinkel, J., Tan, C.P., Goubau, D., Schulz, O., Pichlmair, A., Bier, K., Robb, N., Vreede, F., Barclay, W., Fodor, E., Reis e Sousa, C., 2010. RIG-I detects viral genomic RNA during negative-strand RNA virus infection. Cell 140, 397–408.

Rehwinkel, J., Maelfait, J., Bridgeman, A., Rigby, R., Hayward, B., Liberatore, R.A., Bieniasz, P.D., Towers, G.J., Moita, L.F., Crow, Y.J., Bonthron, D.T., Reis e Sousa, C., 2013. SAMHD1-dependent retroviral control and escape in mice. EMBO J. 32, 2454–2462.

Reinert, L.S., Lopusna, K., Winther, H., Sun, C., Thomsen, M.K., Nandakumar, R., Mogensen, T.H., Meyer, M., Vaegter, C., Nyengaard, J.R., Fitzgerald, K.A., Paludan, S.R., 2016. Sensing of HSV-1 by the cGAS-STING pathway in microglia orchestrates antiviral defence in the CNS. Nat. Commun. 7, 13348.

Reinholz, M., Kawakami, Y., Salzer, S., Kreuter, A., Dombrowski, Y., Koglin, S., Kresse, S., Ruzicka, T., Schauber, J., 2013. HPV16 activates the AIM2 inflammasome in keratinocytes. Arch. Dermatol. Res. 305, 723–732.

Rice, G.I., Bond, J., Asipu, A., Brunette, R.L., Manfield, I.W., Carr, I.M., Fuller, J.C., Jackson, R.M., Lamb, T., Briggs, T.A., Ali, M., Gornall, H., Couthard, L.R., Aeby, A., Attard-Montalto, S.P., Bertini, E., Bodemer, C., Brockmann, K., Brueton, L.A., Corry, P.C., Desguerre, I., Fazzi, E., Cazorla, A.G., Gener, B., Hamel, B.C., Heiberg, A., Hunter, M., Van der Knaap, M.S., Kumar, R., Lagae, L., Landrieu, P.G., Lourenco, C.M., Marom, D., Mcdermott, M.F., Van Der Merwe, W., Orcesi, S., Prendiville, J.S., Rasmussen, M., Shalev, S.A., Soler, D.M., Shinawi, M., Spiegel, R., Tan, T.Y., Vanderver, A., Wakeling, E.L., Wassmer, E., Whittaker, E., Lebon, P., Stetson, D.B., Bonthron, D.T., Crow, Y.J., 2009. Mutations involved in aicardi-goutieres syndrome implicate SAMHD1 as regulator of the innate immune response. Nat. Genet. 41, 829–832.

Rice, G.I., Kasher, P.R., Forte, G.M., Mannion, N.M., Greenwood, S.M., Szynkiewicz, M., Dickerson, J.E., Bhaskar, S.S., Zampini, M., Briggs, T.A., Jenkinson, E.M., Bacino, C.A., Battini, R., Bertini, E., Brogan, P.A., Brueton, L.A., Carpanelli, M., De Laet, C., De Lonlay, P., Del Toro, M., Desguerre, I., Fazzi, E., Garcia-Cazorla, A., Heiberg, A., Kawaguchi, M., Kumar, R., Lin, J.P., Lourenco, C.M., Male, A.M., Marques Jr., W., Mignot, C., Olivieri, I., Orcesi, S., Prabhakar, P., Rasmussen, M., Robinson, R.A., Rozenberg, F., Schmidt, J.L., Steindl, K., Tan, T.Y., Van Der Merwe, W.G., Vanderver, A., Vassallo, G., Wakeling, E.L., Wassmer, E., Whittaker, E., Livingston, J.H., Lebon, P., Suzuki, T., McLaughlin, P.J., Keegan, L.P., O'connell, M.A., Lovell, S.C., Crow, Y.J., 2012. Mutations in ADAR1 cause aicardi-goutieres syndrome associated with a type I interferon signature. Nat. Genet. 44, 1243–1248.

Roberts, T.L., Idris, A., Dunn, J.A., Kelly, G.M., Burnton, C.M., Hodgson, S., Hardy, L.L., Garceau, V., Sweet, M.J., Ross, I.L., Hume, D.A., Stacey, K.J., 2009. HIN-200 proteins regulate caspase activation in response to foreign cytoplasmic DNA. Science 323, 1057–1060.

Rodero, M.P., Tesser, A., Bartok, E., Rice, G.I., Della Mina, E., Depp, M., Beitz, B., Bondet, V., Cagnard, N., Duffy, D., Dussiot, M., Fremond, M.L., Gattorno, M., Guillem, F., Kitabayashi, N., Porcheray, F., Rieux-Laucat, F., Seabra, L., Uggenti, C., Volpi, S., Zeef, L.A.H., Alyanakian, M.A., Beltrand, J., Bianco, A.M., Boddaert, N., Brouzes, C., Candon, S., Caorsi, R., Charbit, M., Fabre, M., Faletra, F., Girard, M., Harroche, A., Hartmann, E., Lasne, D., Marcuzzi, A., Neven, B., Nitschke, P., Pascreau, T., Pastore, S., Picard, C., Picco, P., Piscianz, E., Polak, M., Quartier, P., Rabant, M., Stocco, G., Taddio, A., Uettwiller, F., Valencic, E., Vozzi, D., Hartmann, G., Barchet, W., Hermine, O., Bader-Meunier, B., Tommasini, A., Crow, Y.J., 2017. Type I interferon-mediated autoinflammation due to DNase II deficiency. Nat. Commun. 8, 2176.

Roulois, D., Loo Yau, H., Singhania, R., Wang, Y., Danesh, A., Shen, S.Y., Han, H., Liang, G., Jones, P.A., Pugh, T.J., O'brien, C., De Carvalho, D.D., 2015. DNA-demethylating agents target colorectal cancer cells by inducing viral mimicry by endogenous transcripts. Cell 162, 961–973.

Schlee, M., Roth, A., Hornung, V., Hagmann, C.A., Wimmenauer, V., Barchet, W., Coch, C., Janke, M., Mihailovic, A., Wardle, G., Juranek, S., Kato, H., Kawai, T., Poeck, H., Fitzgerald, K.A., Takeuchi, O., Akira, S., Tuschl, T., Latz, E., Ludwig, J., Hartmann, G., 2009. Recognition of 5′ triphosphate by RIG-I helicase requires short blunt double-stranded RNA as contained in panhandle of negative-strand virus. Immunity 31, 25–34.

Schnieke, A., Stuhlmann, H., Harbers, K., Chumakov, I., Jaenisch, R., 1983. Endogenous Moloney leukemia virus in nonviremic Mov substrains of mice carries defects in the proviral genome. J. Virol. 45, 505–513.

Schoggins, J.W., Wilson, S.J., Panis, M., Murphy, M.Y., Jones, C.T., Bieniasz, P., Rice, C.M., 2011. A diverse range of gene products are effectors of the type I interferon antiviral response. Nature 472, 481–485.

Schulz, O., Diebold, S.S., Chen, M., Naslund, T.I., Nolte, M.A., Alexopoulou, L., Azuma, Y.T., Flavell, R.A., Liljestrom, P., Reis e Sousa, C., 2005. Toll-like receptor 3 promotes cross-priming to virus-infected cells. Nature 433, 887–892.

Sharova, N., Wu, Y., Zhu, X., Stranska, R., Kaushik, R., Sharkey, M., Stevenson, M., 2008. Primate lentiviral Vpx commandeers DDB1 to counteract a macrophage restriction. PLoS Pathog. 4, e1000057.

Sheedy, F.J., Palsson-McDermott, E., Hennessy, E.J., Martin, C., O'leary, J.J., Ruan, Q., Johnson, D.S., Chen, Y., O'neill, L.A., 2010. Negative regulation of TLR4 via targeting of the proinflammatory tumor suppressor PDCD4 by the microRNA miR-21. Nat. Immunol. 11, 141–147.

Stavrou, S., Blouch, K., Kotla, S., Bass, A., Ross, S.R., 2015. Nucleic acid recognition orchestrates the anti-viral response to retroviruses. Cell Host Microbe 17, 478–488.

Stetson, D.B., ko, J.S., Heidmann, T., Medzhitov, R., 2008. Trex1 prevents cell-intrinsic initiation of autoimmunity. Cell 134, 587–598.

Sun, L., Wu, J., Du, F., Chen, X., Chen, Z.J., 2013. Cyclic GMP-AMP synthase is a cytosolic DNA sensor that activates the type I interferon pathway. Science 339, 786–791.

Tabeta, K., Georgel, P., Janssen, E., Du, X., Hoebe, K., Crozat, K., Mudd, S., Shamel, L., Sovath, S., Goode, J., Alexopoulou, L., Flavell, R.A., Beutler, B., 2004. Toll-like receptors 9 and 3 as essential components of innate immune defense against mouse cytomegalovirus infection. Proc. Natl. Acad. Sci. U.S.A. 101, 3516–3521.

Takemura, N., Kawasaki, T., Kunisawa, J., Sato, S., Lamichhane, A., Kobiyama, K., Aoshi, T., Ito, J., Mizuguchi, K., Karuppuchamy, T., Matsunaga, K., Miyatake, S., Mori, N., Tsujimura, T., Satoh, T., Kumagai, Y., Kawai, T., Standley, D.M., Ishii, K.J., Kiyono, H., Akira, S., Uematsu, S., 2014. Blockade of TLR3 protects mice from lethal radiation-induced gastrointestinal syndrome. Nat. Commun. 5, 3492.

Takemura, N., Kurashima, Y., Mori, Y., Okada, K., Ogino, T., Osawa, H., Matsuno, H., Aayam, L., Kaneto, S., Park, E.J., Sato, S., Matsunaga, K., Tamura, Y., Ouchi, Y., Kumagai, Y., Kobayashi, D., Suzuki, Y., Yoshioka, Y., Nishimura, J., Mori, M., Ishii, K.J., Rothenberg, M.E., Kiyono, H., Akira, S., Uematsu, S., 2018. Eosinophil depletion suppresses radiation-induced small intestinal fibrosis. Sci. Transl. Med. 10, eaan0333.

Tang, B., Li, X., Ren, Y., Wang, J., Xu, D., Hang, Y., Zhou, T., Li, F., Wang, L., 2017. MicroRNA-29a regulates lipopolysaccharide (LPS)-induced inflammatory responses in murine macrophages through the Akt1/NF-kappaB pathway. Exp. Cell Res. 360, 74–80.

Tanji, H., Ohto, U., Shibata, T., Miyake, K., Shimizu, T., 2013. Structural reorganization of the Toll-like receptor 8 dimer induced by agonistic ligands. Science 339, 1426–1429.

Wang, Q., Khillan, J., Gadue, P., Nishikura, K., 2000. Requirement of the RNA editing deaminase ADAR1 gene for embryonic erythropoiesis. Science 290, 1765–1768.

Wang, X., Han, Y., Dang, Y., Fu, W., Zhou, T., Ptak, R.G., Zheng, Y.H., 2010. Moloney leukemia virus 10 (MOV10) protein inhibits retrovirus replication. J. Biol. Chem. 285, 14346–14355.

Wassermann, R., Gulen, M.F., Sala, C., Perin, S.G., Lou, Y., Rybniker, J., Schmid-Burgk, J.L., Schmidt, T., Hornung, V., Cole, S.T., Ablasser, A., 2015. Mycobacterium tuberculosis differentially activates cGAS- and inflammasome-dependent intracellular immune responses through ESX-1. Cell Host Microbe 17, 799–810.

Waterston, R.H., Lindblad-Toh, K., Birney, E., Rogers, J., Abril, J.F., Agarwal, P., Agarwala, R., Ainscough, R., Alexandersson, M., An, P., Antonarakis, S.E., Attwood, J., Baertsch, R., Bailey, J., Barlow, K., Beck, S., Berry, E., Birren, B.,

Bloom, T., Bork, P., Botcherby, M., Bray, N., Brent, M.R., Brown, D.G., Brown, S.D., Bult, C., Burton, J., Butler, J., Campbell, R.D., Carninci, P., Cawley, S., Chiaromonte, F., Chinwalla, A.T., Church, D.M., Clamp, M., Clee, C., Collins, F.S., Cook, L.L., Copley, R.R., Coulson, A., Couronne, O., Cuff, J., Curwen, V., Cutts, T., Daly, M., David, R., Davies, J., Delehaunty, K.D., Deri, J., Dermitzakis, E.T., Dewey, C., Dickens, N.J., Diekhans, M., Dodge, S., Dubchak, I., Dunn, D.M., Eddy, S.R., Elnitski, L., Emes, R.D., Eswara, P., Eyras, E., Felsenfeld, A., Fewell, G.A., Flicek, P., Foley, K., Frankel, W.N., Fulton, L.A., Fulton, R.S., Furey, T.S., Gage, D., Gibbs, R.A., Glusman, G., Gnerre, S., Goldman, N., Goodstadt, L., Grafham, D., Graves, T.A., Green, E.D., Gregory, S., Guigo, R., Guyer, M., Hardison, R.C., Haussler, D., Hayashizaki, Y., Hillier, L.W., Hinrichs, A., Hlavina, W., Holzer, T., Hsu, F., Hua, A., Hubbard, T., Hunt, A., Jackson, I., Jaffe, D.B., Johnson, L.S., Jones, M., Jones, T.A., Joy, A., Kamal, M., Karlsson, E.K., et al., 2002. Initial sequencing and comparative analysis of the mouse genome. Nature 420, 520–562.

Watson, R.O., Bell, S.L., Macduff, D.A., Kimmey, J.M., Diner, E.J., Olivas, J., Vance, R.E., Stallings, C.L., Virgin, H.W., Cox, J.S., 2015. The cytosolic sensor cGAS detects mycobacterium tuberculosis DNA to induce type I interferons and activate autophagy. Cell Host Microbe 17, 811–819.

Werner, M., Purta, E., Kaminska, K.H., Cymerman, I.A., Campbell, D.A., Mittra, B., Zamudio, J.R., Sturm, N.R., Jaworski, J., Bujnicki, J.M., 2011. 2′-O-ribose methylation of cap2 in human: function and evolution in a horizontally mobile family. Nucleic Acids Res. 39, 4756–4768.

West, A.P., Khoury-Hanold, W., Staron, M., Tal, M.C., Pineda, C.M., Lang, S.M., Bestwick, M., Duguay, B.A., Raimundo, N., Macduff, D.A., Kaech, S.M., Smiley, J.R., Means, R.E., Iwasaki, A., Shadel, G.S., 2015. Mitochondrial DNA stress primes the antiviral innate immune response. Nature 520, 553–557.

Woo, S.R., Fuertes, M.B., Corrales, L., Spranger, S., Furdyna, M.J., Leung, M.Y., Duggan, R., Wang, Y., Barber, G.N., Fitzgerald, K.A., Alegre, M.L., Gajewski, T.F., 2014. STING-dependent cytosolic DNA sensing mediates innate immune recognition of immunogenic tumors. Immunity 41, 830–842.

Wu, J., Chen, Z.J., 2014. Innate immune sensing and signaling of cytosolic nucleic acids. Annu. Rev. Immunol. 32, 461–488.

Wu, B., Peisley, A., Richards, C., Yao, H., Zeng, X., Lin, C., Chu, F., Walz, T., Hur, S., 2013. Structural basis for dsRNA recognition, filament formation, and antiviral signal activation by MDA5. Cell 152, 276–289.

Wu, B., Peisley, A., Tetrault, D., Li, Z., Egelman, E.H., Magor, K.E., Walz, T., Penczek, P.A., Hur, S., 2014. Molecular imprinting as a signal-activation mechanism of the viral RNA sensor RIG-I. Mol. Cell 55, 511–523.

Xu, S., Ducroux, A., Ponnurangam, A., Vieyres, G., Franz, S., Musken, M., Zillinger, T., Malassa, A., Ewald, E., Hornung, V., Barchet, W., Haussler, S., Pietschmann, T., Goffinet, C., 2016. cGAS-mediated innate immunity spreads intercellularly through HIV-1 env-induced membrane fusion sites. Cell Host Microbe 20, 443–457.

Yanai, H., Ban, T., Wang, Z., Choi, M.K., Kawamura, T., Negishi, H., Nakasato, M., Lu, Y., Hangai, S., Koshiba, R., Savitsky, D., Ronfani, L., Akira, S., Bianchi, M.E., Honda, K., Tamura, T., Kodama, T., Taniguchi, T., 2009. HMGB proteins function as universal sentinels for nucleic-acid-mediated innate immune responses. Nature 462, 99–103.

Yoneyama, M., Fujita, T., 2008. Structural mechanism of RNA recognition by the RIG-I-like receptors. Immunity 29, 178–181.

Zeng, C., Wu, A., Wang, Y., Xu, S., Tang, Y., Jin, X., Wang, S., Qin, L., Sun, Y., Fan, C., Snijder, E.J., Neuman, B.W., Chen, Y., Ahola, T., Guo, D., 2016. Identification and characterization of a ribose 2′-O-methyltransferase encoded by the ronivirus branch of nidovirales. J. Virol. 90, 6675–6685.

Zhang, F., Romano, P.R., Nagamura-Inoue, T., Tian, B., Dever, T.E., Mathews, M.B., Ozato, K., Hinnebusch, A.G., 2001. Binding of double-stranded RNA to protein kinase PKR is required for dimerization and promotes critical autophosphorylation events in the activation loop. J. Biol. Chem. 276, 24946–24958.

Zhang, S.Y., Jouanguy, E., Ugolini, S., Smahi, A., Elain, G., Romero, P., Segal, D., Sancho-Shimizu, V., Lorenzo, L., Puel, A., Picard, C., Chapgier, A., Plancoulaine, S., Titeux, M., Cognet, C., Von Bernuth, H., Ku, C.L., Casrouge, A., Zhang, X.X., Barreiro, L., Leonard, J., Hamilton, C., Lebon, P., Heron, B., Vallee, L., Quintana-Murci, L., Hovnanian, A., Rozenberg, F., Vivier, E., Geissmann, F., Tardieu, M., Abel, L., Casanova, J.L., 2007. TLR3 deficiency in patients with herpes simplex encephalitis. Science 317, 1522–1527.

Zhang, X., Shi, H., Wu, J., Sun, L., Chen, C., Chen, Z.J., 2013. Cyclic GMP-AMP containing mixed phosphodiester linkages is an endogenous high-affinity ligand for STING. Mol. Cell 51, 226–235.

Zhang, L., Huang, C., Guo, Y., Gou, X., Hinsdale, M., Lloyd, P., Liu, L., 2015. MicroRNA-26b modulates the NF-kappaB pathway in alveolar macrophages by regulating PTEN. J. Immunol. 195, 5404–5414.

Zhang, J., Huang, F., Tan, L., Bai, C., Chen, B., Liu, J., Liang, J., Liu, C., Zhang, S., Lu, G., Chen, Y., Zhang, H., 2016a. Host protein moloney leukemia virus 10 (MOV10) acts as a restriction factor of influenza a virus by inhibiting the nuclear import of the viral nucleoprotein. J. Virol. 90, 3966–3980.

Zhang, Z., Ohto, U., Shibata, T., Krayukhina, E., Taoka, M., Yamauchi, Y., Tanji, H., Isobe, T., Uchiyama, S., Miyake, K., Shimizu, T., 2016b. Structural analysis reveals that toll-like receptor 7 is a dual receptor for guanosine and single-stranded RNA. Immunity 45, 737–748.

Zhu, J., Huang, X., Yang, Y., 2007. Innate immune response to adenoviral vectors is mediated by both Toll-like receptor-dependent and -independent pathways. J. Virol. 81, 3170–3180.

Zhu, J., Zhang, Y., Ghosh, A., Cuevas, R.A., Forero, A., Dhar, J., Ibsen, M.S., Schmid-Burgk, J.L., Schmidt, T., Ganapathiraju, M.K., Fujita, T., Hartmann, R., Barik, S., Hornung, V., Coyne, C.B., Sarkar, S.N., 2014. Antiviral activity of human OASL protein is mediated by enhancing signaling of the RIG-I RNA sensor. Immunity 40, 936–948.

Zust, R., Cervantes-Barragan, L., Habjan, M., Maier, R., Neuman, B.W., Ziebuhr, J., Szretter, K.J., Baker, S.C., Barchet, W., Diamond, M.S., Siddell, S.G., Ludewig, B., Thiel, V., 2011. Ribose 2′-*O*-methylation provides a molecular signature for the distinction of self and non-self mRNA dependent on the RNA sensor Mda5. Nat. Immunol. 12, 137–143.

Zust, R., Li, S.H., Xie, X., Velumani, S., Chng, M., Toh, Y.X., Zou, J., Dong, H., Shan, C., Pang, J., Qin, C.F., Newell, E.W., Shi, P.Y., Fink, K., 2018. Characterization of a candidate tetravalent vaccine based on 2′-*O*-methyltransferase mutants. PLoS One 13, e0189262.

CHAPTER TWO

Intracellular RNA Sensing in Mammalian Cells: Role in Stress Response and Cancer Therapies

Nikolai N. Khodarev[1]
Department of Radiation and Cellular Oncology and Ludwig Center for Metastasis Research, The University of Chicago, Chicago, IL, United States
[1]Corresponding author: e-mail address: n-khodarev@uchicago.edu

Contents

International Review of Cell and Molecular Biology, Volume 344
ISSN 1937-6448
https://doi.org/10.1016/bs.ircmb.2018.08.005

Abstract

DNA damage has been considered the primary action of ionizing radiation (IR) in normal and tumor cells resulting in cell autonomous death. However, recent findings have elucidated novel local and systemic effects of IR mediated by the induction of Type I interferons (IFN) and activation of adaptive immune responses. These responses are initiated by DNA- and RNA-dependent activation of pattern recognition receptors (PRR) which comprise an innate immunity system responsible for detection of exogenous pathogens, mounting of an anti-viral response, and activation of adaptive immunity. Activation of cytoplasmic RNA sensors by endogenous dsRNAs has emerged as an important mechanism of tumor cell-intrinsic Type I IFN induction in the response to genotoxic therapies. The fact that an effective anti-tumor effect of IR and chemotherapy response is dependent on the RNA sensing pathways indicates that tumor cells are equipped by functional innate immunity system which recognizes genotoxic stress insult as an intrusion of exogenous RNA pathogens. This indicates that DNA damage leads to the RNA stress, which is recognized as danger-associated molecular patterns (DAMPs) presented by aberrant RNA molecules interacting with cytoplasmic RNA sensors. In the current review, we will discuss the mechanistic bases for RNA-dependent innate responses in tumor cells, RNA ligands recognized as DAMPs, and the potential implications of these findings to improve cancer therapies.

ABBREVIATIONS

ADAR RNA adenosine deaminase
APC antigen-presenting cells
ASC apoptosis-associated speck-like protein containing CARD
CARD caspase activation and recruitment domain
cGAMP cyclic GMP-AMP
cGAS cyclic GMP-AMP synthase (MB21D1)
CKII casein kinase II
CMRT1/2 cap methyltransferase 1/2
CTLA4 cytotoxic T lymphocyte-associated 4 (CD152)
DAMP danger-associated molecular patterns
DC dendritic cells
dsRNAs double-stranded RNA
ERV endogenous retroviruses
GIS gastro-intestinal syndrome
HDAC histone deacetylase
ICT immune checkpoint therapy
IFN interferon
IR ionizing radiation
IRAK interleukin-1 receptor-associated kinase
IRF interferon regulatory factor

ISG	interferon-stimulated genes
LGP2	laboratory of genetics and physiology 2 (RNA helicase DHX58)
LINEs	long interspersed nuclear elements
LRR	leucine-rich repeats
LTR	long terminal repeats
MAPK	mitogen-activated protein kinase
MAVS	mitochondrial anti-viral signaling protein (IPS1 or VISA)
MDA5	melanoma differentiation-associated protein 5 (IFIH1 or AGS7)
MLKL	mixed lineage kinase domain like pseudokinase
MyD88	myeloid differentiation marker 88
NACHT	domain present in NAIP, CIITA, HET-E, and TP1
ndsRNAs	natural double-stranded small RNAs
NLR	NOD-like receptors
NLRP3	nucleotide-binding domain and leucine-rich repeat containing family pyrin 3
NTase	nucleotidyltransferase
OAS	2′,5′-oligoadenylate synthetases
OLR	OAS-like receptors
PAMP	pathogen-associated molecular patterns
PB	processing bodies
PD1	programmed cell death 1 PDCD1 (CD279)
PD-L1	programmed cell death 1 ligand 1 (B7H1)
PKR	RNA-dependent protein kinase
PML-NB	PML nuclear body (also known as ND10)
PP1	protein phosphatase 1
PRR	pattern recognition receptors
RIG-I	retinoic acid inducible gene 1 (RNA helicase DDX58)
RIPK	receptor-interacting serine/threonine protein kinases
RLR	RIG-I-like receptors
RNase L	ribonuclease L
RNMT	RNA guanine-7-methyltransferase
ROS	reactive oxygen species
SG	stress granules
SINEs	short interspersed nuclear elements
snoRNAs	small nucleolar RNAs
snRNAs	small nuclear RNAs
ssRNA	single-stranded RNA
STAT	signal transducer and activator of transcription
STING	stimulator of interferon genes (transmembrane protein 173; TMEM173)
TBI	total body irradiation
TBK1	tank-binding kinase 1
TIM	TRAF-interacting motif
TIR	Toll/IL-1R domain
TLR	Toll-like receptors
TME	tumor microenvironment
TRAF	TNF receptor-associated factor
TRIF	TIR-domain containing protein inducing IFN-beta
XIAP	X-linked inhibitor of apoptosis

1. INTRODUCTION

RNA sensors are presented by several distinct families of cytosolic pattern recognition receptors (PRRs) that are activated through recognition of exogenous (non-self) viral and microbial RNA molecules (Akira et al., 2006; Roers et al., 2016; Schlee and Hartmann, 2016). Binding of these RNA ligands to their cognate receptors leads to complex downstream events which converge on interferon response and inflammatory cytokine production. IFN response leads to the formation of an anti-viral state in infected cells, predominantly associated with induction of various interferon-stimulated genes (ISGs) involved in anti-viral defense and elimination of infected cells (Sadler and Williams, 2008; Samarajiwa et al., 2009; Samuel, 2001). As well, activation of Type I IFN signaling and cytokine production leads to the induction of adaptive immune responses through activation of antigen-presenting cells, cross-priming of CD8+ T cells and recruitment of immune cells to the regions of primary pathogen invasion (Corrales et al., 2017; Fuertes et al., 2013; Yang et al., 2014). Different types of RNA sensors and enzymes involved in controlling the pool of endogenous RNAs are associated with distinct types of autoimmune diseases indicating that alterations in regulation of RNA sensors and their interactions with endogenous self-RNAs can lead to chronic sterile inflammation (Kato and Fujita, 2015; Lassig and Hopfner, 2017; Roers et al., 2016; Wu and Chen, 2014). As well, certain types of genotoxic stress (ionizing radiation (IR), chemotherapy drugs, epigenetic modifiers, and reactive oxygen species (ROS)) can lead to the abnormal accumulation of some endogenous RNA and their interactions with cytoplasmic RNA sensors (Chiappinelli et al., 2015; Engel et al., 2017; Leonova et al., 2013; Ranoa et al., 2016; Rudin and Thompson, 2001). These observations indicate that stress-induced alterations in the quantities or types of endogenous RNAs and their intracellular distribution may be recognized as intrusion of non-self RNA pathogens and create a state of "viral mimicry" (Roulois et al., 2015). These changes in the pool of endogenous RNAs will be further defined here as "RNA stress" considering that such changes can lead to the activation of cytoplasmic RNA sensors and eventually to detrimental effects on cell populations.

The association between RNA stress, immune activation, and cytotoxicity raises an important question about the role of RNA stress in tumor development and response to conventional cancer therapies and immune checkpoint therapy (ICT). It is likely that some key players of these sensing pathways can act as tumor suppressors as activation of such pathways might be beneficial in cancer therapies. Indeed, tumor-suppressive functions were demonstrated for RIG-I, PKR, and RNase L (Carpten et al., 2002; Hou et al., 2014; Koromilas et al., 1992). For radiotherapy the role of nucleic acid sensing is well described for DNA-dependent cGAS/cGAMP/STING signaling in host immune cells (Ahn and Barber, 2014; Deng et al., 2014, 2016; Weichselbaum et al., 2017; Woo et al., 2015). However, the role of RNA sensing in tumor cells is less appreciated in the context of cancer therapy. Accumulating literature indicate importance of RNA sensing in cancer cells subjected to radio/chemotherapy and treatments by epigenetic modifiers (Benci et al., 2016; Boelens et al., 2014; Chiappinelli et al., 2015; Ranoa et al., 2016; Roulois et al., 2015; Widau et al., 2014). The ability of IR to induce interferon and TNF-alpha signaling in cancer cells was previously described and indicated the existence of specific mechanisms connecting IR-induced injury and cytokine production (Amundson et al., 2004; Gorbunova et al., 2012; Hallahan et al., 1990; Khodarev et al., 2004, 2007; Tsai et al., 2007; Weichselbaum et al., 2002; Widau et al., 2014). Recent studies have demonstrated these mechanisms rely on activation of nucleic acid sensors, including RNA sensors, which warrant further investigations in this field. Identification of endogenous ligands involved in the stress response to cancer therapies leading to activation of RNA sensors will be of critical importance. A significant group of these ligands includes dsRNAs produced by stress-activated endogenous retrotransposable elements like LINEs, SINEs, and LTR/ERVs. How genotoxic stress activates these endogenous retrotransposons and how we can use this knowledge to improve tumor response to conventional and immunotherapy are challenging directions of investigations.

In the current review, we provide an outline of known families of RNA sensors and mechanisms of induction and intracellular redistribution of their endogenous RNA ligands by stress stimuli. In addition, we discuss recent data regarding the role for retrotransposable genes and small non-coding RNAs as activators of RNA sensors. We also discuss interactions between RNA sensing systems in cancer cells and DNA sensing systems in host immune cells and role of RNA sensors in conventional therapy and immunotherapy of tumors.

2. OVERVIEW OF RNA SENSING SYSTEMS IN MAMMALIAN CELLS

There are five major types of RNA sensing pattern recognition receptors in mammalian cells, which are Toll-like receptors (TLRs), RIG-I-like receptors (RLRs), OAS-like receptors (OLRs), NOD-like receptors

(NLRs), and RNA-dependent protein kinase PKR. In this section, we provide a brief overview of these PRRs and discuss their ability to recognize stress-induced endogenous RNAs in normal and tumor cells.

2.1 Toll-Like Receptors (TLRs)

TLRs are Type I transmembrane glycoproteins responsible for binding with ligands, which represent molecular structures common to different types of microbes and viruses. Such common structures are named pathogen-associated molecular patterns (PAMPs) (Akira et al., 2006). The first discovered member of this family, Toll, was initially described in *Drosophila melanogaster* as a regulator of dorsoventral patterning and response to fungal infection (Lemaitre et al., 1996). Its human homolog was cloned in 1997 and was the first member of mammalian pattern recognition receptors (PRRs) demonstrated to be evolutionary conserved between flies and mammals (Medzhitov et al., 1997). There are 10 TLRs in human and 13 in mouse (Wu and Chen, 2014). The N-terminal extracellular region of TLRs contains leucine-rich repeats (LRR) which form horseshoe-like domain, responsible for interaction with activating PAMPs. The C-terminal region is similar to interleukin-1 receptor (IL-1R), is referred to as Toll/IL-1R domain (TIR domain), and is responsible for activation of signaling events after binding of LRR domain with appropriate ligands (see Fig. 1). Several TLRs are responsible for binding with RNA ligands. TLR3, TLR8 and TLR7 bind dsRNA and ssRNA, respectively (see Fig. 1). Importantly all TLRs that are responsible for recognition and binding with RNA ligands are located in endosome/lysosome compartments, which separate them from cell-intrinsic nuclear and cytosolic NA but expose to viral and microbial compounds which enter cells through endocytosis (see Fig. 2). Upon ligand binding, TLRs homodimerize and recruit the adaptor proteins myeloid differentiation marker 88 (MyD88) for TLR7/8 or TIR-domain containing protein inducing IFN-beta (TRIF) for TLR3. Further downstream signaling from TLR7/8 involves activation of three major signaling pathways: NFkB pathway, IRF3/IRF7 pathway, and mitogen-activated protein kinase (MAPK) pathway. MyD88 recruited by activated TLR7/8 forms activating complexes with interleukin-1 receptor-associated kinases IRAK-1/4 (see Fig. 2). Activated IRAK-1 binds TNF receptor-associated factor 6 (TRAF6) which is an E3 ligase responsible for activation of IkB kinase and TAK1 kinase through K63 polyubiquitination (Jensen and Thomsen, 2012 and Fig. 2). In the result IKK complex phosphorylates

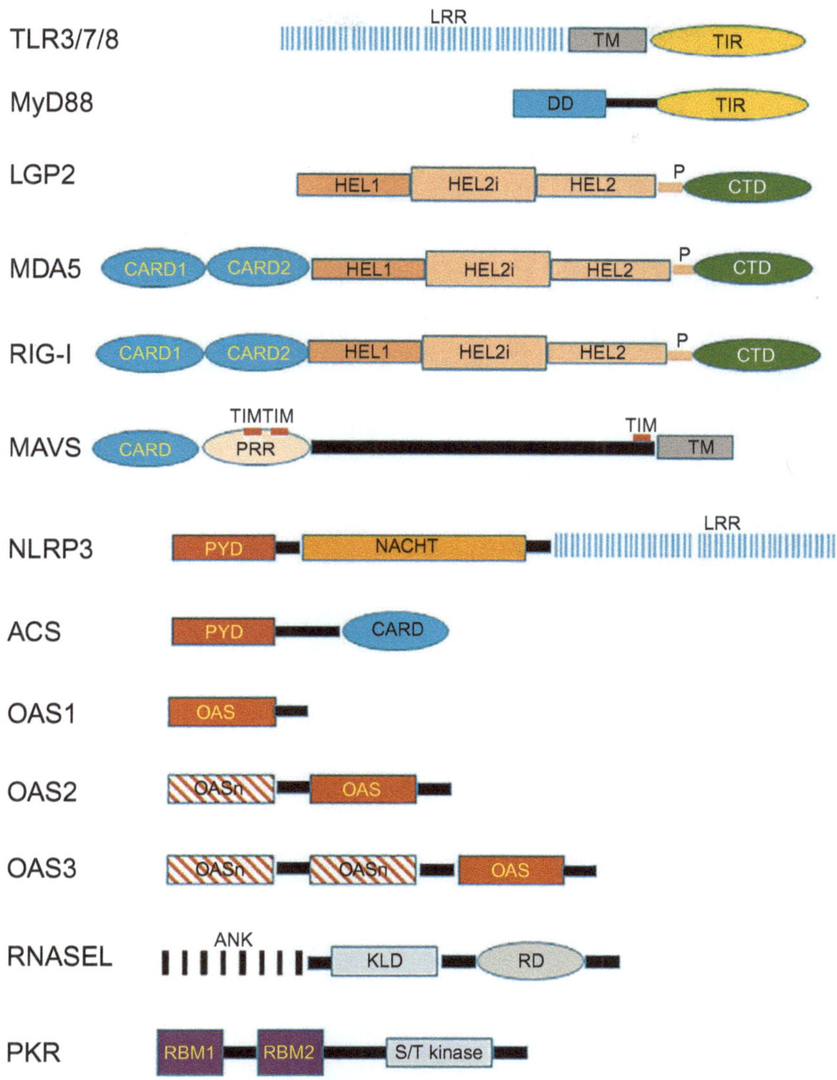

Fig. 1 Domain structure of RNA sensors. LRR, leucine-rich repeats; TM, transmembrane domain; TIR, Toll/IL-1R domain; DD, death domain; HEL-RNA-helicase domains; CTD, C-terminal domains; CARD, caspase activation and recruitment domain; PRR, proline-rich repeats; TIM, TRAF-interacting motive; PYD, pyrin domain; NACHT, domain present in NAIP, CIITA, HET-E, and TP1; ASC, apoptosis-associated speck-like protein containing CARD; OAS, oligoadenylate synthetase domain; OASn, catalytically inactive OAS domain; ANK, ankyrin repeats; KLD, catalytically inactive kinase-like domain; RD, ribonuclease domain (Dong et al., 2001); RBM1, RBM2, RNA-binding domains; S/K, serine/threonine protein kinase domain.

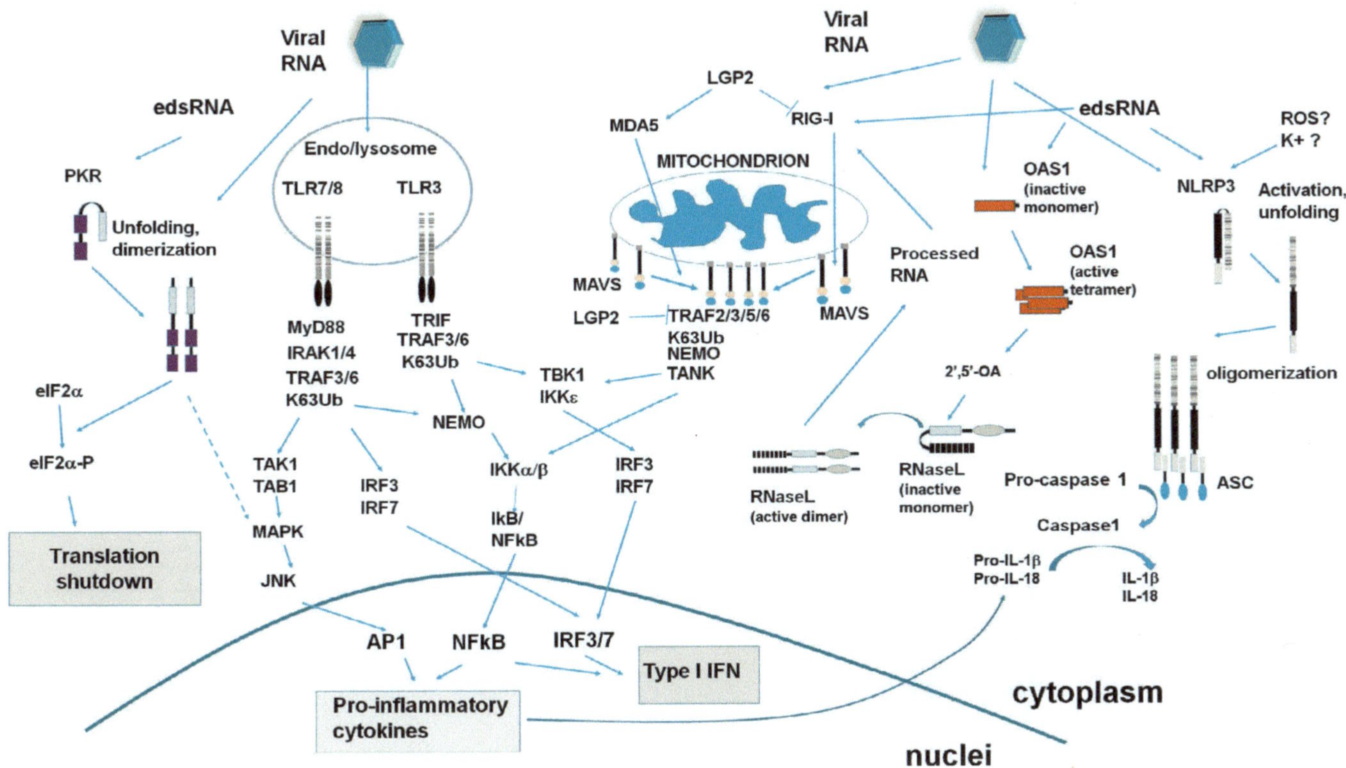

Fig. 2 See legend on opposite page.

IkB suppressor of NFkB which leads to its degradation and nuclear translocation of NFkB; this also leads to the activation of MAPK pathway (see Fig. 2). Signaling from TLR3 is different from other TLRs in that TLR3 uses TRIF, but not MyD88, as adapter protein. TRIF activates NFkB-dependent pathway through TRAF6, as TLR7/8. Additionally, TRIF binds with TRAF3 which leads to activation of Tank-binding kinase 1 (TBK1) and further activation of IRF3. These complex signaling events lead to NFkB/AP1-dependent activation of pro-inflammatory cytokines and IRF3/IRF7/NFkB-dependent activation of Type I IFN response comprising IFN-beta and different isoforms of IFN-alpha.

TLRs have significant differences in cell-specific expression (Mohammad Hosseini et al., 2015). In host immune cells, TLR3 is expressed in NK, DC, and T and B cells, while TLR8 is expressed in monocytes, macrophages, mast cells, and dendritic cells. Data from transgenic mice indicate that the major cellular type, which mostly relies on TLR, but not RLR-based recognition of intracellular viral RNA, are dendritic cells (Kato et al., 2005). TLRs can be expressed in intestinal epithelial cells, where they participate in inflammatory processes associated with interaction with gut microbiome; however, some of them, as TLR4 may be aberrantly overexpressed in colorectal cancer and correlate with poor survival (Luddy et al., 2014; Sato et al., 2009). How common is the expression of different TLRs in various types of tumor cells is still unclear.

Fig. 2 Signaling pathways associated with cytoplasmic RNA sensors. Major cytoplasmic RNA sensors are PKR, TLR3, and TLR7/8 (located in endo/lysosomal compartment), RLRs (LGP2, RIG-I, and MDA5), and their adapter protein MAVS, located in outer membrane of mitochondria and in peroxisomes, inflammasome NLRP3, and OAS/RNase L system. edsRNA, endogenous double-stranded RNAs. Activation of TLRs and RLRs leads to cytoplasmic signaling cascades, which converge on activation of three major transcriptional factors—NFkB, AP1, and IRF3/IRF7. Major groups of genes, transcriptionally activated by these factors in the context of TLR/RLR response are pro-inflammatory cytokines and Type I IFNs (in humans corresponding to IFN-beta and 13 isotypes of IFN-alpha genes; less understood are functions of IFN-kappa and IFN-omega-1). MyD88, adapter protein for TLR7/8; TRIF (TIR-domain containing protein inducing IFN-beta), adapter protein for TLR3. IRAK1/4, interleukin-1 receptor-associated kinases 1 and 4, initiating TLR7/8 signaling cascade; TRAF (TNF-receptor associated factors 2–6), E3 ubiquitin ligase, responsible for K63-linked polyubiquitination of target proteins and itself; K63Ub, K63-linked polyubiquitin chains; NEMO or IKKγ, the regulatory subunit of the inhibitor of IκB kinase (IKK) complex; IKKα/β, serine/threonine protein kinases, responsible for phosphorylation of NFkB repressor IkB. TAK1 (MAP3K7, MEKK7), serine/threonine protein kinase (TGF-beta activated kinase 1); TAB1, TAK1-binding protein 1; IRF3/7-interferon regulatory factors 3/7; TBK1-TANK-binding kinase 1. See text for further explanations.

2.2 RIG-I-Like Receptors (RLRs)

Contrary to TLRs, RLRs are located in the cytosolic fraction of mammalian cells and are not separated from its molecular context by any membrane barriers. RLRs are ubiquitously expressed in different types of mammalian cells, including epithelial tumor cells (Jensen and Thomsen, 2012; Kato et al., 2005; Ranoa et al., 2016). Active function of RLRs in tumor cells and their recruitment in stress response was first demonstrated in 2014, when Widau et al. used an siRNA screen to identify LGP2 as a critical suppressor of lethal response of tumor cells to ionizing radiation (Widau et al., 2014). Independently, Boelens et al. demonstrated that RNA-containing exosomes, released by stromal fibroblasts activate RIG-I in tumor cells, and modulate their response to IR (Boelens et al., 2014).

Several excellent in-depth reviews describe biological, structural, and biochemical properties of nucleic acid PRRs, including the RLR system (Akira et al., 2006; Dixit and Kagan, 2013; Palm and Medzhitov, 2009; Roers et al., 2016; Schlee and Hartmann, 2016; Wu and Chen, 2014). As such, we provide only a brief outline of RLR signaling. RLRs include three DExD/H box RNA helicases (RNA-binding enzymes with conserved motif Asp-Glu-any amino acid-Asp/His (DExD/H)), which include LGP2 (Laboratory of Genetics and Physiology 2) or DHX58 (Cui et al., 2001), RIG-I (retinoic acid inducible gene 1 or DDX58), and MDA5 (melanoma differentiation-associated protein 5, or IFIH1, or AGS7), which belong to the Superfamily 2 (SF2) of RNA helicases (Singleton et al., 2007). All three proteins contain a C-terminal domain (CTD) and the RNA helicase/ATPase domains Hel1, Hel2i, and Hel2 (see Fig. 1). RIG-I and MDA5 also contain N-terminal caspase activation and recruitment domains (CARD domains) responsible for protein–protein interactions. CARD domains are utilized by activated RIG-I and MDA5 to activate mitochondrial anti-viral signaling protein (MAVS, or IPS1, or VISA), which acts as an adapter protein for RLR sensors and initiates downstream signaling (Xu et al., 2014). LGP2 which does not contain a CARD domain is thought to perform regulatory functions by modulating the activity of RIG-I and MDA5. The regulatory properties of LGP2 are associated with its ability to interact with TRAF proteins downstream from MAVS, thereby suppressing RLR/MAVS signaling (Parisien et al., 2018). Simultaneously, LGP2 can positively regulate MDA5 signaling through initiation of RNA-MDA5 complexes (Bruns et al., 2014). Overall LGP2 suppresses RIG-I but in the context of viral infection can activate MDA5 depending on the type of viruses

(Jensen and Thomsen, 2012; Schlee, 2013). Therefore, RIG-I and MDA5 act as primary sensors of cytosolic RNA ligands and LGP2 controls their activity.

To a large degree recruitment of RIG-I or MDA5 in the response to viral or endogenous RNAs is determined by the type and structure of activating RNA ligands and accessibility of the free 5′ ends of RNA. RIG-I is activated by relatively short double-stranded RNA molecules (a minimum length of 10–19 bp and up to 150–300 bp) with triphosphate groups at the 5′ end (5′ppp) of the molecule. Preferable substrates have 5′ blunt ends. Based on the crystal structure of the RNA-binding pocket in RIG-I C-terminal domain (CTD), the lysine-rich cleft at amino acid positions 858–861 provides a docking site to the alpha, beta, and gamma phosphate groups at the 5′ end of dsRNA ligand (Kolakofsky et al., 2012, Wang et al., 2010). Indeed, RIG-I K858A-K861A mutations diminished RIG-I binding to short non-coding endogenous RNAs and consequently abolished the RIG-I-dependent induction of IFN-beta response post-IR (Ranoa et al., 2016). However, several reports indicate that removal of 5′-phosphate groups by phosphatases reduced but not completely abolished IFN-beta inducing activity of different types of RNAs (Boelens et al., 2014, Chiappinelli et al., 2015, Ranoa et al., 2016). Further reports demonstrated that RIG-I can also bind 7mGppp modified RNAs with capped 5′ppp end (cap 0) and can be activated by RNase L cleaved RNAs which contain 5′OH ends (Malathi et al., 2007). Overall these data indicate that RIG-I primarily binds free 5′ ends of relatively short (19–300 bp) dsRNAs. The presence of triphosphorylated 5′ blunt ends can increase the affinity of binding but is not an absolute requirement for binding of RNA ligands to RIG-I. Interesting that some RNA ligands can suppress RIG-I activity. Jiang et al. described VSV-induced IFN-dependent activation of transcription of lncRNA lnc-Lsm3b on the late stages of viral infection. This RNA ligand contains multiple internal stem-loop structures which bind to multiple RIG-I molecules and therefore compete with VSV RNA suppressing RIG-I-induced IFN response (Jiang et al., 2018).

Contrary to RIG-I, MDA5 interacts internally with relatively long dsRNA molecules (>2000 nt) independently of the structure of free 5′ ends or with 5′ ends bound by other RNA-binding proteins. Perhaps, the lack of specificity to terminal structures in MDA5 is connected with the absence of an ordered capping loop that exists in the RIG-I CTD (Yin et al., 2015). The appearance of short cytoplasmic dsRNAs with free 5′ ends can be recognized by RIG-I, while long loops of RNA molecules, enriched by ds structures and potentially associated with other RNA-binding proteins at 5′ ends can be recognized by MDA5 (Bruns et al., 2014; Schlee and Hartmann, 2016).

Post-translational modifications of RNA sensors can suppress their activity at the basal level but activate them in response to infection or stress stimuli. Such modifications can act as an important mechanism to prevent interaction of RNA sensors with endogenous RNAs (Lassig and Hopfner, 2017; Xu et al., 2017b; Wu and Chen, 2014 and Fig. 3). These post-translational modifications (PTMs) include serine/threonine and tyrosine phosphorylation, polyubiquitination, and acetylation. At the basal level both RIG-I and MDA5 are maintained in non-active state through serine/threonine phosphorylation of CARD domains and CTD. For RIG-I this phosphorylation is performed by protein kinases PKCα/β and CKII (Gack et al., 2010; Maharaj et al., 2012). Kinases responsible for phosphorylation of MDA5 have not been identified. Phosphorylation protects RIG-I/MDA5 from K63-linked polyubiquitination which is necessary for activation of these sensors and their binding to MAVS (Gack, 2014; Xu et al., 2017b; Wu and Chen, 2014). As well, acetylation of Lys-909 in CTD of RIG-I

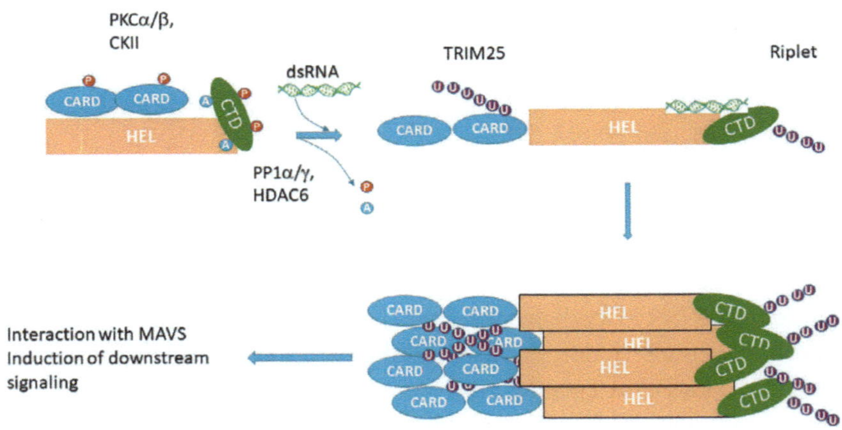

Fig. 3 Post-translational modifications of *RIG-I*. Activation of RIG-I is initiated by dephosphorylation of CARD domains and CTD domain by protein phosphatase 1 (PP1α/γ) and deacetylation of CTD domain by histone deacetylase 6 (HDAC6). Further activation steps are connected with K63-linked polyubiquitination of CTD by E3 ubiquitin ligase Riplet which induces "open" conformation allowing binding with RNA and further K63-linked polyubiquitination of CARD2 domain by tripartite motif containing E3 ubiquitin ligase TRIM25 (Gack et al., 2007). The later leads to the formation of RIG-I tetramers, which interact with MAVS through CARD domains. These interactions induce prion-like oligomerization of MAVS molecules in the outer membrane of mitochondria and activation of MAVS-dependent downstream signaling (see Fig. 2).

prevents interaction of Lys-907 with 5′ppp dsRNA, thereby suppressing activation of RIG-I (Choi et al., 2016). Consistently, activation of RLRs is initiated by dephosphorylation of CARD domains and CTD domain by protein phosphatase 1 and, at least for RIG-I, deacetylation of CTD domain by histone deacetylase 6 (HDAC6; see Fig. 3 and Choi et al., 2016; Gack, 2014; Lassig and Hopfner, 2017; Maharaj et al., 2012; Wies et al., 2013). Further activation steps are connected with K63-linked polyubiquitination of CTD by E3 ubiquitin ligase Riplet (Oshiumi et al., 2009, 2010) which induces an "open" conformation allowing binding with RNA and further K63-linked polyubiquitination of CARD2 domain by tripartite motif containing E3 ubiquitin ligase TRIM25 (Gack et al., 2007). The later leads to the formation of RIG-I tetramers, which interact with MAVS through CARD domains. These interactions induce prion-like oligomerization of MAVS molecules in the outer membrane of mitochondria and activation of MAVS-dependent downstream signaling (see Fig. 2). Such a robust system of PTMs preventing RIG-I/MDA5 activation serves as a control mechanism which protect cells from erroneous recognition of self-RNAs by RLRs. Autoinhibitory conformations which prevent activation of RNA sensors at the basal level seems to be a common rule for different RNA sensors. NLRP3, PKR, and OAS/RNASEL exist in self-inhibitory monomeric conformations, and the initial steps of their activation are connected with unfolding and oligomerization (see below and Fig. 2).

Downstream signaling from MAVS overlaps with signaling pathways, activated by RNA-dependent TLRs (see Fig. 2) but mechanistically is initiated by different events (Vazquez and Horner, 2015). MAVS contains three TRAF-interacting motifs (TIMs) in its PRR domain and close to TM domains (see Fig. 1) (Liu et al., 2013). Initial activation of MAVS-dependent signaling is associated with recruitment of the E3 ubiquitin ligases TRAF2/5 and TRAF6 which interact with MAVS through TIMs and initiate K63-linked polyubiquitination of recruited TRAFs (see Fig. 2 and Paz et al., 2011). This K63-linked polyubiquitination further recruits NEMO/IKKγ through its ubiquitin-binding domains (Liu et al., 2013). In turn, immobilized NEMO leads to the recruitment and activation of TBK1, IKKα, and IKKβ, which eventually leads to activation of NFkB- and IRF3-dependent pathways (see Fig. 2). Similar to TLR signaling MAVS-dependent pathways lead to activation of Type I IFN signaling and NFkB-induced pro-inflammatory cytokines. Contrary to TLR signaling, MAVS signaling does not recruit MAPK/JNK pathway.

2.3 NOD-Like Receptors (NLRs)

NOD-like receptors or nucleotide-binding domain, leucine-rich repeat containing proteins (NLRs) are cytosolic proteins with a wide range of upstream ligands, which upon activation induce production and extracellular secretion of pro-inflammatory cytokines IL-1β and IL-18 as well as activation of MAPK pathway and induction of apoptosis and pyroptosis. Eventually cell death pathways lead to the elimination of infected cells (Jensen and Thomsen, 2012; Mankan et al., 2012; Martinon et al., 2009; Schroder and Tschopp, 2010; Tschopp and Schroder, 2010). Nucleotide-binding domain and leucine-rich repeat containing family pyrin 3 (NLRP3) is a member of the inflammasome family involved in the sensing of intracellular RNA molecules and ROS. NLRP3 contains an N-terminal pyrin domain responsible for interaction with the adaptor ASC (apoptosis-associated speck-like protein containing CARD, see Fig. 1). NACHT domain is involved in oligomerization and activation, while the C-terminal LRR domain is associated with sensing of upstream ligands and autoinhibition of NLRP3. This structure is highly evolutionarily conserved and is homologous to the resistance (R) genes in plants (Tschopp and Schroder, 2010). Details of activation are not fully understood yet, but it involves a combination of aberrant cytoplasmic accumulation of ds and ss RNA molecules and generation of ROS. Activation leads to the unfolding and oligomerization of NLRP3 and initiation of interaction of NLRP3 with ASC through Pyr domains (see Figs. 1 and 2). ASC further recruits pro-caspase 1 through its CARD domain and autoactivation of pro-caspase 1 leads to the release of active caspase-1. Active caspase-1 processes pro-IL-1β and pro-IL-18 into the mature secreted forms, which initiate an inflammatory response. Interestingly, NLRP3 does not activate transcriptional production of pro-IL-1β and pro-IL-18; these are NFkB-dependent events associated with co-activation of TLRs and RLRs, which demonstrates interaction between different families of cytoplasmic RNA sensors (see Fig. 2). Active caspase-1 also induces a specific mode of cell death, namely pyroptosis, which is characterized by the disintegration of cellular membrane and release of pro-inflammatory cytokines and cytosolic context into extracellular space (Cookson and Brennan, 2001). NLRP3 can be activated by influenza A virus and Sendai virus (see Jensen and Thomsen, 2012). NLRP3 is also involved in the sensing of endogenous RNA DAMPs, such as dsRNAs formed by Pol II transcribed Alu1 repeats, and associated with pathogenesis of age-related macular degeneration (Kaneko et al., 2011; Tarallo et al., 2012).

2.4 RNA-Dependent Protein Kinase (PKR)

RNA-dependent protein kinase PKR and 2′,5′-oligoadenylate synthases were the first enzymatic activities associated with intracellular accumulation of dsRNA and later considered as RNA sensors (Farrell et al., 1978; Wu and Chen, 2014). PKR contains two N-terminal RNA-binding domains and C-terminal serine/threonine protein kinase domain (see Fig. 1). PKR is constitutively expressed in all tissues at the basal level. In an inactive form PKR exist as monomers; upon binding with dsRNA PKR autophosphorylates itself, dimerizes and phosphorylates downstream targets (see Fig. 2). PKR is activated by a broad range of viruses; it is also directly activated by Type I IFNs, as a part of an anti-viral response (Sadler and Williams, 2008), to prevent translation of viral mRNAs and amplification of active virions. The best characterized downstream target of PKR is translational initiation factor eIF2-alpha. Upon phosphorylation of eIF2-alpha, it suppresses the ability of eIF2 complex to initiate translation of mRNA in ribosome. Therefore, activation of PKR leads to the global shutdown of translation and often to cell death (Mu et al., 2016). To evade PKR-dependent inhibition, HSV-1 uses its $\gamma_1$34.5 gene, which recruits cellular protein phosphatase 1 to dephosphorylate eIF2-alpha and restore normal translation (He et al., 1997). In addition to eIF2-alpha phosphorylation, PKR was implicated in phosphorylation of IkB, JNK, and MAPK and control of apoptosis, autophagy, and metabolic stress (reviewed in Marchal et al., 2014; Watanabe et al., 2018). Importantly PKR is associated with control of cell cycle; constitutively active PKR led to G2/M arrest in CHO cells (Dagon et al., 2001). Similar to NLRP3 and RLRs, PKR was recently implicated not only in recognition of viral RNA but also in recognition of endogenous RNAs. Elbarbary et al. described an interaction of PKR with dsRNA structures formed by inverted Alu1 repeats in 3′UTR of different human mRNAs (Elbarbary et al., 2013). Similar to recognition of viral dsRNA, these interactions led to phosphorylation of eIF2-alpha and translational repression. Binding of another dsRNA-binding protein, Staufen1, to the same repeats prevented their interactions with PKR and restored normal translation. PKR was also reported to interact with small nucleolar RNAs (snoRNAs) but the functional consequences of these interactions remain unclear (Youssef et al., 2015). In addition to its anti-viral functions, PKR was considered as tumor suppressor gene (Meurs et al., 1990, 1993). Dominant negative constructs of PKR induced malignant transformation (Koromilas et al., 1992). These data indicate a link between RNA sensing system and malignant transformation. Tumor-suppressive functions of RNA sensors are associated with their ability to activate downstream IFN signaling in association with pro-death pathways and directed toward elimination of infected or endogenously activated cells; this may explain how PKR suppression activates malignant transformation and tumor escape from immunologic surveillance.

2.5 OAS-Like Receptors (OLRs)

2′,5′-Oligoadenylate synthetases (OAS) represent a group of template-independent nucleotidyltransferases. They belong to a large superfamily of nucleotidyltransferase (NTase) fold proteins, recently characterized based on structural and enzymatic properties of hundreds of different proteins with NTase activity (Kuchta et al., 2009). The same superfamily contains cGAS (also known as MB21D1 member of MB21 family). Recent structural insights allowed to combine OAS and cGAS in one family of cytoplasmic OAS-like receptors based on the common 3D structure of catalytic core, common enzymatic activity, mechanism of catalysis in the active center, and mechanism of allosteric regulation by nucleic acid binding (Hornung et al., 2014; Lohofener et al., 2015). Together with PKR, OAS proteins were among first enzymes described as activated in mammalian cells by dsRNA associated with viral infection and interferon treatment (Williams et al., 1979). OAS proteins are responsible for dsRNA-induced synthesis of 2′-5′-linked oligomers of adenosine from ATP, with the general structure pppA(2′p5′A)n, where $n \geq 2$. These oligomers act as secondary messengers to activate RNase L which in turn leads to the degradation of viral RNA (Hovanessian and Justesen, 2007; Kristiansen et al., 2011; Malathi et al., 2007, 2010; Sadler and Williams, 2008). Human OASs are presented by three enzymatically active proteins: OAS1 (40/46 kDa), OAS2 (69/71 kDa), and OAS3 (100 kDa). They correspond to proteins with one, two, and three OAS domains, respectively (see Fig. 1). However in each of these proteins only one domain is catalytically active. The additional domains are inactive and most likely were generated by duplication of the initial OAS domain (Hovanessian and Justesen, 2007). An additional OAS-related gene, OAS-like or OASL, corresponds to an enzymatically inactive protein with inactivated OAS domain and an ISG15-like domain. Different OAS proteins have unique intracellular locations, with OAS1 and OAS2 located in cytoplasmic mitochondrial and microsomal fractions, while OAS3 is associated with ribosomes and nuclear fractions (Besse et al., 1998; Chebath et al., 1987). Initially it was thought that OAS3 form only 2′-5′A dimers with lower affinity for RNase L and is a weak activator of RNase L (Hovanessian and Justesen, 2007). However recently Ibsen et al. purified this enzyme to homogeneity and demonstrated that it had higher affinity for dsRNA as compared with OAS1 and produce long oligomers of 2′-5′A with high intracellular activity toward RNase L (Ibsen et al., 2014), RNase L is an 80 kDa protein with N-terminal ankyrin repeats and C-terminal serine/threonine kinase-like domain and ribonuclease domain (see Fig. 1). It resides in the cytoplasm in inactive self-inhibitory monomeric form (see Fig. 2). Upon binding of ankyrin repeats with 2′,5′-oligoA RNase L forms catalytically active homodimers and degrades viral RNA, preventing translation and amplification of virions (Dong and Silverman, 1995). Importantly, RNase L degraded RNA molecules are able to activate cytoplasmic RLR system in a RIG-I-dependent way (Malathi et al., 2007) which in turn leads to the activation of Type I IFN signaling and pro-inflammatory cytokine response through MAVS-associated downstream signaling cascades (see Fig. 2). Thus outcome of OAS/RNase L signaling partially overlaps with PKR and partially with RLR signaling pathways. Similar to PKR, RNase L has properties of a tumor suppressor gene and has a linkage with hereditary prostate cancer (Carpten et al., 2002; Lengyel, 1993). Thus, at least some members of cytoplasmic RNA sensing pathways which are involved in initiation of the downstream signaling pathways associated with IFN signaling, cytotoxic response, and innate immunity activation also act as tumor suppressors. This suggests that the homeostatic balance of RNA sensors and RNA ligands in cytoplasm may be important factor of regulation of anti-tumor innate and adaptive immunity and suppression of malignant transformation.

3. DISCRIMINATION OF SELF FROM NON-SELF RNA. RNA STRESS AS AN ABERRANT RECOGNITION OF SELF-RNAs

Mammalian cells are equipped with extensive and potentially redundant systems of cytoplasmic RNA sensors. The existence of multiple types of RNA sensors associated with different downstream signaling pathways indicate the importance of cellular protection from exogenous RNAs which can accumulate in the cytoplasm during pathogen intrusion. However at the same time the cytoplasm contains multiple types of endogenous RNAs. It is clear that cells should evolve mechanisms that discriminate self and non-self RNAs to avoid activation of RNA sensors by different types of endogenous

RNAs. Indeed, several types of such discriminative mechanisms have been described in the literature. One important mechanism is based on RNA modifications which are present in mammalian cells but absent in viral/microbial RNAs. In eukaryotes Pol II transcribed pre-mRNAs in the nucleus are modified by 7-methylguanosine cap which is added to the first transcribed nucleotide by enzymes triphosphatase-guanylyltransferase (RNGTT) and RNA guanine-7-methyltransferase (RNMT) (cap 0, see Fig. 4). Further modifications are introduced in the first and second transcribed nucleotides by cap methyltransferase 1 (CMRT1 or ISG95) and 2 (CMRT2) (Aregger and Cowling, 2017; Belanger et al., 2010). These enzymes methylate O-2 residues of ribose at the first and second nucleotides, leading to the formation of cap 1 and cap 2 modified mRNAs (Fig. 4). These modifications are absent in viral and microbial RNAs and can be used by RNA sensors for discrimination between endogenous and exogenous non-self RNAs. RIG-I can accommodate cap 0-modified RNA molecules but cannot bind cap 1-modified RNA (Schuberth-Wagner et al., 2015; Werner et al., 2011). Cap 1 modification also prevents RNA binding with TLR7 and TLR8 (Kariko et al., 2005) and suppresses binding of RNA with another sensor IFIT1 (Habjan et al., 2013). In combination with recognition of the 5′ end structure of RNA molecules (see Section 2), this modification provides a robust mechanism for self vs non-self-recognition (Lassig and Hopfner, 2017; Roers et al., 2016; Schlee and Hartmann, 2016). In addition, the secondary structure of RNA ligands is of critical importance. Long viral RNAs in the cytoplasm form double-stranded structures, which are preferentially recognized by TLR3 and TLR8, RIG-I and MDA5 and PKR. Enzyme adenosine deaminase (ADAR) provides A-to-I editing of dsRNA structure thus disrupting the formation of ds structures in endogenous RNAs. The importance of dsRNA signaling, which is associated with IFN production, was recently demonstrated by Mannion et al., who reported that mutations of ADAR1 led to the accumulation of stable dsRNAs and embryonic lethality in a mouse model, which could be rescued by depletion of MAVS or mutations in the Type I IFN receptor (Mannion et al., 2014). Consistently Liddicoat et al. demonstrated that mutations in ADAR1 can be rescued by depletion of MDA5, another member of the RLR pathway (Liddicoat et al., 2015). Recently it was demonstrated that independently from MDA5/MAVS pathway ADAR1 lethality in human cells can be rescued by RNase L depletion, consistently with functions of OAS/RNase L system as important sensor of RNA signaling (Li et al., 2017b). In humans mutations in ADAR1 are associated with Aicardi–Goutiers syndrome (AGS) and

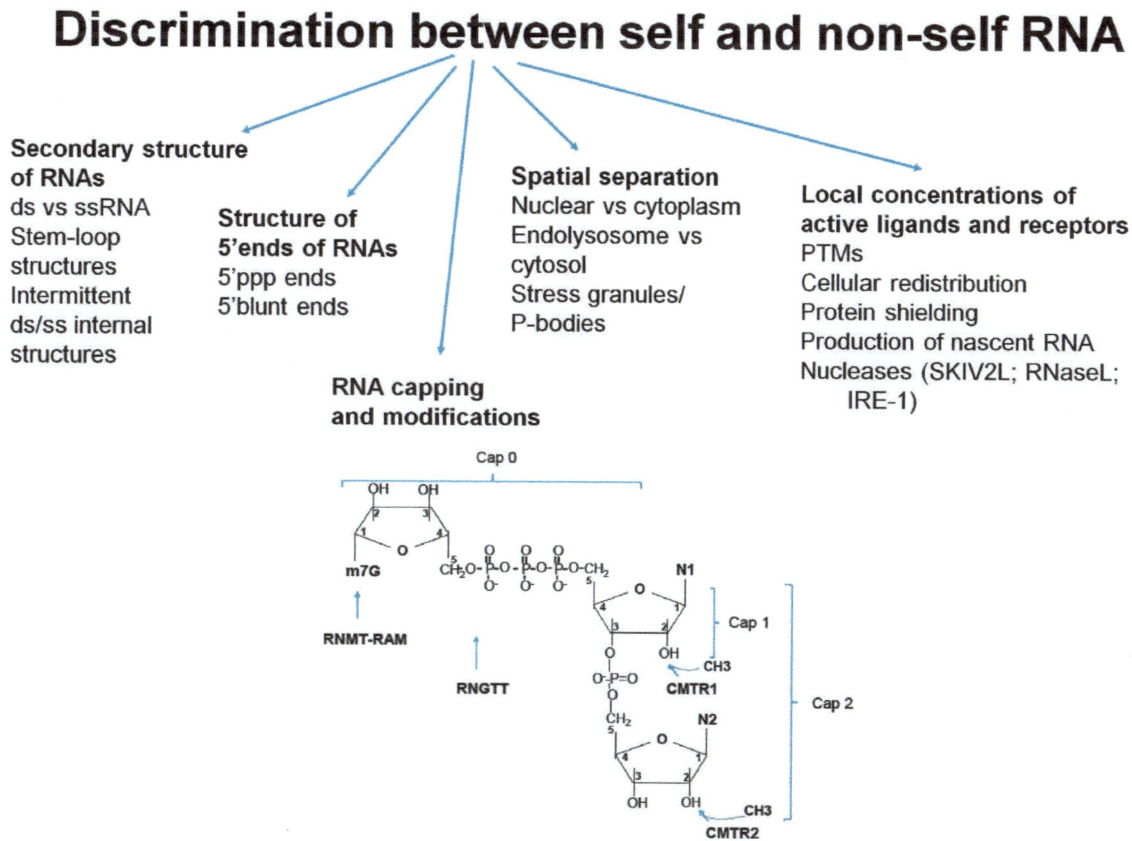

Fig. 4 See legend on next page.

constitutive activation of Type I IFN signaling (Rice et al., 2012). Overall these observations indicate that recognition of dsRNA structures is important for activation of different types of RNA sensors. They also indicate that mammalian cells have intrinsic control mechanisms which scan and remove dsRNA structures from endogenous RNAs where they tend to appear at the basal level. Such dsRNAs exist in the eukaryotic transcriptome more often than it was considered before and have some unusual functionality. A comprehensive description of "dsRNAome" has been provided recently in *Caenorhabditis elegans* (Whipple et al., 2015). In these animals, dsRNAs are mostly represented by intergenic and intronic noncoding RNAs associated with development and response to exogenous pathogens (Whipple et al., 2015). Such a common and evolutionary conserved manifestation of dsRNA signaling in eukaryotes indicates its importance in basic cellular functions. Major mechanisms of self vs non-self RNA discrimination are intensively discussed in literature and summarized in Fig. 4 (see Lassig and Hopfner, 2017; Roers et al., 2016; Schlee and Hartmann, 2016 for reviews). These mechanisms can be separated into four major groups: (1) structure of RNA ligands; (2) modifications of RNA; (3) physical separation of RNA sensors from endogenous RNA ligands; and (4) alterations in local concentrations of RNA ligands in various compartments containing sensing molecules due to increased production, activation through PTMs (see Fig. 3), or degradation of RNA ligands by nucleases.

Fig. 4 Major mechanisms of discrimination between self and non-self RNAs. Discrimination between self and non-self RNAs is based on the multiple mechanisms which include: (1) recognition of the secondary and tertiary structure of RNA ligands; (2) recognition of specific modifications of RNA ligands, primarily RNA methylation; (3) physical separation of RNA sensors from endogenous RNA ligands; and (4) alterations of local concentrations of RNA ligands in the compartments containing sensing molecules due to increased production, activation/suppression through PTMs (see Fig. 3), or suppressed degradation of RNA ligands by nucleases. Scheme illustrates capping of mammalian RNA which serves as important mechanism of discrimination. Cap 0 is represented by 7-methylguanosine, connected with the first nucleotide by unusual 5′-5′-triphosphate bridge. Cap 0 modification is provided by sequential action of two enzymes-triphosphatase and guanylyltransferase RNGTT and subsequent methylation of guanine in N7 position by cap methyltransferase RNMT-RAM (Aregger and Cowling, 2017). RNMT is catalytic subunit and RAM-regulatory subunit (RNMT activating miniprotein). In mammals first two nucleotides of nascent mRNA can be further modified by O-2 methylation by cap methyltransferase 1 (CMTR1) and cap methyltransferase 2 (CMTR2) (Belanger et al., 2010). Cap 1 corresponds to 2-O methylation of the first transcribed nucleotide, cap 2 to the methylation of both first and second transcribed nucleotides.

Alterations of each of these mechanisms can lead to abnormal interactions of RNA sensors with endogenous ligands and initiate cascades of signaling events eventually leading to the induction of cytotoxic/cytostatic signaling and/or formation of chronic inflammatory responses. Genotoxic stress, epigenetic modifiers, and reactive oxygen species (ROS) can activate different RNA sensors in tumor and host cells. One reason for such activation is connected with increased production of dsRNAs in the result of stress stimuli and their cellular redistribution to the compartments where they can interact with RNA sensors mimicking viral state (Roulois et al., 2015; Takemura et al., 2014). However, other mechanisms of such activation should also be considered consisting with complex mechanisms of discrimination of viral and endogenous RNAs (see Fig. 4). We propose such a response to stress stimuli which leads to the acute activation of different RNA sensors in mammalian cells without external pathogens (in aseptic conditions) as "RNA stress." The specificity of this type of stress response is determined by several unusual factors. First, it is induced by genotoxic stress as the response to DNA damage. So RNA stress represents an underestimated pathway of stress response located outside of traditional pathways, controlled by ATR/ATM and p53 (Blackford and Jackson, 2017; Manic et al., 2015). Second, this RNA stress is associated with global changes in the context, processing, and cellular distribution of RNA molecules. Mechanisms of these changes in response to DNA damage just recently became to attract attention and are an important part of understanding of stress response in normal and tumor cells (Giono et al., 2016). And third, this stress is associated with specific downstream signaling, where Type I IFNs and pro-inflammatory cytokines lead to the elimination of responding cells, and development of inflammatory and immune response. Due to these features RNA stress can have a cytotoxic impact on cells, which induces this type of response, but it has potentially stimulatory effect on immune cells and activation of adaptive immune responses. These features make this type of stress an attractive target for cancer therapies, providing opportunities to increase cytotoxic insult to tumor cells while initiating anti-tumor response of host adaptive immunity.

4. ENDOGENOUS LIGANDS OF RNA SENSORS AS STRESS-INDUCED DAMPs

The first experimental evidence for the appearance of endogenous RNA DAMPs in cells subjected to genotoxic stress were provided by RNA transfer experiments, where total RNA was purified from non-treated or treated donor cells and transferred to reporter cells containing constructs for detection of Type I IFN signaling. Leonova et al. demonstrated appearance of IFNβ-activating RNAs in p53-null MEFs treated with 5-Aza-dC (Leonova et al., 2013). Consistent with these findings, Chiappinelli et al. demonstrated that 5-Aza-dC treatment of cancer cell lines Tyk-nu and A2780 led to the accumulation of cytoplasmic RNAs that can induce IFNβ production in recipient HT29 cells (Chiappinelli et al., 2015). In the context of IR response, Ranoa et al. demonstrated that irradiation in therapeutic doses leads to the appearance of IFNβ-activating RNAs in different tumor cell lines and HEK293 cells during first 24–72h post-IR (Ranoa et al., 2016). Importantly, when total RNA from donor cells were treated with nucleases specific for various forms of single- and double-stranded DNA or RNA prior to stimulation of recipient cells, the results suggested that the IFNβ-inducing activity is mostly associated with dsRNAs (Ranoa et al., 2016). Taken together these data indicated that genotoxic stress leads to the appearance of specific RNAs recognized by RNA sensing systems, which induces a detrimental IFN response. Such RNAs act as DAMPs and lead to the activation of innate immune response in different tumor cells in a cell autonomous context, without direct or remote interactions with host immune cells. Current data indicate that a significant fraction of stress-induced RNAs accumulating in cytoplasm and bound to different RNA sensors belong to different families of retrotransposons and repetitive non-coding RNAs, traditionally considered "junk" or "parasitic" DNA. Truncated and full size retrotransposable elements represent approximately 40% of human and mouse genomes and are able to propagate in the host genome through a "copy and paste" mechanism. This mechanism combines transcription of RNA intermediates from internal retroelement (RE) promoters by Pol II or Pol III, generation of DNA copy by intrinsic reverse transcriptase (RT), and genomic reintegration using integrase or endonuclease activity depending on the type of RE (Wicker et al., 2007). Recent data revealed examples of retrotransposons activated by genotoxic stress and detected in complexes with different cytoplasmic RNA sensors. They included long interspersed nuclear elements (LINEs), long terminal repeats (LTRs), including endogenous retroviral elements (ERVs) and several families, related to the short interspersed nuclear elements (SINEs, or Alu1 repeats in human and B1/B2 repeats in mice), including 7SL RNA, tRNA, and 5S RNA (Wicker et al., 2007) (see Fig. 5). Stress-induced transcription of different endogenous RE and direct binding of SINEs, LINEs,

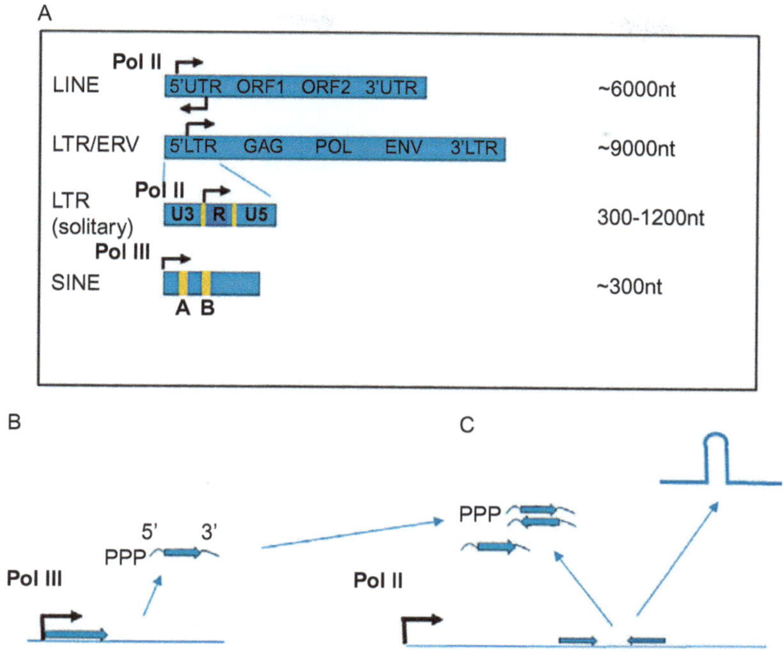

Fig. 5 Structure of major retrotransposable elements which can activate cytoplasmic RNA sensors. (A) General structure of LINEs, LTR/ERV, and SINEs; arrows indicate positions of transcription start sites by Pol II or Pol III. UTR-untranslated regions; LTR-long terminal repeats; A, B-A, and B-boxes in the left monomer of SINE. (B and C) Pol III and Pol II-dependent transcription of Alu1 elements and formation of double-stranded structures from Alu1 repeats. Pol III initiate transcription of solitary full size Alu1 repeats. These transcripts contain 5′ppp and can form internal double-stranded structure (Chung et al., 2018). Alternatively, such Pol III generated transcripts can hybridize with Pol II generated protein-coding RNA transcripts which contain antisense Alu1 repeats in intronic, 5′UTR, or 3′UTR regions. Alu1 repeats also can form intramolecular stem-loop structures if they are presented in the same transcript by direct and reverse copies (C).

and LTRs/ERVs RNAs with RNA sensors have been recently described by several laboratories and may be an essential mechanism of RNA stress induced by cancer therapies (Boelens et al., 2014; Chiang et al., 2018; Chiappinelli et al., 2015; Guler et al., 2017; Leonova et al., 2013; Ranoa et al., 2016; Rudin and Thompson, 2001; Zeng et al., 2014) (see Table 1). Importantly, the highly repetitive nature of these RE (10^5–10^6 copies/genome) and vast abundance of short truncated versions leads to an enrichment of double-stranded RNAs in the pool of RNA molecules transcribed from these genes which partially explains their affinity for RLRs (see Fig. 5).

Table 1 Ligands of RNA Sensors

	Types of RNA	Inducing Agent	RNA Sensor	References
SINEs	Pol III-derived SINEs (AluI, B1/B2)	IR, chemo drugs cisplatin, and etoposide	?	Rudin and Thompson (2001)
	SINEs	Exosomes from stromal cells co-cultured with breast tumor cells	RIG-I	Boelens et al. (2014)
	Alu 1, B1/B2	5-aza-2′-deoxycytidine (5-aza-dC), IR	?	Leonova et al. (2013)
	Alu1	ADAR1 mutations	MDA5	Liddicoat et al. (2015)
	Alu1	Dicer I mutations	NLRP3 inflammasome	Tarallo et al. (2012) and Kaneko et al. (2011)
	Alu1	Cell cycle	PKR	Kim et al. (2014a)
	Transfer RNA (tRNA)	IR	RIG-I	Ranoa et al. (2016)
			IFIT5	Katibah et al. (2014)
	Pol III transcribed RN7SL1	Exosomes from co-cultures	RIG-I	Boelens et al. (2014) and Nabet et al. (2017)
	Pol III transcribed 5SrRNA	HSV-1; EBV, IAV	RIG-I	Chiang et al. (2018)
LINEs	LINEs	Exosomes from stromal cells co-cultured with breast tumor cells	RIG-I	Boelens et al. (2014)
	L1MA2	IR	RIG-I	Ranoa et al. (2016)
	LINEs (L1PA2–6; L1HS)	Erlotinib; vemurafenib	RIG-I	Guler et al. (2017)

LTRs/ERVs	LTR25, LTRGa	IR	RIG-I	Ranoa et al. (2016)
	ERVs	T cell independent type 2 antigen (NP-Ficoll)	MAVS through RIG-I or MDA5	Zeng et al. (2014)
	ERVs	DNMTi, colorectal cancer	MDA5/MAVS	Roulois et al. (2015)
	ERVs	DNMTi, ovarian cancer	MDA5/MAVS; TLR?	Chiappinelli et al. (2015)
	Solitary LTR12E, 12D, 12C	DNMTi, HDACi, lung cancer	?	Brocks et al. (2017)
snRNAs/ scaRNAs	U1, U2 snRNA	UVB, IR, etoposide, cisplatin, and doxorubicin	TLR3; RIG-I	Bernard et al. (2012), Borkowski et al. (2015), and Ranoa et al. (2016)
	U4, U6, U12 snRNA; scaRNA9; scaRNA18	UVB	TLR3	Bernard et al. (2012) and Borkowski et al. (2015)
lncRNA	lnc-Lsm3	VSV	RIG-I	Jiang et al. (2018)

4.1 SINEs/Alu1 Elements

The best studied examples of retrotransposable elements activated by genotoxic stimuli and responsible for activation of different types of RNA sensors are Alu1 elements or SINEs. SINEs are AT-rich repetitive sequences that are ~300 bp in length (or shorter for truncated Alu1 versions), which can integrate in the genome and often form tracts of repetitive elements either in a direct or reverse orientation. Most common sites of Alu1 integration are intergenic regions or intronic regions of protein-coding genes (see Fig. 5). Alu have their own internal promoters recognized by Pol III (Manic et al., 2015). Notably, Pol III is capable of generating RNA transcripts with 5′ppp ends and is functional both in the nucleus and the cytosol (Ablasser et al., 2009). Therefore, Alu1 can be transcribed by Pol II in the context of long mRNA for protein-coding genes or by Pol III as individual short RNA. Alu not only have extensive intramolecular double-stranded structures (Chung et al., 2018) but also can form intermolecular double-stranded structures (see Fig. 5) which can be successfully recognized by RIG-I/MDA5 and PKR (Chung et al., 2018; Hornung et al., 2006; Schlee et al., 2009; Schuberth-Wagner et al., 2015). Rudin and Thompson initially demonstrated that IR, etoposide, and cisplatin increased the levels of Alu1 RNA in human cells and its homologs B1 and B2 RNAs in murine cells (Rudin and Thompson, 2001) (see Table 1). Treatment with demethylating agents lead to the accumulation of Alu1 and B1/B2 in human and mouse cell lines, and this correlated with lethal IFNβ responses (see Leonova et al., 2013 and Table 1). Kim et al. demonstrated that protein kinase R (PKR) can be activated by Alu1 dsRNAs that are released in the cytoplasm due to the transient breakdown of the nuclear envelope during mitosis (Kim et al., 2014a). Alu1 double-stranded structures located in Pol II transcripts represent more than 90% of all cell-intrinsic double-stranded structures recognized and edited by ADAR1 (Chung et al., 2018). Depletion of ADAR1 led to the PKR-dependent shutdown of translation in IFNβ-treated cells and MDA5-dependent activation of Type I IFN response in neuronal progenitor cells. A striking observation regarding the functions of Alu1 dsRNA was provided by Tarallo et al. (2012) and Kaneko et al. (2011). These authors investigated patients with geographic atrophy (GA), an advanced form of age-related macular degeneration resulting from retinal pigmented epithelium (RPE) cell degeneration. They found this disease is associated with the Dicer1-dependent accumulation of Alu1-containing dsRNAs, most likely produced by Pol II in the cytoplasm of RPE cells. Same accumulation of Alu repeats was also observed in mice with inactivated Dicer 1 and was rescued by suppression of Alu1 by antisense oligonucleotides (Kaneko et al., 2011). Further data revealed that Alu1 dsRNA activates NLRP3 inflammasome and MyD88/IL18/caspase 8 signaling associated with developing of RPE degeneration and cell death (Kim et al., 2014b; Tarallo et al., 2012). Finally, recent data indicate that two other members of SINE family closely related to Alu, 7SL RNA, and 5SrRNA pseudogene RNA5SP141 (Wicker et al., 2007) are activated through Pol III transcription in the result of c-Myc/Notch activation or viral infection and bind to RIG-1, inducing IFNβ response (Chiang et al., 2018; Nabet et al., 2017).

4.2 Long Interspersed Elements (LINEs)

Another source of repetitive RNAs is long interspersed elements, or LINEs, which represent ~20% of the human genome (Wicker et al., 2007). LINEs are endogenous retrotransposable elements, which contain functionally active genes for reverse transcription and integration into the genome. They are transcribed by Pol II from internal promoters in 5′ and 3′ UTRs (see Fig. 5). They autonomously encode proteins that allow them to maintain their self-replication and self-dissemination in the genome (Hancks and Kazazian, 2016). LINE1 is active in human genome and induces retrotranspositions in cultured human cells. Full size functionally active copies of LINEs represent only a minor fraction of the genome (~100 copies per human genome), while the vast majority of LINEs in human and mouse genome are truncated, comprising ~1,500,000 copies (Kassiotis and Stoye, 2016). Boelens et al. demonstrated that LINEs were overrepresented in RIG-I-activating exosomes as compared with cellular RNA (Boelens et al., 2014) (see Table 1). Ranoa et al. also demonstrated that LINE RNA L1MA2 co-immunoprecipitated with RIG-I post-IR (Ranoa et al., 2016). Consistent with these observations recent data indicate that cytotoxic responses of cancer cells to the EGFR inhibitor Erlotinib was associated with activation of the LINE1 family and that cells with acquired resistance to Erlotinib repressed LINE1 through epigenetic changes of chromatin (Guler et al., 2017). LINE1-induced cytotoxicity was associated with RIG-I-dependent activation of IFN signaling and was suppressed by siRNA depletion of LINEs and/or RIG-I knockdowns. Remarkably, reverse transcriptase inhibitors (RTi) partially rescued drug-resistant cell population, indicating that cytotoxicity of LINEs is associated not only with their transcription but also with their potential reintegration in genomes (Guler et al., 2017).

Similar data were previously described for SINEs (Rudin and Thompson, 2001), which do not have their own RT but hijack this enzyme from LINEs (Mager and Stoye, 2015).

4.3 Long Terminal Repeats/Endogenous Retroviruses (LTR/ERVs)

Long terminal repeat (LTR) retroelements are another group of endogenous retrotransposable elements that are flanked by LTRs at the ends, and make up about 8% of the human genome (Kassiotis and Stoye, 2016; Wicker et al., 2007). LTRs at the 5′ and 3′ ends serve as transcription initiation and termination sites, respectively (Fig. 5), while internal LTR sequences encode proteins with reverse transcriptase (RT) and integrase activity. Endogenous retroviruses (ERVs) are also classified under LTR retroelements, with the capability of encoding envelope protein (Env) for production of viral particles. A group of ERV Env proteins—Syncytins—is used by the host for formation of syncytium during placental development (Grow et al., 2015; Vargiu et al., 2016). LTR sequences were enriched in RIG-I-activating exosomes (Boelens et al., 2014) (see Table 1). Ionizing radiation induced the binding of LTR25, LTR6A, and MamGypsy2-LTR RNA sequences to RIG-I (Ranoa et al., 2016). Consistent with these findings, ERVs were found to activate MDA5/MAVS signaling in tumor cells after treatment by DNA demethylating agents (Chiappinelli et al., 2015; Roulois et al., 2015). Similarly, endogenous ERVs were also reported to induce T cell-independent B cell activation through RIG-I or MDA5 signaling (Zeng et al., 2014). Importantly, many promoter sequences responsible for transcriptional activation of endogenous retroelements and ERVs are controlled by DNA methylation (Leonova et al., 2013; Rebollo et al., 2012). This may explain their reactivation and subsequent interaction with RLRs after treatment by DNA demethylating agents (Chiappinelli et al., 2015; Leonova et al., 2013; Roulois et al., 2015).

These data indicate that repetitive retrotransposable elements represent an essential group of cell-intrinsic RNA ligands activated by different forms of stress and able to stimulate various types of cytoplasmic RNA sensors in normal and tumor cells. These REs have a significant impact on genome functions and stability. They provide whole genome network of transcriptional regulation of multiple host genes (Chuong et al., 2016, 2017). In addition, they are involved in genome rearrangements due to insertional mutagenesis and induction of different recombinational events (Deininger et al., 2003). The ability of cytoplasmic RNA sensors to interact with

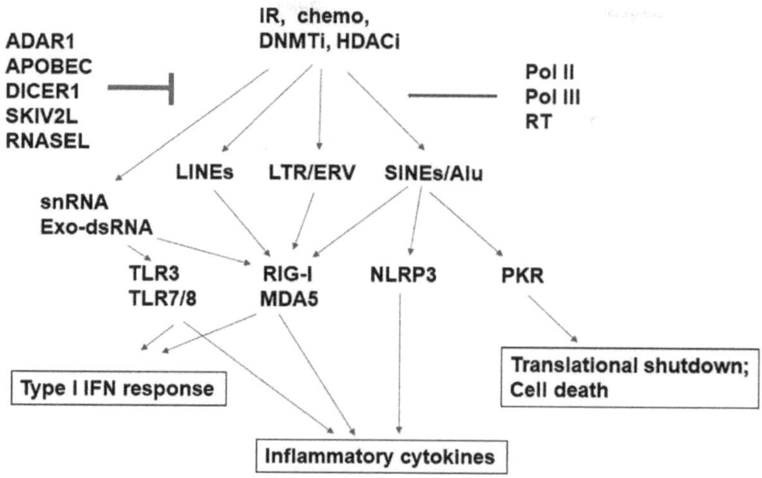

Fig. 6 Major stress-induced RNA ligands and consequences of their interactions with cognate receptors. Conventional cancer therapy and epigenetic inhibitors of DNA methyl transferases (DNMTi) and histone deacetylases (HDACi) lead to the aberrant transcriptional activation (derepression) and reverse transcription of multiple retrotransposable elements. Same stimuli can also lead to the transcription and nuclear-cytoplasmic redistribution of small non-coding RNAs. Several families of enzymes control aberrant accumulation of double-stranded RNA transcripts, which includes (but not limited to) adenosine deaminases ADAR, cytidine deaminases APOBEC, RNAase III Dicer1, and exo-RNAse complex SKIV2L. Interaction of these RNA ligands with cytoplasmic RNA sensors TLR3/7/8, RLRs, NLRP3, and PKR leads to the downstream signaling with three major outcomes: activation of Type I IFN signaling, activation of inflammatory response, and translational shutdown/activation of the cell death in cell subpopulations with unbalanced RNA stress.

transcriptional RNA intermediates of stress-activated RE leads to downstream signaling associated with induction of cell death in activated cells and mobilization of host immune response against such cells (Volkman and Stetson, 2014). This means that RNA sensors act as suppressors of RE-associated genomic instability and participate in the maintenance of homeostatic genetic stability of cell populations (see Fig. 6). Therefore interactions between RE and cytoplasmic RNA sensors reveal new functions of innate immune system in the maintenance of genomic stability in normal and tumor cells.

4.4 Small Non-Coding RNAs (snRNAs and snoRNAs)

Several laboratories also reported involvement of endogenous small nuclear RNAs (snRNAs) in activation of RLR and TLR systems, indicating an ability of snRNAs to act as DAMPs (Bernard et al., 2012; Ranoa et al., 2016;

Takemura et al., 2014) (see Table 1). These non-coding RNAs have sizes between 100 and 300 nucleotides and have a well-characterized secondary structure enriched by double-stranded stem loops (Borkowski et al., 2015). They play important roles in mRNA post-transcriptional processing by forming complexes with several proteins to form spliceosomes (Lee and Rio, 2015). snRNAs are transcribed in the nucleus by RNA Pol II, and rapidly exported to the cytoplasm, where they bind to Sm proteins. The newly assembled small nuclear ribonucleoprotein particles (snRNPs) are then transported back into the nucleus. Eventually, mature snRNPs are incorporated into spliceosomes in the nucleoplasm (Chari and Fischer, 2010). Bernard et al. demonstrated that UV-irradiated keratinocytes activate U1, U2, U4, and U6 RNAs which initiate Type I IFN signaling through binding with RNA sensing receptor TLR3 (Bernard et al., 2012; Borkowski et al., 2015). On the other hand, Ranoa et al. demonstrated that IR and chemotherapy induced the formation of RIG-I complexes with either snRNA U1 or U2 (Ranoa et al., 2016). Interestingly, mapping of U1 sequences involved in RIG-I interaction post-IR revealed that the same region (stem loops 1, 2, and part of 3 of U1) is recognized by TLR3 post-UV (Bernard et al., 2012, Ranoa et al., 2016). Indeed, synthetic constructs of these individual stem-loop regions or their combination are able to elicit IFNβ response in reporter cells (Ranoa et al., 2016). Considering recruitment of multiple DExD/H box RNA helicases in the assembly and functioning of spliceosome and their specific interactions with snRNAs (Chang et al., 2013), these data pose provocative questions about potential interactions of RLR DExD/H box RNA helicases with splicing machinery, which can be mediated by U1/U2 snRNAs. Especially interesting is that U1/U2 initiate splicing catalysis through specific binding with 5′ and 3′ splicing sites in intronic regions and their hijacking by RIG-I can potentially lead to the accumulation of unprocessed transcripts with retained introns (Matera and Wang, 2014).

Another type of small non-coding RNA is snoRNA. These small RNAs are involved in modifications of ribosomal RNA (Falaleeva et al., 2017). snoRNAs were demonstrated to act as ligands of PKR in the condition of metabolic stress but the functional consequences of these interactions remain unclear (Youssef et al., 2015).

Therefore, dsRNAs transcribed from different members of SINE/Alu, LINEs, and LTR/ERV families as well as snRNAs are well-documented cell-intrinsic ligands (RNA DAMPs) for TLR3, PKR, RIG-I/MDA5, and NLRP3 inflammasome (Fig. 6). They can be induced by different stress stimuli including IR, chemotherapy, kinase inhibitors, and epigenetic modifiers and lead to three major downstream events. These are Type I IFN response with cytotoxic and immune-activating functions activated through TLR3 and RIG-I/MDA5, translational shutdown and growth arrest activated through PKR, and inflammatory response/caspase 8 induction activated through NLRP3, TLR3/7/8, and RLRs. Accumulation of these ligands and their interaction with cytoplasmic sensors are positively controlled by Pol II/Pol III transcription and cell cycle-dependent integrity of nuclear membrane and negatively controlled by ADAR1, DICER1, cytoplasmic RNA degrading complex SKIV2L and RNase L (Chung et al., 2018; Kaneko et al., 2011; Kim et al., 2014a; Leonova et al., 2013; Li et al., 2017b; Rudin and Thompson, 2001; Tarallo et al., 2012).

5. EPIGENETIC SILENCING AND CHROMATIN CONFORMATION IN REGULATION OF RNA DAMPs

The ability of RNAs transcribed from repetitive RE genes to activate innate immune response through binding with RNA sensors (Fig. 6) suggest there exists regulatory mechanisms to silence the transcription of such genes. Such silencing is achieved through multiple regulatory mechanisms, including epigenetic modifications of DNA and chromatin, condensation of chromatin domains in heterochromatin structures, and recruitment of transcriptional suppressors in the promoter regions of integrated RE (Schlesinger and Goff, 2015; Thompson et al., 2016). DNA methylation of intrinsic Pol II and Pol III promoters of LINEs, LTR/ERV, and SINEs is well described in literature (see Goodier, 2016 for review). SINEs can also affect DNA methylation of neighboring chromatin regions. Integration of Alu1 in the genome can act as a nucleation event for increased methylation in this region and suppression of transcription of neighboring genes (Arnaud et al., 2000; Bakshi et al., 2016; Carnell and Goodman, 2003). As well, different histone modifications are associated with RE, leading to their transcriptional silencing (Goodier, 2016 and Fig. 7). Specific histone marks, associated with heterochromatic repressed regions of chromatin are often found in RE loci and participate in epigenetic silencing of these RNA DAMPs (Goodier, 2016; Yang and Wang, 2016). This indicates that suppressors of DNA methylation and histone modifications may lead to reactivation of RE and induce their interactions with RNA sensors. Indeed, inhibitors of DNA methylation and HDAC inhibitors can directly reactivate endogenous promoters in many RE.

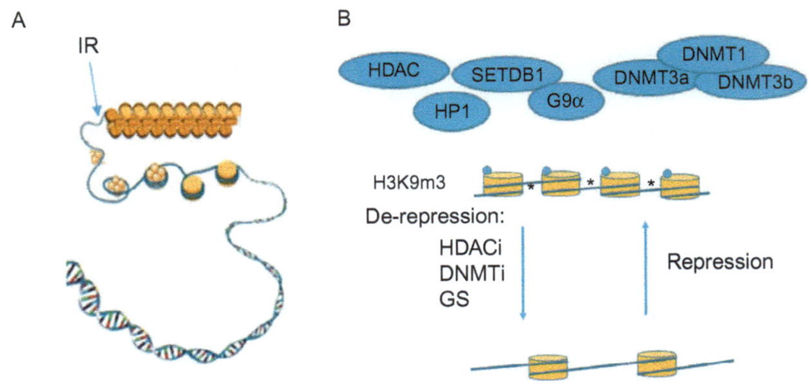

Fig. 7 Conformational and epigenetic changes in chromatin associated with transcriptional repression/activation of retrotransposable elements. (A) IR leads to chromatin decondensation through introduction of single-strand breaks and relaxation of chromatin loops. (B) Modifications of DNA and histones associated with repression/activation of LINEs and LTRs. Asterisks, 5-methylcytosine in DNA; blue dots, histone modifications, associated with inactive chromatin (mostly-H3K9m3); HDACi, inhibitors of histone deacetylases; DNMTi, inhibitors of DNA methyl transferases; GS, genotoxic stress; HP1, heterochromatin protein 1 (also known as chromobox homolog 5 CBX5); SETDB1 (set domain protein, bifurcated, 1), histone H3 methyltransferase; G9a (euchromatic histone-lysine *n*-methyltransferase 2; EHMT2), part of H3 methyltransferase complex; DNMT1, DNMT3a, DNMT3b, DNA methyltransferases.

This reactivation leads to the appearance of treatment-induced non-annotated TSSs (TINATS) which were significantly enriched in solitary LTRs of ERV/LTR12C family (Brocks et al., 2017). This is consistent with observations indicating that epigenetic modifiers activate LTR/ERVs and SINEs and recruit RLR/IFN-dependent pathway through binding with RIG-I (Chiappinelli et al., 2015; Leonova et al., 2013). H3K9me3-mediated chromatin repression of LINE-1 elements was found to be associated with lung cancer cell resistance to erlotinib and vemurafenib; HDAC inhibitors reactivated LINE-1 and led to resensitization of cell population to drugs (Brocks et al., 2017). In another report a member of the LINE1 family, L1MA2, was detected in complexes with RIG-I post-IR (Ranoa et al., 2016) (Table 1). As well genotoxic stress leads to the DNA damage through accumulation of ss and ds breaks; as a consequence supercoiled and highly compact chromatin domains become relaxed and transform into an open loop conformation (see Fig. 7 and Corless and Gilbert, 2016). This open conformation allows RNA polymerases and accessory transcriptional factors to gain access to promoter regions and initiate transcription start sites (TSS) that are otherwise hidden in untreated cells. This mechanism was described previously for ability of IR-induced endonucleases to target and "relax" chromatin regions enriched by LINE1 and Alu1 elements, which can potentially stimulate their transcription (Khodarev et al., 1998, 2000). Another mechanism is associated with stress-induced appearance of aberrant TSS in the sense and antisense strands, leading to the convergent transcription and formation of corresponding dsRNAs (Lin et al., 2014; Nakamori et al., 2011). This mechanism was demonstrated for regions with microsatellite tracts, where convergent transcription led to the induction of apoptotic cell death (Lin et al., 2010). More recently Portal et al. found multiple natural double-stranded small RNAs (ndsRNAs) produced as a result of convergent transcription in retinoic acid modulator region (RAM) of chromosome 8q24.21 with multiple functions including regulation of mitosis (Portal et al., 2015). These ndsRNAs were up-regulated by retinoic acid, indicating their response to external cues. Magnusson et al. demonstrated that TNF-alpha and LPS-treated fibroblasts induce multiple TSS in opposite strands, leading to the appearance of nascent dsRNAs (enhancer RNAs) in intergenic regions (Magnuson et al., 2015). Together these data indicate that stress induces transcriptional activation of retrotransposable elements, repressed on the basal level and also can lead to the aberrant formation of dsRNAs due to convergent transcription in sense and antisense orientation.

6. INTERCELLULAR AND INTRACELLULAR REDISTRIBUTION OF RNA DAMPs

RNA DAMPs can be delivered to RNA sensors through extracellular space (cell-extrinsic pathways) or as a result of intracellular redistribution (cell-intrinsic pathways). Extrinsic pathways are initiated by the release of dsRNAs or RNA-containing exosomes in the extracellular space, which can be induced by genotoxic stress or interactions between tumor and stromal cells (Bernard et al., 2012; Boelens et al., 2014; Borkowski et al., 2015; Takemura et al., 2014). Free ds and ssRNA molecules can further enter the intracellular space through yet unidentified mechanisms and activate Type I IFN signaling through interactions with Toll-like receptors (see Fig. 2A). Exposure of intestinal epithelial cells to gamma irradiation led to the release of dsRNAs into the extracellular space in a p53-dependent manner (Takemura et al., 2014). TLR3−/− animals and TLR antagonists were used to demonstrate that neighboring cells are activated by these dsRNA molecules in a TLR3/p53-dependent fashion (Takemura et al., 2014).

Boelens et al. (2014) described a RAB27B-dependent release of exosomes from stromal cells with their subsequent uptake by neighboring tumor cells. These exosomes were enriched with LINE/LTR sequences and a member of SINE family RN7SL1 RNA and activated Type I IFN signaling in target tumor cells through preferential activation of RIG-I.

In eukaryotes, compartmentalization is an essential determinant in preventing unwanted activation of cytosolic RNA sensors by self-RNAs. Ranoa et al. demonstrated that IR led to the redistribution of RIG-I stimulatory ligands U1 and U2 from the nucleus to the cytoplasm (Ranoa et al., 2016). Kim et al. demonstrated that protein kinase R (PKR) can also be activated by dsRNAs comprised by inverted Alu1 repeats that are released in the cytoplasm due to the transient breakdown of the nuclear envelope during mitosis (Kim et al., 2014a). As a special type of redistribution, we may consider the ability of some RNA ligands to release from complexes with RNA-binding proteins (RBP) and further interact with RLR receptors (process, named "unshielding"). When RIG-I ligand RN7SL1 was complexed with SRP proteins SRP9 and SRP14 it was "shielded" from interaction with RIG-I and did not induce RIG-I-dependent Type I IFN response. However, if Notch/c-Myc signaling was activated in fibroblasts, RN7SL1 was overexpressed and a significant fraction of this RNA remained in the free form (unshielded) and successfully activated RIG-I (Nabet et al., 2017). Similarly, Chiang et al. recently demonstrated that HSV-1 infection leads to nuclear/cytoplasmic relocalization of 5S pseudogene RNA5SP141, its dissociation from 5S binding cellular and mitochondrial ribosomal proteins RLP5 and MRPL18, and subsequent binding with RIG-I (Chiang et al., 2018). Therefore, genotoxic stress can lead to the redistribution of intrinsic dsRNA molecules and their receptors and co-localize them in compartments where they can activate innate immune responses. Further investigation is necessary to characterize these mechanisms and the potential implications of these interactions. Together these data indicate there are various pathways which can lead to the activation of RNA-dependent pattern recognition receptors in tumor cells subjected to genotoxic stress.

7. STRESS GRANULES, P-BODIES, AND PML NUCLEAR BODIES AS CELLULAR COMPARTMENTS OF RNA STRESS RESPONSE AND IFN SIGNALING

Uncontrolled interactions of endogenous RNA ligands with cytoplasmic RNA sensors represent RNA stress, which can be induced by cytotoxic treatments associated with cancer therapy, radiation injury, hypoxic conditions, metabolic alterations, and other conditions which can lead to accumulation and cellular redistribution of different types of RNA. There is a group of cellular structures which contain RNA molecules, different RNA processing enzymes and RNA-binding proteins (RBPs), which are induced by different stress conditions and are associated with elimination of unwanted RNAs. Among them best described are stress granules (SG) and processing bodies (PB). Stress granules are membraneless cytoplasmic structures which contain translation initiation factors, 40S ribosomal subunits, and different RNA molecules, mostly mRNAs directed toward degradation. Formation of SGs is induced by phosphorylation of eIF2, which is provided by activated PKR (see Fig. 2) or several others stress-activated serine/threonine protein kinases (see Anderson et al., 2015 for review). Phosphorylation of eIF2 leads to the suppression of translation initiation and formation of stalled 48S complexes of 40S ribosomal subunits with mRNAs (Kedersha et al., 2002). Nucleation of SGs is initiated by aggregation-prone RNA-binding proteins TIA-1 and G3BP1, which are major markers of these subcellular structures (Anderson et al., 2015). SGs can further fuse/interact with PBs, where latter are specialized bodies for degradation of RNAs. PBs contain exonucleases XPN1 and members of exosome SKIV2L complex, involved in the surveillance of cytoplasmic pool of retroelements RNA (see Fig. 6), enzymes responsible for mRNA decapping, deadenylase complex CCR4/CAF1/NOT, nonsense-mediated mRNA decay proteins (Conti and Izaurralde, 2005), and members of si/microRNA degradation machinery, including Argonautes proteins (Liu et al., 2005). Therefore P-bodies are cytoplasmic sites of RNA processing and degradation, responsible for elimination of unwanted RNA species from cytoplasm. SG/PBs act as negative regulators of RNA-dependent innate immunity system depleting endogenous untranslated RNA. On the other hand, they can act as a cellular compartment for initial interactions of RNA ligands with appropriate RNA sensors. Recent data indicate that recognition of viral RNAs by RIG-I involves sequential interactions of RIG-I, TRIM25, and MAVS with SGs (Sanchez-Aparicio et al., 2017). Primary recognition of RNA by Rig-I occurs in SGs in complexes with TRIM25 (see Fig. 3) while activated Rig-I moves from SG to mitochondrion compartment where it interacts with MAVS. Mengovirus infection led to the PKR-dependent formation of SGs and activation of IFN signaling; MDA5 was found to be localized in these SG, but this localization was not necessary for induction of IFN signaling (Langereis et al., 2013). Perhaps similar to

Rig-I in this system localization of MDA5 in SG was important for initial interactions with ligands while activation of downstream signaling was associated with MAVS interaction in mitochondria (see Sanchez-Aparicio et al., 2017). Another granular compartment-PML nuclear body (PML-NB), or nuclear domain 10 bodies (ND10, see Lopez et al., 2002; Xu et al., 2016; Xu and Roizman, 2017) also attracted attention in the context of innate immunity signaling (see Scherer and Stamminger, 2016 for review). PML-NB comprised several cellular proteins (PML, SP100, ATRX, and hDAXX), which are modified by SUMO, form distinguishable nuclear granules after infection by DNA- and RNA-containing viruses and are involved in the regulation of IFN response induced by interaction of innate immunity nucleic acid sensors with cognate ligands (Bernardi and Pandolfi, 2007; Scherer and Stamminger, 2016). PML-NB can interact and regulate activity of IRF3, MHCII transcriptional co-activator CIITA and Stat1, thereby activating Type I IFN response, tumor neoantigens cross-presentation by APC/DC, and activating ISG response to IFN treatments in cancer and host cells. These data indicate that RNA sensoring pathways can be regulated by specific cytoplasmic and nuclear bodies, which potentially can interact with each other. How these structures are associated with recognition of specific RNA ligands by different types of RNA sensors is important question for further investigations.

8. RELATIONSHIPS BETWEEN RNA-DEPENDENT AND DNA-DEPENDENT PATHWAYS IN TUMOR AND HOST IMMUNE CELLS

Tumor tissue represents a complex mixture of cancer, stromal, and immune cells. Local irradiation or systemic therapy delivers genotoxic injury to both compartments during cancer therapy. Traditionally mechanisms of innate immunity and molecular functions of different PRRs were described in relation to different populations of host immune cells. However, data summarized in this review indicate that tumor cells are also equipped by innate immunity system which can recognize stress-induced RNA DAMPs in a cell autonomous context and activate Type I IFN signaling in cell populations subjected to stress. Considering the potential co-existence of different nucleic acid sensing mechanisms in each tumor and host cell, will these cells respond similarly to the same stress stimuli or their response will be based on different nucleic acid sensing mechanisms? Unequal usage of some DNA and RNA sensing pathways in tumor and host immune cells started to attract attention in recent publications (Lurescia et al., 2018; Seelige et al., 2018). For example, IR leads to the activation of DNA-dependent cGAS/STING pathways in myeloid and dendritic cells (Deng et al., 2014, 2016; Xu et al., 2017a). Why does the same genotoxic insult lead to the preferential activation of RLR sensing system in different types of tumor cells with LGP2 as a common radioprotector (Ranoa et al., 2016; Widau et al., 2014)? One possible explanation may be the ability of given cell types to respond to genotoxic stimuli by the accumulation of DNA fragments in the cytoplasm, as was previously demonstrated in early work using lymphocytes (Wyllie et al., 1984). Cells of mesenchymal and hematopoietic origin appear to be more prone to apoptotic DNA cleavage and accumulation of nucleosomal DNA in the cytoplasm as compared to solid tumors arising from exodermal and endodermal epithelial cells (Green and Scott, 1994; Rich et al., 2000). The data indicate this is due in part to the ability to activate apoptotic nucleases in different cell types (Khodarev et al., 1998, 2000). Our data indicate that epithelial tumors and glial cells accumulate less cytoplasmic DNA as compared with hematopoietic cells in the response to IR (see Fig. 8). We hypothesize that epithelial tumors harbor a restricted capacity to induce apoptotic DNA fragmentation to genotoxic stress and respond primarily by an RNA stress response while host immune cells respond by endonucleolytic cleavage of genomic DNA resulting in an efflux of DNA fragments into the cytoplasm or uptake of fragmented DNA from dying tumor cells and activation of STING-dependent pathways (see Fig. 8). Indeed literature data indicate association of several nucleases with apoptotic functions with activation of DNA signaling pathways in different types of cells. Activity of endonuclease Mus81 is associated with amount of cytosolic DNA in immortalized MEFs, prostate cancer cells, and B cell lymphoma cell lines; AT-rich DNA fragments, produced by Mus81 enter the cytoplasm and activate cGAS/Sting/IFN-beta pathway, consistent with tumor-suppressive functions of this nuclease (Ho et al., 2016). Deoxyribonuclease 1 was associated with apoptotic DNA cleavage (Keyel, 2017; Peitsch et al., 1994). Recent data indicate that DNase 1 and DNase 1L3 can be associated with susceptibility to systemic lupus erythemathosus through activation of cytoplasmic DNA sensoring pathways (Al-Mayouf et al., 2011; Yasutomo et al., 2001). Similarly, DNase II was identified as enzyme responsible for induction of apoptotic DNA fragmentation in some cancer cells (Barry and Eastman, 1993; Keyel, 2017). Recently DNase II was associated with cytoplasmic sensing of mitochondrial DNA, preventing activation of TLR9 and AIM2 (Baum et al., 2015; Oka et al., 2012).

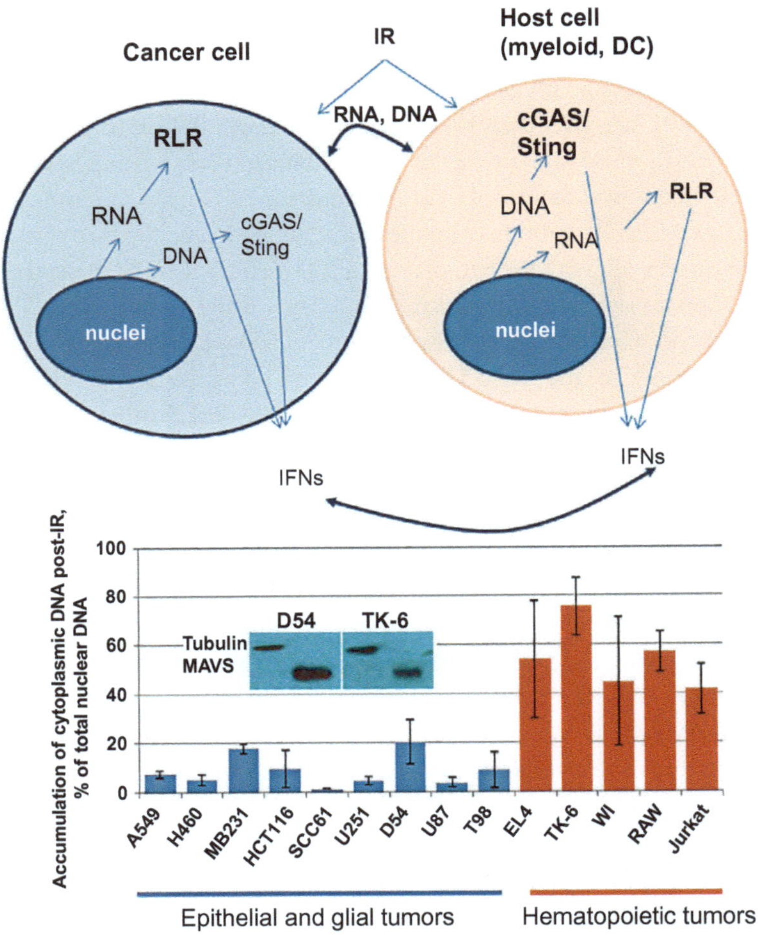

Fig. 8 RNA and DNA sensing pathways in cancer and host immune cells. (A) RNA and DNA sensing pathways co-exist in both cancer and host immune cells but may have different abundance and functional significance dependent on the cell context. IFNs produced in each cell compartment can act on the counterpart cells; similarly, DNA and RNA released in the extracellular space after stress stimuli can activate nucleic acid sensoring pathways both in cancer and host cells. (B) Accumulation of DNA fragments in the cytoplasm of irradiated cells. Different cell lines were preincubated for 48 h with 3H-thymidine, irradiated at 5 Gy and 24 h post-IR cytosolic and nuclear/mitochondrial fractions were separated according to Liu and Fagotto (2011). Cytosolic DNA is presented as a fraction of nuclear/mitochondrial DNA; inset shows distribution of cytoplasmic marker tubulin and mitochondrial marker MAVS in obtained fractions.

This hypothesis is consistent with recent data which demonstrated that cGAS/STING pathway is suppressed in many types of tumor cells. For example, the majority of melanoma and colorectal cancer cell lines tested demonstrated impaired responses and/or absent expression of cGAS,

STING, or both due to epigenetic silencing of these genes (Xia et al., 2016a, b). Recurrent losses of cGAS/STING expression were also demonstrated in clinical samples and thought to be related to epigenetic silencing rather than with mutational events. So at least in some types of cancer cells stress-induced activation of Type I IFN signaling can be achieved through RNA-dependent RLR signaling rather than DNA-dependent cGAS/STING activation. If such observations will be proved clinically this may provide important opportunity to target genotoxic therapy and immunotherapy to the tumor cell compartment through differential activation of PRR pathways preferentially expressed in the tumor cells of the given patient.

9. ROLE OF RNA SENSORS IN CONVENTIONAL THERAPY AND IMMUNOTHERAPY OF CANCER

To a large degree, the functional significance of nucleic acid sensing in cancer therapy is determined by the induction of IFN signaling. IFNs act on several cell types in the tumor microenvironment, including tumor, stromal, and immune cells (see Borden et al., 2007; Cheon et al., 2014; Minn, 2015; Parker et al., 2016). IFN signaling is complex in that it can potentially leads to pro-death or pro-survival effects (Khodarev et al., 2004, 2007, 2012). Multiple clinical trials of IFN administration as anti-tumor therapies were developed since the early 1980s (Borden et al., 2007; Parker et al., 2016). Unfortunately, these trials demonstrated significant dose-limiting adverse effects of systemic IFN administration. Current treatment strategies have focused on the immune-modifying effects of IFNs, which provide a rational for combination of IFN/RNA-based therapy with immune checkpoint therapy (ICT), which recently demonstrated significant clinical success (Herrera et al., 2016; Weichselbaum et al., 2017). In this section, we outline the direct effects of IFNs on tumor cells in the context of tumor responses to conventional genotoxic therapy and the immune effects of IFNs in the context of immunotherapy. We also discuss approaches to target these treatments to tumor cells and to combine the therapeutic effects from cell-intrinsic and immunogenic actions of RNA sensors in tumor cells.

9.1 Action of IFNs on Tumor Cells and Role of RLRs in Conventional Cancer Therapy

In tumor cells, IFNs lead to growth arrest and cell cycle perturbations and also regulate cell death through induction of apoptosis or necroptosis

(Borden and Williams, 2011; Minn and Wherry, 2016). The anti-proliferative action of IFNs has been attributed, in part, to the up-regulation of p21 (CDKN1A), which is a p53-dependent inhibitor of CDK2 activity. Activation of p21 leads to G1/S arrest and to lesser degree G2/M arrest; it also drives tumor cells into a senescent state (Georgakilas et al., 2017; Hobeika et al., 1997; Sharpless and Sherr, 2015). Additionally, IFN-alpha can activate the p38 MAPK pathway leading to growth arrest (Chiang et al., 2006; Zhao et al., 2011). IFNs induce apoptosis through extrinsic pathways by activation of death receptors, and intrinsic pathways through caspase 9 activation that ultimately leads to the activation of effector caspases 3 and 7 (Green and Llambi, 2015). Death receptor signaling can be activated by induction of specific ligands, such as TRAIL or Fas/CD95 (Bernardo et al., 2013) or by direct binding of the transcription factor IRF1 with corresponding promoter elements in caspase 8 (Casciano et al., 2004; Ruiz De Almodovar et al., 2004). The cell-intrinsic caspase 9-dependent pathway can be activated by the IFN-inducible protein XAF-1, which inhibits X-linked inhibitor of apoptosis XIAP and therefore activates caspase 9 and effector caspases 3/7 (Liston et al., 2001). IFNs can also induce necroptosis, a mode of regulated cell death morphologically related to necrosis but specifically induced by the signaling pathways which recruit receptor-interacting serine/threonine protein kinases 1 and 3 (RIPK1 and RIPK3) and MLKL (mixed lineage kinase domain-like pseudokinase) (Galluzzi et al., 2018). Necroptosis can be activated by TNFR1, TLR3/4, and the DNA-interacting sensor DAI. IFNAR1 −/− and IFNGR1 −/− mice are resistant to the induction of necroptosis, indicating that both Type I and Type II IFNs are pivotal for induction of this mode of the cell death (Galluzzi et al., 2018). Overall these selected examples indicate that IFNs have multiple cell-intrinsic effects on tumor cells, which leads to the suppression of proliferation and induction of cell death (see Borden et al., 2007; Chawla-Sarkar et al., 2003; Kotredes and Gamero, 2013 for in-depth reviews). Therefore, the ability of cancer therapies to induce IFN production through activation of RNA or DNA sensing pathways can increase the cytotoxic insult delivered by therapy.

Consistent with these considerations, Takemura et al. tested the response of wild-type and TLR −/− mice to whole-body irradiation at doses leading to gastro-intestinal syndrome (GIS) and lethality as a result of crypt cell death (Takemura et al., 2014). They found that animals with TLR3 knockout were significantly more resistant to total body irradiation (TBI). Death of animals was associated with destruction of intestinal villi and accumulation of TUNEL-positive and caspase-3 positive cells, consistent with apoptotic degradation. Administration of TLR3 ligand poly(I:C) alone or in combination with TBI augmented cell death and suppressed the amount of proliferating cells in TLR3 wild type, but not TLR3 knockout animals. Recruitment of TLRs was initiated by the p53-dependent extracellular accumulation of dsRNA and suppressed by low molecular weight inhibitor of dsRNA/TLR3 binding. Finally, downstream signaling from TLR3 was associated with activation of a TRIF-RIPK1 pathway consistent with necroptosis signaling (Galluzzi et al., 2018; Takemura et al., 2014). Thus IR activates dsRNA signaling pathway which recruits TLR3 and TRIF-RIPK1 signaling and agonist of the TLR3 receptor augmented this cytotoxic pathway and inhibitor mitigated its toxicity (Takemura et al., 2014).

Widau et al. used a targeted siRNA screen against ISGs associated with tumor radio/chemoresistance and found that LGP2 was the major radioprotective gene in 14 cancer cell lines representing seven different cancer histologies (Widau et al., 2014). Knockdown of LGP2 led to 10–100-fold radiosensitization of cancer cells in vitro. In addition, neutralizing antibodies against IFN-beta partially rescued IR-induced LGP2-dependent cell death indicating that the radioprotective functions of LGP2 at least partially are associated with the suppression of the Type I IFN production. These data were the first to identify involvement of RLRs in the cellular response to IR and demonstrated that inhibition of LGP2 leads to radiosensitization of tumor cells. Consistent with these findings, Ranoa et al. demonstrated that LGP2 knockout animals are significantly more sensitive to TBI at 5.5 Gy, with median survivals of 11 days vs 14 days for wild-type animals. LGP2 knockout also led to a significant accumulation of IFN-beta in the serum after IR as compared with wild-type animals (Ranoa et al., 2016). Similar to observations of Takemara et al., lethality was associated with GIS and caspase-3-dependent apoptosis in enterocytes lining the microvilli. Importantly, depletion of RIG-I, but not MDA5, led to significant radioprotection of animals, suppression of IFN-beta accumulation in the serum after TBI, and inhibition of apoptotic cell death of enterocytes (Ranoa et al., 2016). These data revealed that the balance between LGP2 and RIG-I is critical for the cellular response to IR, where RIG-I signaling can increase and LGP2 signaling can suppress IR delivered cytotoxicity. Similarly various DNA-damaging chemotherapies, including cisplatin, doxorubicin, and etoposide, induced apoptosis in normal and tumor cells with wild-type RIG-I. Suppression of RIG-I abrogated the apoptotic response and IFN signaling (Ranoa et al., 2016). Furthermore, RIG-I suppression in xenografted tumors rendered them radiosensitive. Together these data indicate that RNA signaling in normal and tumor cells is involved in the response to IR and chemotherapy, and activation of RNA signaling leads to increased cytotoxic responses while suppression of RNA signaling leads to reduced cytotoxicity and decreased therapeutic effects.

9.2 Action of IFNs on Immune Cells and Role of RNA Sensors in Cancer Immunotherapy

Activation of anti-tumor immunity is one of the major breakthroughs in cancer therapy during the last decade. The primary mechanism for immune checkpoint therapy involves elimination or suppression of the feedback inhibition of adaptive immune response (see Herrera et al., 2016; Minn, 2015; Weichselbaum et al., 2017 for in-depth reviews). Immune checkpoints are presented by specialized receptors expressed on the cell membrane of T cells and involved in the negative regulation of T cell functions in the adaptive immune response. The best studied examples of such receptors are programmed cell death 1(PD1)-PD1 ligands1/2 (PD-L1/PD-L2) system and CTL antigen 4 (CTLA4). The PD1 receptor is up-regulated on the surface of effector T cells upon antigen-specific activation. Its cognate ligand PD-L1 can be expressed on many cell types including tumor cells, APC/DC, and tumor-associated macrophages (TAMs) (Herrera et al., 2016; Minn and Wherry, 2016). Interaction of PD1 and PD-L1 leads to the suppression of T cell function. Consistently, neutralizing antibodies to PD1 or PD-L1 can eliminate this feedback inhibition, reactivate anti-tumor immune response, and lead to the robust tumor regression (Borghaei et al., 2015; Brahmer et al., 2015). Similarly, CTLA4 is expressed on effector T cells and can interact with CD80/CD86 ligands on APC/DC cells thereby attenuating T cell activation. Neutralizing antibodies to CTLA4 can reactivate T cells and have demonstrated synergistic effects with radiotherapy, both for local control and for control of distant metastases (Herrera et al., 2016; Ruocco et al., 2012; Twyman-Saint Victor et al., 2015).

IFN signaling is critical for both regulation of immune checkpoint receptors and activation of adaptive immune responses. PD-L1 is regulated through IFN/JAK signaling, which can potentially induce different outcomes depending on the timing and intensity of exposure to IFNs (Cheon et al., 2014; Lee et al., 2006; Mostafavi et al., 2016). Recent data show that the PD-L1 promoter region contains IRF1, STAT1/STAT3, and STAT2/STAT5 binding sites, which control the expression of PD-L1 mRNA

(Garcia-Diaz et al., 2017). Activation of IFN signaling in tumor cells can lead to the expression of PD-L1 and therefore render them resistant to T cell-dependent adaptive immune response. On the other hand, IFN signaling is necessary for anti-tumor immunity by activating APCs and CD8+ T cell cytotoxic responses (Burnette et al., 2011; Deng et al., 2014; Yang et al., 2014). IFNs can also increase the efficacy of antigen presentation in the context of MHC-I in APCs/DCs through regulation of MHC and proteasome expression (Kessler et al., 2002; York et al., 1999). They also participate in the formation of chemokine networks responsible for the recruitment of APCs and T cells from draining lymph nodes to the tumor microenvironment (Bauer et al., 2006; Corrales et al., 2017). Such combination of pro-survival and pro-death features is typical for IFN signaling and should be considered in further strategies exploiting this signaling in cancer therapy (Cheon et al., 2014; Khodarev et al., 2007, 2012; Minn and Wherry, 2016).

Consistent with these considerations, activation of RNA sensors associated with induction of IFN signaling (see Fig. 2) is important for adaptive immune response against tumor cells. Early studies demonstrated that synthetic polyribonucleotides can activate IFN signaling in tumor cells and exert anti-tumor action in clinical settings. Multiple clinical trials demonstrated that administration of poly(I:C) or poly(A:U) alone or in combination with genotoxic therapy led to improved clinical responses in patients with bladder cancer, gastric cancer, low grade gliomas, and breast cancer (Aranda et al., 2014; Hartman et al., 2014; Parker et al., 2016; Qin et al., 2014). Importantly, these responses were not accompanied by the detrimental side effects associated with administration of purified or recombinant interferons (reviewed in Borden et al., 2007; Parker et al., 2016). Currently it is well established that synthetic polyribonucleotides are the ligands for several RNA sensors, including TLR3 and TLR8; they also can activate MDA5 and PKR (see Fig. 2 and Karpus et al., 2012; Swetha and Ramaiah, 2015). This notion provides an opportunity to combine RNA sensor activation by conventional cancer therapies with current immune checkpoint therapies. In this regard, it was demonstrated that suppression of IFN signaling reduces the efficacy of anti-CTLA-4 therapy in preclinical models (Benci et al., 2016) and is associated with resistance to anti-PD-1 therapy in melanoma patients (Zaretsky et al., 2016). Different endogenous retroelements can be transcriptionally activated in cancer cells and activate RNA sensing innate immunity pathways (see Figs. 6 and 7). Recent data indicate that expression of ERVs in tumor samples was associated with markers of histone modifications and immune checkpoint markers in multiple cancer types. These data suggested that ERV transcription can be potentially associated with efficacy of ICT which relies on the expression of immune checkpoint receptors (Panda et al., 2018). Expression of ERV3.2 and ERVK.2 was also significantly associated with response to PD-1 blockade in patients with renal cell carcinoma (Panda et al., 2018). These indicate that endogenous ligands of RNA sensors can act as potential predictors of individual sensitivity to immune checkpoint therapy. Further investigations of ICT combinations with RLR activation are on-going in clinical trials (Barsoum et al., 2017; Panda et al., 2018).

9.3 Dualism of IFN Signaling and Potential Strategies to Activate Cytotoxic Functions in Tumor Cells

One of the most enigmatic features of IFN signaling is its ability to simultaneously activate pro-death and pro-survival pathways (Khodarev et al., 2004, 2007; Minn and Wherry, 2016; Parker et al., 2016; Platanias, 2005). IFNs alpha, beta, and gamma and their master transcriptional factors Stat1, Stat2, IRF9, and IRF3/7 can act as tumor suppressors and oncogenes depending on the cellular context, timing of treatment, and conditions of experiments. ISGs associated with induction of IFN signaling can act as biomarkers of cytotoxic response or biomarkers of tumor resistance to genotoxic therapy (Berry et al., 2012; Cheon et al., 2014; Khodarev et al., 2004, 2007; Minn, 2015; Weichselbaum et al., 2008). To explain this apparent paradox, we suggested a selection hypothesis where tumor-initiating clones are initially sensitive to direct action of IFNs and genotoxic stress, whereas their selection in the tumor microenvironment which is saturated by cytotoxic IFNs leads to the formation of cross-resistant clones that are resistant to IFNs and radio/chemotherapy (Khodarev et al., 2007, 2009, 2012). Recent clinical data provide direct proof of this hypothesis. Zaretsky et al. demonstrated that tumors which relapsed after failure of anti-PD-1 therapy in melanoma patients contain loss-of-function mutations in Jak1 and Jak2 kinases. Using patient-derived cell lines they demonstrated that such mutations abrogate Stat1 phosphorylation in response to IFN gamma and/or IFN-alpha/beta and make tumor cells resistant to anti-proliferative/pro-death signaling of IFNs (Zaretsky et al., 2016). This means that the direct action of IFNs on tumor cells indeed is responsible for the selection against tumor clones in cancer patients and that such selection leads to the formation of IFN resistant aggressive tumor clones, as predicted by selection hypothesis (Khodarev et al., 2012). Therefore optimization and targeting of therapies which involve IFN signaling must be based on understanding of this bi-modal action of IFNs on tumor and host immune cells. Diversified system of RNA sensors may provide an excellent opportunity to flexibly regulate IFN signaling in individual tumor microenvironments (TME). Overall this direction of investigations is at the very initial stages and, other than synthetic polyribonucleotides poly(I:C) and poly(A:U), little is known regarding pharmacological agonists and antagonists of different RNA sensors.

One of the attractive targets is RIG-I. Contrary to cGAS/STING, RIG-I is ubiquitously expressed in tumor cells and induces cell death preferentially in tumor cells (Barsoum et al., 2017; Ranoa et al., 2016). This indicates that activation of RIG-I can target cytotoxicity to tumor cell compartment, rather than induce detrimental effects in normal or immune cells. Some evidence suggests that RIG-I activation can occur through application of all-trans-retinoic derivatives and antioxidants. *Acitretin* is a potent agonist of the RLR pathway and induces Type I IFN expression in HIV-infected cells (Li et al., 2016). *Tretinoin*, another all-trans-retinoic acid derivative activates RIG-I and an IRF3/IFN-beta pathway in BRAF-mutated melanoma cells (Szabo et al., 2016). Engel et al. found that tumor hypoxia suppresses RIG-I expression in HIF1-alpha- and ROS-dependent manner while administration of vitamin C restores RIG-I expression and functionality through ROS scavenging (Engel et al., 2017). Identification of tumor-derived RNA ligands of RIG-I and/or MDA5 could be used to sensitize tumors to radiation therapy and certain chemotherapies. Recent data indicate that the stem-loop regions of snRNAs can specifically activate RLR signaling through selective activation of RIG-I (Ranoa et al., 2016). Chiang et al. described optimization of VSV-derived 5′ppp dsRNA, which led to synthesis of M5 and M8 oligoribonucleotides with strong specificity to RIG-I, as compared with MDA5 and TLR3 (Chiang et al., 2015). M5 and M8 (59 and 99 nucleotides, respectively) formed stem-loop structures with just single-nucleotide loops and demonstrated high activity in induction of Type I IFN signaling and ability to suppress replication of Dengue and Influenza viruses. Barsoum et al. described fully synthetic RNA derivative designated as RGT100, which was specific to RIG-I (Barsoum et al., 2017). Intratumoral or systemic injections of RGT100 in mice bearing B16 melanoma or CT26 colorectal carcinoma tumors led to local and systemic suppression of contralateral tumors and lung metastases. Application of RGT100 induced robust and prolonged Type I IFN response and strong activation of NK-mediated anti-tumor activity. RGT100 is currently in phase I/II clinical trial (NCT03065023). Another approach to target RIG-I in tumor cells came from observations that addition of 5′ppp groups to siRNA mimics did not change their targeting specificity. Therefore "bifunctional" RNA agonists were designed which combine ability to target preferentially RIG-I due to preferred 2D structure and structure of 5′ end of RNA ligand and suppress important tumor genes with oncogenic functions. The validity of this approach was demonstrated using 5′ppp siRNAs against TGF-beta (Ellermeier et al., 2013), Bcl2 (Poeck et al., 2008), glutaminase (Meng et al., 2014), and MDR1 (Li et al., 2017a). In all cases, suppression of targeted genes was accompanied by the activation of RIG-I, IFN signaling, and apoptotic responses in tumor cells. Therefore, further detection and characterization of structural and functional properties of RNA ligands of different RNA sensors could provide new approaches to the rational design of drugs with abilities to preferentially activate cytotoxic pathways in tumor cells with simultaneous activation of IFN-dependent innate and adaptive anti-tumor immunity.

10. CONCLUDING REMARKS AND FUTURE DIRECTIONS

IFN signaling is unusual in its ability to mediate both pro-death and pro-survival signals and to act simultaneously on host and tumor cells. The ability of tumor cells to induce IFN signaling is an important factor modulating tumor development, response to therapy, and recruitment of the host immune system (Cheon et al., 2014; Khodarev et al., 2012; Minn, 2015). IR and chemotherapy induce IFN signaling in tumor cells through recruitment of RNA sensors of innate immunity and recognition of stress-induced RNA molecules as danger-associated molecular patterns (DAMPs). Accumulation of RNA DAMPs after genotoxic insults is considered an RNA stress. A significant fraction of RNA DAMPs is presented by transcripts from human retrotransposable elements and endogenous retroviruses, thus revealing new and unusual functions of these genomic parasites in the response of cancer cells to genotoxic stress. Another common group of RNA DAMPs is small non-coding RNAs, including snRNAs that are involved in splicing. The potential link between stress response, RNA sensors, and splicing may be an important direction for further investigations. Several agonists and antagonists of different RNA sensors were investigated as potential drugs in anti-tumor therapy with encouraging results. Overall the data presented in the current review indicate that tumor cells are equipped with cell-intrinsic RNA-dependent innate immune systems, which could be used to improve tumor responses to anti-cancer genotoxic and immune checkpoint therapies.

ACKNOWLEDGMENTS

Author is grateful to Dr. Ralph R. Weichselbaum for fruitful discussions of the manuscript; Dr. Sean P. Pitroda for discussions and help in preparation of manuscript; and Xiaona Huang and Tara P. Mojtahed for technical assistance with experiments described in Fig. 8. This work was partially supported by the Lung Cancer Research Foundation (LCRF) and by the Virginia and D.K. Ludwig Fund for Cancer Research (to R.R.W.).

REFERENCES

Ablasser, A., Bauernfeind, F., Hartmann, G., Latz, E., Fitzgerald, K.A., Hornung, V., 2009. RIG-I-dependent sensing of poly(dA:dT) through the induction of an RNA polymerase III-transcribed RNA intermediate. Nat Immunol. 10, 1065–1067.

Ahn, J., Barber, G.N., 2014. Self-DNA, STING-dependent signaling and the origins of autoinflammatory disease. Curr. Opin. Immunol. 31, 121–126.

Akira, S., Uematsu, S., Takeuchi, O., 2006. Pathogen recognition and innate immunity. Cell 124, 783–801.

Al-Mayouf, S.M., Sunker, A., Abdwani, R., Abrawi, S.A., Almurshedi, F., Alhashmi, N., et al., 2011. Loss-of-function variant in DNASE1L3 causes a familial form of systemic lupus erythematosus. Nat. Genet. 43, 1186–1188.

Amundson, S.A., Grace, M.B., Mcleland, C.B., Epperly, M.W., Yeager, A., Zhan, Q., et al., 2004. Human in vivo radiation-induced biomarkers: gene expression changes in radiotherapy patients. Cancer Res. 64, 6368–6371.

Anderson, P., Kedersha, N., Ivanov, P., 2015. Stress granules, P-bodies and cancer. Biochim. Biophys. Acta 1849, 861–870.

Aranda, F., Vacchelli, E., Obrist, F., Eggermont, A., Galon, J., Sautes-Fridman, C., et al., 2014. Trial watch: Toll-like receptor agonists in oncological indications. Oncoimmunology 3, e29179.

Aregger, M., Cowling, V.H., 2017. Regulation of mRNA capping in the cell cycle. RNA Biol. 14, 11–14.

Arnaud, P., Goubely, C., Pelissier, T., Deragon, J.M., 2000. SINE retroposons can be used in vivo as nucleation centers for de novo methylation. Mol. Cell. Biol. 20, 3434–3441.

Bakshi, A., Herke, S.W., Batzer, M.A., Kim, J., 2016. DNA methylation variation of human-specific Alu repeats. Epigenetics 11, 163–173.

Barry, M.A., Eastman, A., 1993. Identification of deoxyribonuclease II as an endonuclease involved in apoptosis. Arch. Biochem. Biophys. 300, 440–450.

Barsoum, J., Renn, M., Schuberth, C., Jakobs, C., Schwickart, A., Schlee, M., et al., 2017. Selective stimulation of Rig-I with a novel synthetic RNA includes strong anti-tumor immunity in mouse tumor models. Cancer Immunol. Res. 5, B44.

Bauer, J.W., Baechler, E.C., Petri, M., Batliwalla, F.M., Crawford, D., Ortmann, W.A., et al., 2006. Elevated serum levels of interferon-regulated chemokines are biomarkers for active human systemic lupus erythematosus. PLoS Med. 3, e491.

Baum, R., Sharma, S., Carpenter, S., Li, Q.Z., Busto, P., Fitzgerald, K.A., et al., 2015. Cutting edge: AIM2 and endosomal TLRs differentially regulate arthritis and autoantibody production in DNase II-deficient mice. J. Immunol. 194, 873–877.

Belanger, F., Stepinski, J., Darzynkiewicz, E., Pelletier, J., 2010. Characterization of hMTr1, a human Cap1 2'-O-ribose methyltransferase. J. Biol. Chem. 285, 33037–33044.

Benci, J.L., Xu, B., Qiu, Y., Wu, T.J., Dada, H., Twyman-Saint Victor, C., et al., 2016. Tumor interferon signaling regulates a multigenic resistance program to immune checkpoint blockade. Cell 167, 1540–1554.e12.

Bernard, J.J., Cowing-Zitron, C., Nakatsuji, T., Muehleisen, B., Muto, J., Borkowski, A.W., et al., 2012. Ultraviolet radiation damages self noncoding RNA and is detected by TLR3. Nat. Med. 18, 1286–1290.

Bernardi, R., Pandolfi, P.P., 2007. Structure, dynamics and functions of promyelocytic leukaemia nuclear bodies. Nat. Rev. Mol. Cell Biol. 8, 1006–1016.

Bernardo, A.R., Cosgaya, J.M., Aranda, A., Jimenez-Lara, A.M., 2013. Synergy between RA and TLR3 promotes type I IFN-dependent apoptosis through upregulation of TRAIL pathway in breast cancer cells. Cell Death Dis. 4, e479.

Berry, C.M., Hertzog, P.J., Mangan, N.E., 2012. Interferons as biomarkers and effectors: lessons learned from animal models. Biomark. Med. 6, 159–176.

Besse, S., Rebouillat, D., Marie, I., Puvion-Dutilleul, F., Hovanessian, A.G., 1998. Ultrastructural localization of interferon-inducible double-stranded RNA-activated enzymes in human cells. Exp. Cell Res. 239, 379–392.

Blackford, A.N., Jackson, S.P., 2017. ATM, ATR, and DNA-PK: the trinity at the heart of the DNA damage response. Mol. Cell 66, 801–817.

Boelens, M.C., Wu, T.J., Nabet, B.Y., Xu, B., Qiu, Y., Yoon, T., et al., 2014. Exosome transfer from stromal to breast cancer cells regulates therapy resistance pathways. Cell 159, 499–513.

Borden, E.C., Williams, B.R., 2011. Interferon-stimulated genes and their protein products: what and how? J. Interf. Cytokine Res. 31, 1–4.

Borden, E.C., Sen, G.C., Uze, G., Silverman, R.H., Ransohoff, R.M., Foster, G.R., et al., 2007. Interferons at age 50: past, current and future impact on biomedicine. Nat. Rev. Drug Discov. 6, 975–990.

Borghaei, H., Paz-Ares, L., Horn, L., Spigel, D.R., Steins, M., Ready, N.E., et al., 2015. Nivolumab versus docetaxel in advanced nonsquamous non-small-cell lung cancer. N. Engl. J. Med. 373, 1627–1639.

Borkowski, A.W., Kuo, I.H., Bernard, J.J., Yoshida, T., Williams, M.R., Hung, N.J., et al., 2015. Toll-like receptor 3 activation is required for normal skin barrier repair following UV damage. J. Invest. Dermatol. 135, 569–578.

Brahmer, J., Reckamp, K.L., Baas, P., Crino, L., Eberhardt, W.E., Poddubskaya, E., et al., 2015. Nivolumab versus docetaxel in advanced squamous-cell non-small-cell lung cancer. N. Engl. J. Med. 373, 123–135.

Brocks, D., Schmidt, C.R., Daskalakis, M., Jang, H.S., Shah, N.M., Li, D., et al., 2017. DNMT and HDAC inhibitors induce cryptic transcription start sites encoded in long terminal repeats. Nat. Genet.

Bruns, A.M., Leser, G.P., Lamb, R.A., Horvath, C.M., 2014. The innate immune sensor LGP2 activates antiviral signaling by regulating MDA5-RNA interaction and filament assembly. Mol. Cell 55, 771–781.

Burnette, B.C., Liang, H., Lee, Y., Chlewicki, L., Khodarev, N.N., Weichselbaum, R.R., et al., 2011. The efficacy of radiotherapy relies upon induction of type I interferon-dependent innate and adaptive immunity. Cancer Res. 71, 2488–2496.

Carnell, A.N., Goodman, J.I., 2003. The long (LINEs) and the short (SINEs) of it: altered methylation as a precursor to toxicity. Toxicol. Sci. 75, 229–235.

Carpten, J., Nupponen, N., Isaacs, S., Sood, R., Robbins, C., Xu, J., et al., 2002. Germline mutations in the ribonuclease L gene in families showing linkage with HPC1. Nat. Genet. 30, 181–184.

Casciano, I., De Ambrosis, A., Croce, M., Pagnan, G., Di Vinci, A., Allemanni, G., et al., 2004. Expression of the caspase-8 gene in neuroblastoma cells is regulated through an essential interferon-sensitive response element (ISRE). Cell Death Differ. 11, 131–134.

Chang, T.H., Tung, L., Yeh, F.L., Chen, J.H., Chang, S.L., 2013. Functions of the DExD/H-box proteins in nuclear pre-mRNA splicing. Biochim. Biophys. Acta 1829, 764–774.

Chari, A., Fischer, U., 2010. Cellular strategies for the assembly of molecular machines. Trends Biochem. Sci. 35, 676–683.
Chawla-Sarkar, M., Lindner, D.J., Liu, Y.F., Williams, B.R., Sen, G.C., Silverman, R.H., et al., 2003. Apoptosis and interferons: role of interferon-stimulated genes as mediators of apoptosis. Apoptosis 8, 237–249.
Chebath, J., Benech, P., Hovanessian, A., Galabru, J., Revel, M., 1987. Four different forms of interferon-induced 2',5'-oligo(A) synthetase identified by immunoblotting in human cells. J. Biol. Chem. 262, 3852–3857.
Cheon, H., Borden, E.C., Stark, G.R., 2014. Interferons and their stimulated genes in the tumor microenvironment. Semin. Oncol. 41, 156–173.
Chiang, E., Dang, O., Anderson, K., Matsuzawa, A., Ichijo, H., David, M., 2006. Cutting edge: apoptosis-regulating signal kinase 1 is required for reactive oxygen species-mediated activation of IFN regulatory factor 3 by lipopolysaccharide. J. Immunol. 176, 5720–5724.
Chiang, C., Beljanski, V., Yin, K., Olagnier, D., Ben Yebdri, F., Steel, C., et al., 2015. Sequence-specific modifications enhance the broad spectrum antiviral response activated by RIG-I agonists. J. Virol. 89, 8011–8025.
Chiang, J.J., Sparrer, K.M.J., Van Gent, M., Lassig, C., Huang, T., Osterrieder, N., et al., 2018. Viral unmasking of cellular 5S rRNA pseudogene transcripts induces RIG-I-mediated immunity. Nat. Immunol. 19, 53–62.
Chiappinelli, K.B., Strissel, P.L., Desrichard, A., Li, H., Henke, C., Akman, B., et al., 2015. Inhibiting DNA methylation causes an interferon response in cancer via dsRNA including endogenous retroviruses. Cell 162, 974–986.
Choi, S.J., Lee, H.C., Kim, J.H., Park, S.Y., Kim, T.H., Lee, W.K., et al., 2016. HDAC6 regulates cellular viral RNA sensing by deacetylation of RIG-I. EMBO J. 35, 429–442.
Chung, H., Calis, J.J.A., Wu, X., Sun, T., Yu, Y., Sarbanes, S.L., et al., 2018. Human ADAR1 prevents endogenous RNA from triggering translational shutdown. Cell 172, 811–824.e14.
Chuong, E.B., Elde, N.C., Feschotte, C., 2016. Regulatory evolution of innate immunity through co-option of endogenous retroviruses. Science 351, 1083–1087.
Chuong, E.B., Elde, N.C., Feschotte, C., 2017. Regulatory activities of transposable elements: from conflicts to benefits. Nat. Rev. Genet. 18, 71–86.
Conti, E., Izaurralde, E., 2005. Nonsense-mediated mRNA decay: molecular insights and mechanistic variations across species. Curr. Opin. Cell Biol. 17, 316–325.
Cookson, B.T., Brennan, M.A., 2001. Pro-inflammatory programmed cell death. Trends Microbiol. 9, 113–114.
Corless, S., Gilbert, N., 2016. Effects of DNA supercoiling on chromatin architecture. Biophys. Rev. 8, 245–258.
Corrales, L., Matson, V., Flood, B., Spranger, S., Gajewski, T.F., 2017. Innate immune signaling and regulation in cancer immunotherapy. Cell Res. 27, 96–108.
Cui, Y., Li, M., Walton, K.D., Sun, K., Hanover, J.A., Furth, P.A., et al., 2001. The Stat3/5 locus encodes novel endoplasmic reticulum and helicase-like proteins that are preferentially expressed in normal and neoplastic mammary tissue. Genomics 78, 129–134.
Dagon, Y., Dovrat, S., Vilchik, S., Hacohen, D., Shlomo, G., Sredni, B., et al., 2001. Double-stranded RNA-dependent protein kinase, PKR, down-regulates CDC2/cyclin B1 and induces apoptosis in non-transformed but not in v-mos transformed cells. Oncogene 20, 8045–8056.
Deininger, P.L., Moran, J.V., Batzer, M.A., Kazazian Jr., H.H., 2003. Mobile elements and mammalian genome evolution. Curr. Opin. Genet. Dev. 13, 651–658.
Deng, L., Liang, H., Xu, M., Yang, X., Burnette, B., Arina, A., et al., 2014. STING-dependent cytosolic DNA sensing promotes radiation-induced type I interferon-dependent antitumor immunity in immunogenic tumors. Immunity 41, 843–852.

Deng, L., Liang, H., Fu, S., Weichselbaum, R.R., Fu, Y.X., 2016. From DNA damage to nucleic acid sensing: a strategy to enhance radiation therapy. Clin. Cancer Res. 22, 20–25.

Dixit, E., Kagan, J.C., 2013. Intracellular pathogen detection by RIG-I-like receptors. Adv. Immunol. 117, 99–125.

Dong, B., Silverman, R.H., 1995. 2-5A-dependent RNase molecules dimerize during activation by 2–5A. J. Biol. Chem. 270, 4133–4137.

Dong, B., Niwa, M., Walter, P., Silverman, R.H., 2001. Basis for regulated RNA cleavage by functional analysis of RNase L and Ire1p. RNA 7, 361–373.

Elbarbary, R.A., Li, W., Tian, B., Maquat, L.E., 2013. STAU1 binding 3' UTR IRAlus complements nuclear retention to protect cells from PKR-mediated translational shutdown. Genes Dev. 27, 1495–1510.

Ellermeier, J., Wei, J., Duewell, P., Hoves, S., Stieg, M.R., Adunka, T., et al., 2013. Therapeutic efficacy of bifunctional siRNA combining TGF-beta1 silencing with RIG-I activation in pancreatic cancer. Cancer Res. 73, 1709–1720.

Engel, C., Brugmann, G., Lambing, S., Muhlenbeck, L.H., Marx, S., Hagen, C., et al., 2017. RIG-I resists hypoxia-induced immunosuppression and dedifferentiation. Cancer Immunol. Res. 5, 455–467.

Falaleeva, M., Welden, J.R., Duncan, M.J., Stamm, S., 2017. C/D-box snoRNAs form methylating and non-methylating ribonucleoprotein complexes: old dogs show new tricks. Bioessays 39. https://doi.org/10.1002/bies.201600264.

Farrell, P.J., Sen, G.C., Dubois, M.F., Ratner, L., Slattery, E., Lengyel, P., 1978. Interferon action: two distinct pathways for inhibition of protein synthesis by double-stranded RNA. Proc. Natl. Acad. Sci. U. S. A. 75, 5893–5897.

Fuertes, M.B., Woo, S.R., Burnett, B., Fu, Y.X., Gajewski, T.F., 2013. Type I interferon response and innate immune sensing of cancer. Trends Immunol. 34, 67–73.

Gack, M.U., 2014. Mechanisms of RIG-I-like receptor activation and manipulation by viral pathogens. J. Virol. 88, 5213–5216.

Gack, M.U., Shin, Y.C., Joo, C.H., Urano, T., Liang, C., Sun, L., et al., 2007. TRIM25 RING-finger E3 ubiquitin ligase is essential for RIG-I-mediated antiviral activity. Nature 446, 916–920.

Gack, M.U., Nistal-Villan, E., Inn, K.S., Garcia-Sastre, A., Jung, J.U., 2010. Phosphorylation-mediated negative regulation of RIG-I antiviral activity. J. Virol. 84, 3220–3229.

Galluzzi, L., Vitale, I., Aaronson, S.A., Abrams, J.M., Adam, D., Agostinis, P., et al., 2018. Molecular mechanisms of cell death: recommendations of the Nomenclature Committee on Cell Death 2018. Cell Death Differ. 25, 486–541.

Garcia-Diaz, A., Shin, D.S., Moreno, B.H., Saco, J., Escuin-Ordinas, H., Rodriguez, G.A., et al., 2017. Interferon receptor signaling pathways regulating PD-L1 and PD-L2 expression. Cell Rep. 19, 1189–1201.

Georgakilas, A.G., Martin, O.A., Bonner, W.M., 2017. p21: a two-faced genome guardian. Trends Mol. Med. 23, 310–319.

Giono, L.E., Nieto Moreno, N., Cambindo Botto, A.E., Dujardin, G., Munoz, M.J., Kornblihtt, A.R., 2016. The RNA response to DNA damage. J. Mol. Biol. 428, 2636–2651.

Goodier, J.L., 2016. Restricting retrotransposons: a review. Mob. DNA 7, 16.

Gorbunova, V., Hine, C., Tian, X., Ablaeva, J., Gudkov, A.V., Nevo, E., et al., 2012. Cancer resistance in the blind mole rat is mediated by concerted necrotic cell death mechanism. Proc. Natl. Acad. Sci. U. S. A. 109, 19392–19396.

Green, D.R., Llambi, F., 2015. Cell death signaling. Cold Spring Harb. Perspect. Biol. 7, a006080.

Green, D.R., Scott, D.W., 1994. Activation-induced apoptosis in lymphocytes. Curr. Opin. Immunol. 6, 476–487.

Grow, E.J., Flynn, R.A., Chavez, S.L., Bayless, N.L., Wossidlo, M., Wesche, D.J., et al., 2015. Intrinsic retroviral reactivation in human preimplantation embryos and pluripotent cells. Nature 522, 221–225.

Guler, G.D., Tindell, C.A., Pitti, R., Wilson, C., Nichols, K., Kaiwai Cheung, T., et al., 2017. Repression of stress-induced LINE-1 expression protects cancer cell subpopulations from lethal drug exposure. Cancer Cell 32, 221–237.e13.

Habjan, M., Hubel, P., Lacerda, L., Benda, C., Holze, C., Eberl, C.H., et al., 2013. Sequestration by IFIT1 impairs translation of 2'O-unmethylated capped RNA. PLoS Pathog. 9, e1003663.

Hallahan, D.E., Beckett, M.A., Kufe, D., Weichselbaum, R.R., 1990. The interaction between recombinant human tumor necrosis factor and radiation in 13 human tumor cell lines. Int. J. Radiat. Oncol. Biol. Phys. 19, 69–74.

Hancks, D.C., Kazazian Jr., H.H., 2016. Roles for retrotransposon insertions in human disease. Mob. DNA 7, 9.

Hartman, L.L., Crawford, J.R., Makale, M.T., Milburn, M., Joshi, S., Salazar, A.M., et al., 2014. Pediatric phase II trials of poly-ICLC in the management of newly diagnosed and recurrent brain tumors. J. Pediatr. Hematol. Oncol. 36, 451–457.

He, B., Gross, M., Roizman, B., 1997. The gamma(1)34.5 protein of herpes simplex virus 1 complexes with protein phosphatase 1alpha to dephosphorylate the alpha subunit of the eukaryotic translation initiation factor 2 and preclude the shutoff of protein synthesis by double-stranded RNA-activated protein kinase. Proc. Natl. Acad. Sci. U. S. A. 94, 843–848.

Herrera, F.G., Bourhis, J., Coukos, G., 2016. Radiotherapy combination opportunities leveraging immunity for the next oncology practice. CA Cancer J. Clin. 67, 65–85.

Ho, S.S., Zhang, W.Y., Tan, N.Y., Khatoo, M., Suter, M.A., Tripathi, S., et al., 2016. The DNA structure-specific endonuclease MUS81 mediates DNA sensor STING-dependent host rejection of prostate cancer cells. Immunity 44, 1177–1189.

Hobeika, A.C., Subramaniam, P.S., Johnson, H.M., 1997. IFNalpha induces the expression of the cyclin-dependent kinase inhibitor p21 in human prostate cancer cells. Oncogene 14, 1165–1170.

Hornung, V., Ellegast, J., Kim, S., Brzozka, K., Jung, A., Kato, H., et al., 2006. 5'-Triphosphate RNA is the ligand for RIG-I. Science 314, 994–997.

Hornung, V., Hartmann, R., Ablasser, A., Hopfner, K.P., 2014. OAS proteins and cGAS: unifying concepts in sensing and responding to cytosolic nucleic acids. Nat. Rev. Immunol. 14, 521–528.

Hou, J., Zhou, Y., Zheng, Y., Fan, J., Zhou, W., Ng, I.O., et al., 2014. Hepatic RIG-I predicts survival and interferon-alpha therapeutic response in hepatocellular carcinoma. Cancer Cell 25, 49–63.

Hovanessian, A.G., Justesen, J., 2007. The human 2'-5'oligoadenylate synthetase family: unique interferon-inducible enzymes catalyzing 2'-5' instead of 3'-5' phosphodiester bond formation. Biochimie 89, 779–788.

Ibsen, M.S., Gad, H.H., Thavachelvam, K., Boesen, T., Despres, P., Hartmann, R., 2014. The 2'-5'-oligoadenylate synthetase 3 enzyme potently synthesizes the 2'-5'-oligoadenylates required for RNase L activation. J. Virol. 88, 14222–14231.

Jensen, S., Thomsen, A.R., 2012. Sensing of RNA viruses: a review of innate immune receptors involved in recognizing RNA virus invasion. J. Virol. 86, 2900–2910.

Jiang, M., Zhang, S., Yang, Z., Lin, H., Zhu, J., Liu, L., et al., 2018. Self-recognition of an inducible host lncRNA by RIG-I feedback restricts innate immune response. Cell 173, 906–919.e13.

Kaneko, H., Dridi, S., Tarallo, V., Gelfand, B.D., Fowler, B.J., Cho, W.G., et al., 2011. DICER1 deficit induces Alu RNA toxicity in age-related macular degeneration. Nature 471, 325–330.

Kariko, K., Buckstein, M., Ni, H., Weissman, D., 2005. Suppression of RNA recognition by Toll-like receptors: the impact of nucleoside modification and the evolutionary origin of RNA. Immunity 23, 165–175.
Karpus, O.N., Heutinck, K.M., Wijnker, P.J., Tak, P.P., Hamann, J., 2012. Triggering of the dsRNA sensors TLR3, MDA5, and RIG-I induces CD55 expression in synovial fibroblasts. PLoS One 7, e35606.
Kassiotis, G., Stoye, J.P., 2016. Immune responses to endogenous retroelements: taking the bad with the good. Nat. Rev. Immunol. 16, 207–219.
Katibah, G.E., Qin, Y., Sidote, D.J., Yao, J., Lambowitz, A.M., Collins, K., 2014. Broad and adaptable RNA structure recognition by the human interferon-induced tetratricopeptide repeat protein IFIT5. Proc. Natl. Acad. Sci. U. S. A. 111, 12025–12030.
Kato, H., Fujita, T., 2015. RIG-I-like receptors and autoimmune diseases. Curr. Opin. Immunol. 37, 40–45.
Kato, H., Sato, S., Yoneyama, M., Yamamoto, M., Uematsu, S., Matsui, K., et al., 2005. Cell type-specific involvement of RIG-I in antiviral response. Immunity 23, 19–28.
Kedersha, N., Chen, S., Gilks, N., Li, W., Miller, I.J., Stahl, J., et al., 2002. Evidence that ternary complex (eIF2-GTP-tRNA(i)(Met))-deficient preinitiation complexes are core constituents of mammalian stress granules. Mol. Biol. Cell 13, 195–210.
Kessler, B.M., Glas, R., Ploegh, H.L., 2002. MHC class I antigen processing regulated by cytosolic proteolysis-short cuts that alter peptide generation. Mol. Immunol. 39, 171–179.
Keyel, P.A., 2017. Dnases in health and disease. Dev. Biol. 429, 1–11.
Khodarev, N.N., Sokolova, I.A., Vaughan, A.T., 1998. Mechanisms of induction of apoptotic DNA fragmentation. Int. J. Radiat. Biol. 73, 455–467.
Khodarev, N.N., Bennett, T., Shearing, N., Sokolova, I., Koudelik, J., Walter, S., et al., 2000. LINE L1 retrotransposable element is targeted during the initial stages of apoptotic DNA fragmentation. J. Cell. Biochem. 79, 486–495.
Khodarev, N.N., Beckett, M., Labay, E., Darga, T., Roizman, B., Weichselbaum, R.R., 2004. STAT1 is overexpressed in tumors selected for radioresistance and confers protection from radiation in transduced sensitive cells. Proc. Natl. Acad. Sci. U. S. A. 101, 1714–1719.
Khodarev, N.N., Minn, A.J., Efimova, E.V., Darga, T.E., Labay, E., Beckett, M., et al., 2007. Signal transducer and activator of transcription 1 regulates both cytotoxic and prosurvival functions in tumor cells. Cancer Res. 67, 9214–9220.
Khodarev, N.N., Roach, P., Pitroda, S.P., Golden, D.W., Bhayani, M., Shao, M.Y., et al., 2009. STAT1 pathway mediates amplification of metastatic potential and resistance to therapy. PLoS One 4, e5821.
Khodarev, N.N., Roizman, B., Weichselbaum, R.R., 2012. Molecular pathways: interferon/Stat1 pathway: role in the tumor resistance to genotoxic stress and aggressive growth. Clin. Cancer Res. 18, 1–7.
Kim, Y., Lee, J.H., Park, J.E., Cho, J., Yi, H., Kim, V.N., 2014a. PKR is activated by cellular dsRNAs during mitosis and acts as a mitotic regulator. Genes Dev. 28, 1310–1322.
Kim, Y., Tarallo, V., Kerur, N., Yasuma, T., Gelfand, B.D., Bastos-Carvalho, A., et al., 2014b. DICER1/Alu RNA dysmetabolism induces caspase-8-mediated cell death in age-related macular degeneration. Proc. Natl. Acad. Sci. U. S. A. 111, 16082–16087.
Kolakofsky, D., Kowalinski, E., Cusack, S., 2012. A structure-based model of RIG-I activation. RNA 18, 2118–2127.
Koromilas, A.E., Roy, S., Barber, G.N., Katze, M.G., Sonenberg, N., 1992. Malignant transformation by a mutant of the IFN-inducible dsRNA-dependent protein kinase. Science 257, 1685–1689.

Kotredes, K.P., Gamero, A.M., 2013. Interferons as inducers of apoptosis in malignant cells. J. Interf. Cytokine Res. 33, 162–170.
Kristiansen, H., Gad, H.H., Eskildsen-Larsen, S., Despres, P., Hartmann, R., 2011. The oligoadenylate synthetase family: an ancient protein family with multiple antiviral activities. J. Interf. Cytokine Res. 31, 41–47.
Kuchta, K., Knizewski, L., Wyrwicz, L.S., Rychlewski, L., Ginalski, K., 2009. Comprehensive classification of nucleotidyltransferase fold proteins: identification of novel families and their representatives in human. Nucleic Acids Res. 37, 7701–7714.
Langereis, M.A., Feng, Q., Van Kuppeveld, F.J., 2013. MDA5 localizes to stress granules, but this localization is not required for the induction of type I interferon. J. Virol. 87, 6314–6325.
Lassig, C., Hopfner, K.P., 2017. Discrimination of cytosolic self and non-self RNA by RIG-I-like receptors. J. Biol. Chem. 292, 9000–9009.
Lee, Y., Rio, D.C., 2015. Mechanisms and regulation of alternative pre-mRNA splicing. Annu. Rev. Biochem. 84, 291–323.
Lee, S.J., Jang, B.C., Lee, S.W., Yang, Y.I., Suh, S.I., Park, Y.M., et al., 2006. Interferon regulatory factor-1 is prerequisite to the constitutive expression and IFN-gamma-induced upregulation of B7-H1 (CD274). FEBS Lett. 580, 755–762.
Lemaitre, B., Nicolas, E., Michaut, L., Reichhart, J.M., Hoffmann, J.A., 1996. The dorsoventral regulatory gene cassette spatzle/Toll/cactus controls the potent antifungal response in Drosophila adults. Cell 86, 973–983.
Lengyel, P., 1993. Tumor-suppressor genes: news about the interferon connection. Proc. Natl. Acad. Sci. U. S. A. 90, 5893–5895.
Leonova, K.I., Brodsky, L., Lipchick, B., Pal, M., Novototskaya, L., Chenchik, A.A., et al., 2013. p53 cooperates with DNA methylation and a suicidal interferon response to maintain epigenetic silencing of repeats and noncoding RNAs. Proc. Natl. Acad. Sci. U. S. A. 110, E89–E98.
Li, P., Kaiser, P., Lampiris, H.W., Kim, P., Yukl, S.A., Havlir, D.V., et al., 2016. Stimulating the RIG-I pathway to kill cells in the latent HIV reservoir following viral reactivation. Nat. Med. 22, 807–811.
Li, D., Gale, R.P., Liu, Y., Lei, B., Wang, Y., Diao, D., et al., 2017a. 5'-Triphosphate siRNA targeting MDR1 reverses multi-drug resistance and activates RIG-I-induced immune-stimulatory and apoptotic effects against human myeloid leukaemia cells. Leuk. Res. 58, 23–30.
Li, Y., Banerjee, S., Goldstein, S.A., Dong, B., Gaughan, C., Rath, S., et al., 2017b. Ribonuclease L mediates the cell-lethal phenotype of double-stranded RNA editing enzyme ADAR1 deficiency in a human cell line, eLife 6, e25687.
Liddicoat, B.J., Piskol, R., Chalk, A.M., Ramaswami, G., Higuchi, M., Hartner, J.C., et al., 2015. RNA editing by ADAR1 prevents MDA5 sensing of endogenous dsRNA as nonself. Science 349, 1115–1120.
Lin, Y., Leng, M., Wan, M., Wilson, J.H., 2010. Convergent transcription through a long CAG tract destabilizes repeats and induces apoptosis. Mol. Cell. Biol. 30, 4435–4451.
Lin, W.Y., Lin, Y., Wilson, J.H., 2014. Convergent transcription through microsatellite repeat tracts induces cell death. Mol. Biol. Rep. 41, 5627–5634.
Liston, P., Fong, W.G., Kelly, N.L., Toji, S., Miyazaki, T., Conte, D., et al., 2001. Identification of XAF1 as an antagonist of XIAP anti-caspase activity. Nat. Cell Biol. 3, 128–133.
Liu, X., Fagotto, F., 2011. A method to separate nuclear, cytosolic, and membrane-associated signaling molecules in cultured cells. Sci. Signal. 4, pl2.
Liu, J., Valencia-Sanchez, M.A., Hannon, G.J., Parker, R., 2005. MicroRNA-dependent localization of targeted mRNAs to mammalian P-bodies. Nat. Cell Biol. 7, 719–723.

Liu, S., Chen, J., Cai, X., Wu, J., Chen, X., Wu, Y.T., et al., 2013. MAVS recruits multiple ubiquitin E3 ligases to activate antiviral signaling cascades. elife 2, e00785.

Lohofener, J., Steinke, N., Kay-Fedorov, P., Baruch, P., Nikulin, A., Tishchenko, S., et al., 2015. The Activation mechanism of 2'-5'-oligoadenylate synthetase gives new insights into OAS/cGAS triggers of innate immunity. Structure 23, 851–862.

Lopez, P., Jacob, R.J., Roizman, B., 2002. Overexpression of promyelocytic leukemia protein precludes the dispersal of ND10 structures and has no effect on accumulation of infectious herpes simplex virus 1 or its proteins. J. Virol. 76, 9355–9367.

Luddy, K.A., Robertson-Tessi, M., Tafreshi, N.K., Soliman, H., Morse, D.L., 2014. The role of Toll-like receptors in colorectal cancer progression: evidence for epithelial to leucocytic transition. Front. Immunol. 5, 429.

Lurescia, S., Fioretti, D., Rinaldi, M., 2018. Targeting cytosolic nucleic acid-sensing pathways for cancer immunotherapies. Front. Immunol. 9, 1–9.

Mager, D.L., Stoye, J.P., 2015. Mammalian endogenous retroviruses. Microbiol. Spectr. 3, MDNA3-0009-2014.

Magnuson, B., Veloso, A., Kirkconnell, K.S., De Andrade Lima, L.C., Paulsen, M.T., Ljungman, E.A., et al., 2015. Identifying transcription start sites and active enhancer elements using BruUV-seq. Sci. Rep. 5, 17978.

Maharaj, N.P., Wies, E., Stoll, A., Gack, M.U., 2012. Conventional protein kinase C-alpha (PKC-alpha) and PKC-beta negatively regulate RIG-I antiviral signal transduction. J. Virol. 86, 1358–1371.

Malathi, K., Dong, B., Gale Jr., M., Silverman, R.H., 2007. Small self-RNA generated by RNase L amplifies antiviral innate immunity. Nature 448, 816–819.

Malathi, K., Saito, T., Crochet, N., Barton, D.J., Gale Jr., M., Silverman, R.H., 2010. RNase L releases a small RNA from HCV RNA that refolds into a potent PAMP. RNA 16, 2108–2119.

Manic, G., Obrist, F., Sistigu, A., Vitale, I., 2015. Trial watch: targeting ATM-CHK2 and ATR-CHK1 pathways for anticancer therapy. Mol. Cell. Oncol. 2, e1012976.

Mankan, A.K., Kubarenko, A., Hornung, V., 2012. Immunology in clinic review series; focus on autoinflammatory diseases: inflammasomes: mechanisms of activation. Clin. Exp. Immunol. 167, 369–381.

Mannion, N.M., Greenwood, S.M., Young, R., Cox, S., Brindle, J., Read, D., et al., 2014. The RNA-editing enzyme ADAR1 controls innate immune responses to RNA. Cell Rep. 9, 1482–1494.

Marchal, J.A., Lopez, G.J., Peran, M., Comino, A., Delgado, J.R., Garcia-Garcia, J.A., et al., 2014. The impact of PKR activation: from neurodegeneration to cancer. FASEB J. 28, 1965–1974.

Martinon, F., Mayor, A., Tschopp, J., 2009. The inflammasomes: guardians of the body. Annu. Rev. Immunol. 27, 229–265.

Matera, A.G., Wang, Z., 2014. A day in the life of the spliceosome. Nat. Rev. Mol. Cell Biol. 15, 108–121.

Medzhitov, R., Preston-Hurlburt, P., Janeway Jr., C.A., 1997. A human homologue of the Drosophila Toll protein signals activation of adaptive immunity. Nature 388, 394–397.

Meng, G., Xia, M., Xu, C., Yuan, D., Schnurr, M., Wei, J., 2014. Multifunctional antitumor molecule 5'-triphosphate siRNA combining glutaminase silencing and RIG-I activation. Int. J. Cancer 134, 1958–1971.

Meurs, E., Chong, K., Galabru, J., Thomas, N.S., Kerr, I.M., Williams, B.R., et al., 1990. Molecular cloning and characterization of the human double-stranded RNA-activated protein kinase induced by interferon. Cell 62, 379–390.

Meurs, E.F., Galabru, J., Barber, G.N., Katze, M.G., Hovanessian, A.G., 1993. Tumor suppressor function of the interferon-induced double-stranded RNA-activated protein kinase. Proc. Natl. Acad. Sci. U. S. A. 90, 232–236.

Minn, A.J., 2015. Interferons and the immunogenic effects of cancer therapy. Trends Immunol. 36, 725–737.

Minn, A.J., Wherry, E.J., 2016. Combination cancer therapies with immune checkpoint blockade: convergence on interferon signaling. Cell 165, 272–275.

Mohammad Hosseini, A., Majidi, J., Baradaran, B., Yousefi, M., 2015. Toll-like receptors in the pathogenesis of autoimmune diseases. Adv. Pharm. Bull. 5, 605–614.

Mostafavi, S., Yoshida, H., Moodley, D., Leboite, H., Rothamel, K., Raj, T., et al., 2016. Parsing the interferon transcriptional network and its disease associations. Cell 164, 564–578.

Mu, X., Ahmad, S., Hur, S., 2016. Endogenous retroelements and the host innate immune sensors. Adv. Immunol. 132, 47–69.

Nabet, B.Y., Qiu, Y., Shabason, J.E., Wu, T.J., Yoon, T., Kim, B.C., et al., 2017. Exosome RNA unshielding couples stromal activation to pattern recognition receptor signaling in cancer. Cell 170, 352–366.e13.

Nakamori, M., Pearson, C.E., Thornton, C.A., 2011. Bidirectional transcription stimulates expansion and contraction of expanded (CTG)*(CAG) repeats. Hum. Mol. Genet. 20, 580–588.

Oka, T., Hikoso, S., Yamaguchi, O., Taneike, M., Takeda, T., Tamai, T., et al., 2012. Mitochondrial DNA that escapes from autophagy causes inflammation and heart failure. Nature 485, 251–255.

Oshiumi, H., Matsumoto, M., Hatakeyama, S., Seya, T., 2009. Riplet/RNF135, a RING finger protein, ubiquitinates RIG-I to promote interferon-beta induction during the early phase of viral infection. J. Biol. Chem. 284, 807–817.

Oshiumi, H., Miyashita, M., Inoue, N., Okabe, M., Matsumoto, M., Seya, T., 2010. The ubiquitin ligase Riplet is essential for RIG-I-dependent innate immune responses to RNA virus infection. Cell Host Microbe 8, 496–509.

Palm, N.W., Medzhitov, R., 2009. Pattern recognition receptors and control of adaptive immunity. Immunol. Rev. 227, 221–233.

Panda, A., De Cubas, A., Beckermann, K., Riedlinger, G., Stein, M.N., Mayer, T.M., 2018. Expression of endogenous retroviruses and response to immune checkpoint therapy in renal cell cancer. J. Clin. Oncol. 36, 104.

Parisien, J.P., Lenoir, J.J., Mandhana, R., Rodriguez, K.R., Qian, K., Bruns, A.M., et al., 2018. RNA sensor LGP2 inhibits TRAF ubiquitin ligase to negatively regulate innate immune signaling. EMBO Rep. 19, e45176.

Parker, B.S., Rautela, J., Hertzog, P.J., 2016. Antitumour actions of interferons: implications for cancer therapy. Nat. Rev. Cancer 16, 131–144.

Paz, S., Vilasco, M., Werden, S.J., Arguello, M., Joseph-Pillai, D., Zhao, T., et al., 2011. A functional C-terminal TRAF3-binding site in MAVS participates in positive and negative regulation of the IFN antiviral response. Cell Res. 21, 895–910.

Peitsch, M.C., Polzar, B., Tschopp, J., Mannherz, H.G., 1994. About the involvement of deoxyribonuclease I in apoptosis. Cell Death Differ. 1, 1–6.

Platanias, L.C., 2005. Mechanisms of type-I- and type-II-interferon-mediated signalling. Nat. Rev. Immunol. 5, 375–386.

Poeck, H., Besch, R., Maihoefer, C., Renn, M., Tormo, D., Morskaya, S.S., et al., 2008. 5'-Triphosphate-siRNA: turning gene silencing and Rig-I activation against melanoma. Nat. Med. 14, 1256–1263.

Portal, M.M., Pavet, V., Erb, C., Gronemeyer, H., 2015. Human cells contain natural double-stranded RNAs with potential regulatory functions. Nat. Struct. Mol. Biol. 22, 89–97.

Qin, M., Li, Y., Yang, X., Wu, H., 2014. Safety of Toll-like receptor 9 agonists: a systematic review and meta-analysis. Immunopharmacol. Immunotoxicol. 36, 251–260.

Ranoa, D.R., Parekh, A.D., Pitroda, S.P., Huang, X., Darga, T., Wong, A.C., et al., 2016. Cancer therapies activate RIG-I-like receptor pathway through endogenous non-coding RNAs. Oncotarget 7, 26496–26515.

Rebollo, R., Miceli-Royer, K., Zhang, Y., Farivar, S., Gagnier, L., Mager, D.L., 2012. Epigenetic interplay between mouse endogenous retroviruses and host genes. Genome Biol. 13, R89.

Rice, G.I., Kasher, P.R., Forte, G.M., Mannion, N.M., Greenwood, S.M., Szynkiewicz, M., et al., 2012. Mutations in ADAR1 cause Aicardi–Goutieres syndrome associated with a type I interferon signature. Nat. Genet. 44, 1243–1248.

Rich, T., Allen, R.L., Wyllie, A.H., 2000. Defying death after DNA damage. Nature 407, 777–783.

Roers, A., Hiller, B., Hornung, V., 2016. Recognition of endogenous nucleic acids by the innate immune system. Immunity 44, 739–754.

Roulois, D., Loo Yau, H., Singhania, R., Wang, Y., Danesh, A., Shen, S.Y., et al., 2015. DNA-demethylating agents target colorectal cancer cells by inducing viral mimicry by endogenous transcripts. Cell 162, 961–973.

Rudin, C.M., Thompson, C.B., 2001. Transcriptional activation of short interspersed elements by DNA-damaging agents. Genes Chromosom. Cancer 30, 64–71.

Ruiz De Almodovar, C., Ruiz-Ruiz, C., Rodriguez, A., Ortiz-Ferron, G., Redondo, J.M., Lopez-Rivas, A., 2004. Tumor necrosis factor-related apoptosis-inducing ligand (TRAIL) decoy receptor TRAIL-R3 is up-regulated by p53 in breast tumor cells through a mechanism involving an intronic p53-binding site. J. Biol. Chem. 279, 4093–4101.

Ruocco, M.G., Pilones, K.A., Kawashima, N., Cammer, M., Huang, J., Babb, J.S., et al., 2012. Suppressing T cell motility induced by anti-CTLA-4 monotherapy improves antitumor effects. J. Clin. Invest. 122, 3718–3730.

Sadler, A.J., Williams, B.R., 2008. Interferon-inducible antiviral effectors. Nat. Rev. Immunol. 8, 559–568.

Samarajiwa, S.A., Forster, S., Auchettl, K., Hertzog, P.J., 2009. INTERFEROME: the database of interferon regulated genes. Nucleic Acids Res. 37, D852–D857.

Samuel, C.E., 2001. Antiviral actions of interferons. Clin. Microbiol. Rev. 14, 778–809. table of contents.

Sanchez-Aparicio, M.T., Ayllon, J., Leo-Macias, A., Wolff, T., Garcia-Sastre, A., 2017. Subcellular localizations of RIG-I, TRIM25, and MAVS complexes. J. Virol. 91, e01155-16.

Sato, Y., Goto, Y., Narita, N., Hoon, D.S., 2009. Cancer cells expressing toll-like receptors and the tumor microenvironment. Cancer Microenviron. 2 (Suppl. 1), 205–214.

Scherer, M., Stamminger, T., 2016. Emerging role of PML nuclear bodies in innate immune signaling. J. Virol. 90, 5850–5854.

Schlee, M., 2013. Master sensors of pathogenic RNA–RIG-I like receptors. Immunobiology 218, 1322–1335.

Schlee, M., Hartmann, G., 2016. Discriminating self from non-self in nucleic acid sensing. Nat. Rev. Immunol. 16, 566–580.

Schlee, M., Roth, A., Hornung, V., Hagmann, C.A., Wimmenauer, V., Barchet, W., et al., 2009. Recognition of 5' triphosphate by RIG-I helicase requires short blunt double-stranded RNA as contained in panhandle of negative-strand virus. Immunity 31, 25–34.

Schlesinger, S., Goff, S.P., 2015. Retroviral transcriptional regulation and embryonic stem cells: war and peace. Mol. Cell. Biol. 35, 770–777.

Schroder, K., Tschopp, J., 2010. The inflammasomes. Cell 140, 821–832.

Schuberth-Wagner, C., Ludwig, J., Bruder, A.K., Herzner, A.M., Zillinger, T., Goldeck, M., et al., 2015. A conserved histidine in the RNA sensor RIG-I controls immune tolerance to N1-2'O-methylated self RNA. Immunity 43, 41–51.

Seelige, R., Searles, S., Bui, J.D., 2018. Innate sensing of cancer's non-immunologic hallmarks. Curr. Opin. Immunol. 50, 1–8.

Sharpless, N.E., Sherr, C.J., 2015. Forging a signature of in vivo senescence. Nat. Rev. Cancer 15, 397–408.
Singleton, M.R., Dillingham, M.S., Wigley, D.B., 2007. Structure and mechanism of helicases and nucleic acid translocases. Annu. Rev. Biochem. 76, 23–50.
Swetha, M., Ramaiah, K.V.A., 2015. Insulin treatment promotes tyrosine phosphorylation of PKR and inhibits polyIC induced PKR threonine phosphorylation. Arch. Biochem. Biophys. 585, 98–108.
Szabo, A., Fekete, T., Koncz, G., Kumar, B.V., Pazmandi, K., Foldvari, Z., et al., 2016. RIG-I inhibits the MAPK-dependent proliferation of BRAF mutant melanoma cells via MKP-1. Cell. Signal. 28, 335–347.
Takemura, N., Kawasaki, T., Kunisawa, J., Sato, S., Lamichhane, A., Kobiyama, K., et al., 2014. Blockade of TLR3 protects mice from lethal radiation-induced gastrointestinal syndrome. Nat. Commun. 5, 3492.
Tarallo, V., Hirano, Y., Gelfand, B.D., Dridi, S., Kerur, N., Kim, Y., et al., 2012. DICER1 loss and Alu RNA induce age-related macular degeneration via the NLRP3 inflammasome and MyD88. Cell 149, 847–859.
Thompson, P.J., Macfarlan, T.S., Lorincz, M.C., 2016. Long terminal repeats: from parasitic elements to building blocks of the transcriptional regulatory repertoire. Mol. Cell 62, 766–776.
Tsai, M.H., Cook, J.A., Chandramouli, G.V., Degraff, W., Yan, H., Zhao, S., et al., 2007. Gene expression profiling of breast, prostate, and glioma cells following single versus fractionated doses of radiation. Cancer Res. 67, 3845–3852.
Tschopp, J., Schroder, K., 2010. NLRP3 inflammasome activation: the convergence of multiple signalling pathways on ROS production? Nat. Rev. Immunol. 10, 210–215.
Twyman-Saint Victor, C., Rech, A.J., Maity, A., Rengan, R., Pauken, K.E., Stelekati, E., et al., 2015. Radiation and dual checkpoint blockade activate non-redundant immune mechanisms in cancer. Nature 520, 373–377.
Vargiu, L., Rodriguez-Tome, P., Sperber, G.O., Cadeddu, M., Grandi, N., Blikstad, V., et al., 2016. Classification and characterization of human endogenous retroviruses; mosaic forms are common. Retrovirology 13, 7.
Vazquez, C., Horner, S.M., 2015. MAVS coordination of antiviral innate immunity. J. Virol. 89, 6974–6977.
Volkman, H.E., Stetson, D.B., 2014. The enemy within: endogenous retroelements and autoimmune disease. Nat. Immunol. 15, 415–422.
Wang, Y., Ludwig, J., Schuberth, C., Goldeck, M., Schlee, M., Li, H., et al., 2010. Structural and functional insights into 5'-ppp RNA pattern recognition by the innate immune receptor RIG-I. Nat. Struct. Mol. Biol. 17, 781–787.
Watanabe, T., Imamura, T., Hiasa, Y., 2018. Roles of protein kinase R in cancer: potential as a therapeutic target. Cancer Sci. 109, 919–925.
Weichselbaum, R.R., Kufe, D.W., Hellman, S., Rasmussen, H.S., King, C.R., Fischer, P.H., et al., 2002. Radiation-induced tumour necrosis factor-alpha expression: clinical application of transcriptional and physical targeting of gene therapy. Lancet Oncol. 3, 665–671.
Weichselbaum, R.R., Ishwaran, H., Yoon, T., Nuyten, D.S., Baker, S.W., Khodarev, N., et al., 2008. An interferon-related gene signature for DNA damage resistance is a predictive marker for chemotherapy and radiation for breast cancer. Proc. Natl. Acad. Sci. U. S. A. 105, 18490–18495.
Weichselbaum, R.R., Liang, H., Deng, L., Fu, Y.X., 2017. Radiotherapy and immunotherapy: a beneficial liaison? Nat. Rev. Clin. Oncol. 14, 365–379.
Werner, M., Purta, E., Kaminska, K.H., Cymerman, I.A., Campbell, D.A., Mittra, B., et al., 2011. 2'-O-ribose methylation of cap2 in human: function and evolution in a horizontally mobile family. Nucleic Acids Res. 39, 4756–4768.

Whipple, J.M., Youssef, O.A., Aruscavage, P.J., Nix, D.A., Hong, C., Johnson, W.E., et al., 2015. Genome-wide profiling of the C. elegans dsRNAome. RNA 21, 786–800.
Wicker, T., Sabot, F., Hua-Van, A., Bennetzen, J.L., Capy, P., Chalhoub, B., et al., 2007. A unified classification system for eukaryotic transposable elements. Nat. Rev. Genet. 8, 973–982.
Widau, R.C., Parekh, A.D., Ranck, M.C., Golden, D.W., Kumar, K.A., Sood, R.F., et al., 2014. RIG-I-like receptor LGP2 protects tumor cells from ionizing radiation. Proc. Natl. Acad. Sci. U. S. A. 111, E484–E491.
Wies, E., Wang, M.K., Maharaj, N.P., Chen, K., Zhou, S., Finberg, R.W., et al., 2013. Dephosphorylation of the RNA sensors RIG-I and MDA5 by the phosphatase PP1 is essential for innate immune signaling. Immunity 38, 437–449.
Williams, B.R., Golgher, R.R., Brown, R.E., Gilbert, C.S., Kerr, I.M., 1979. Natural occurrence of 2-5A in interferon-treated EMC virus-infected L cells. Nature 282, 582–586.
Woo, S.R., Corrales, L., Gajewski, T.F., 2015. The STING pathway and the T cell-inflamed tumor microenvironment. Trends Immunol. 36, 250–256.
Wu, J., Chen, Z.J., 2014. Innate immune sensing and signaling of cytosolic nucleic acids. Annu. Rev. Immunol. 32, 461–488.
Wyllie, A.H., Morris, R.G., Smith, A.L., Dunlop, D., 1984. Chromatin cleavage in apoptosis: association with condensed chromatin morphology and dependence on macromolecular synthesis. J. Pathol. 142, 67–77.
Xia, T., Konno, H., Ahn, J., Barber, G.N., 2016a. Deregulation of STING signaling in colorectal carcinoma constrains DNA damage responses and correlates with tumorigenesis. Cell Rep. 14, 282–297.
Xia, T., Konno, H., Barber, G.N., 2016b. Recurrent loss of STING signaling in melanoma correlates with susceptibility to viral oncolysis. Cancer Res. 76, 6747–6759.
Xu, P., Roizman, B., 2017. The SP100 component of ND10 enhances accumulation of PML and suppresses replication and the assembly of HSV replication compartments. Proc. Natl. Acad. Sci. U. S. A. 114, E3823–E3829.
Xu, H., He, X., Zheng, H., Huang, L.J., Hou, F., Yu, Z., et al., 2014. Structural basis for the prion-like MAVS filaments in antiviral innate immunity. elife 3, e01489.
Xu, P., Mallon, S., Roizman, B., 2016. PML plays both inimical and beneficial roles in HSV-1 replication. Proc. Natl. Acad. Sci. U. S. A. 113, E3022–E3028.
Xu, M.M., Pu, Y., Han, D., Shi, Y., Cao, X., Liang, H., et al., 2017a. Dendritic cells but not macrophages sense tumor mitochondrial DNA for cross-priming through signal regulatory protein alpha signaling. Immunity 47, 363–373.e5.
Xu, X.X., Wan, H., Nie, L., Shao, T., Xiang, L.X., Shao, J.Z., 2017b. RIG-I: a multifunctional protein beyond a pattern recognition receptor. Protein Cell 9, 246–253.
Yang, F., Wang, P.J., 2016. Multiple LINEs of retrotransposon silencing mechanisms in the mammalian germline. Semin. Cell Dev. Biol. 59, 118–125.
Yang, X., Zhang, X., Fu, M.L., Weichselbaum, R.R., Gajewski, T.F., Guo, Y., et al., 2014. Targeting the tumor microenvironment with interferon-beta bridges innate and adaptive immune responses. Cancer Cell 25, 37–48.
Yasutomo, K., Horiuchi, T., Kagami, S., Tsukamoto, H., Hashimura, C., Urushihara, M., et al., 2001. Mutation of DNASE1 in people with systemic lupus erythematosus. Nat. Genet. 28, 313–314.
Yin, Q., Fu, T.M., Li, J., Wu, H., 2015. Structural biology of innate immunity. Annu. Rev. Immunol. 33, 393–416.
York, I.A., Goldberg, A.L., Mo, X.Y., Rock, K.L., 1999. Proteolysis and class I major histocompatibility complex antigen presentation. Immunol. Rev. 172, 49–66.
Youssef, O.A., Safran, S.A., Nakamura, T., Nix, D.A., Hotamisligil, G.S., Bass, B.L., 2015. Potential role for snoRNAs in PKR activation during metabolic stress. Proc. Natl. Acad. Sci. U. S. A. 112, 5023–5028.

Zaretsky, J.M., Garcia-Diaz, A., et al., 2016. Mutations associated with acquired resistance to PD-1 blockade in melanoma. N. Engl. J. Med. 375, 819–829.
Zeng, M., Hu, Z., Shi, X., Li, X., Zhan, X., Li, X.D., et al., 2014. MAVS, cGAS, and endogenous retroviruses in T-independent B cell responses. Science 346, 1486–1492.
Zhao, L.J., Hua, X., He, S.F., Ren, H., Qi, Z.T., 2011. Interferon alpha regulates MAPK and STAT1 pathways in human hepatoma cells. Virol. J. 8, 157.

CHAPTER THREE

Negative Regulation of Cytosolic Sensing of DNA

Takayuki Abe*,†, Sagi D. Shapira*,†,1

*Department of Systems Biology, Columbia University, New York, NY, United States
†Department of Microbiology and Immunology, Columbia University, New York, NY, United States
[1]Corresponding author: e-mail address: ss4197@columbia.edu

Contents

Abstract

In mammals, cytosolic detection of nucleic acids is critical in initiating innate antiviral responses against invading pathogens (like bacteria, viruses, fungi and parasites). These programs are mediated by multiple cytosolic and endosomal sensors and adaptor molecules (c-GAS/STING axis and TLR9/MyD88 axis, respectively) and lead to the production of type I interferons (IFNs), pro-inflammatory cytokines, and chemokines. While the identity and role of multiple pattern recognition receptors (PRRs) have been elucidated, such immune surveillance systems must be tightly regulated to limit collateral damage and prevent aberrant responses to self- and non-self-nucleic acids. In this review, we discuss recent advances in our understanding of how cytosolic sensing of DNA is controlled during inflammatory immune responses.

1. INTRODUCTION

The innate immune system, which is conserved across virtually all multi-cellular organisms on the planet (from sponges to insects, plants, and vertebrates), includes a diverse set of receptors capable of detecting

ISSN 1937-6448
https://doi.org/10.1016/bs.ircmb.2018.09.002

molecular patterns like sugars, lipids, polymers, and nucleic acids that are principal in prompting protective responses (Broz and Monack, 2013; Kieser and Kagan, 2017; Kumar et al., 2011a). In vertebrates, cytosolic detection of nucleic acids (derived from microbes like viruses, intracellular bacteria, fungi and parasites) is critical in initiating innate (characterized by production of type I IFNs) and adaptive (characterized by T and B cell responses) immune responses (Wu and Chen, 2014). Recent efforts have focused on identifying relevant immune surveillance sensors and components of downstream signaling—Toll-like receptors (TLRs) and their cognate ligands, cytosolic sensing of RNA (primary mediated by the RIG-I/IPS-1 axis), cytosolic sensing of DNA (primary mediated by the cGAS/STING axis), and the inflammasome pathway (primary mediated by NOD-like receptors; NLRs) (Broz and Monack, 2013; Kieser and Kagan, 2017; Kumar et al., 2011a). Yet, while the identities of immune surveillance systems have been revealed, inflammatory programs mediated by these sensors must be tightly regulated to prevent aberrant and inappropriate responses to self-derived ligands (like self RNA and DNA) that may be released from damaged cells, senescent cells, apoptotic cells, or during fertilization (Barber, 2017; de Oliveira Mann and Kranzusch, 2017). Indeed, RIG-I, the cytosolic receptor for RNA distinguishes self from non-self RNA through interaction with a 5′-triphosphate that is unique to viral RNA (Wu and Chen, 2014). Though exceptions to this requirement have been reported, the ability of RIG-I to distinguish self from non-self in this way ensures that anomalous immune responses to cellular RNA do not occur. Responses to DNA are far more agnostic. cGAS, in collaboration with STING, does not distinguish between cellular and foreign DNA (Barber, 2015; Crowl et al., 2017). Indeed, HT-DNA (Herring Testes-DNA), as well as ISD (Interferon-Stimulated DNA) oligonucleotides, is known to act as potent stimulators of cGAS/STING in multiple mammalian cell types. It has also been suggested that chronic cGAS/STING activation induced by self DNA may be responsible for induction of aberrant inflammatory diseases like systemic lupus erythematosus (SLE), Aicardi-Goutières syndrome (AGS), and polyarthritis (Barber, 2015; Crowl et al., 2017). While apoptotic cells represent a possible source of DNA, the existence of DNases in cytoplasmic (e.g., DNase-III, also known as TREX1) and lysosomal compartments (e.g., DNase-II) can clear potential ligands and ensure that inappropriate responses are not initiated (Barber, 2015; Crowl et al., 2017). In the event of mitochondrial DNA (mtDNA) release into the cytoplasm (which occurs following mitochondrial damage), intracellular caspase activation controls the aberrant immune response (McArthur et al., 2018; Rongvaux et al., 2014; White et al., 2014). Similarly, cGAS/STING and the necessary cofactors, and cellular DNA are compartmentalized such that sensing of self-DNA is avoided; the receptor in the cytosol and the ligand (DNA) in the nucleus (Barber, 2015). However, recent publications have illustrated that cell cycle progression in the context of DNA-damage may lead to the formation of micronuclei which elicit cGAS/STING-mediated DNA sensing (Harding et al., 2017; Mackenzie et al., 2017). In addition, cellular DNA can serve as a cGAS/STING ligand following cellular senescence (defined as the senescence-associated secretory phenotype; SASP) (Gluck et al., 2017; Yang et al., 2017). Induction of SASP factors, like inflammatory cytokines and chemokines, may then reinforce senescent cells via autocrine and paracrine routes. Similarly, during fertilization, sperm cell-derived DNA can be found in oocyte cytoplasm may serve to activate cytoplasmic nucleic acid sensing (NAS) pathway to induce inflammatory responses. Thus, there are at least two contexts unrelated to infection where cytosolic responses must be repressed: during mitosis when chromosomal DNA naturally exists in cytoplasm, and during fertilization when sperm-derived DNA enters oocyte cytoplasm. Such control can be achieved through the downregulation of molecules necessary for prompting responses to nucleic acids, and/or the upregulation of a negative regulators that act as molecular safeguards. In this review, we provide an update on cellular machinery that negative controls cytosolic sensing of DNA and discuss therapeutic opportunities for the germ-cell specific NLR family member, NLRP14, a recently identified inhibitor of DNA-sensing during fertilization.

2. CANDIDATE SENSORS OF INTRACELLULAR DNA

In contrast to the identification of molecular machinery responsible for TLR and RNA-sensing RIG-like receptor (RLR) pathways, identification of a universally accepted cytosolic DNA sensor and its related signaling pathway involved a more untidy trajectory (Unterholzner, 2013). Since the existence of a cytosolic sensor for DNA was first postulated (one that could elicit TLR-independent production of IFN to transfected double-stranded DNA; dsDNA) (Ishii et al., 2006; Stetson and Medzhitov, 2006), a great deal of effort has gone into revealing its identity. First, DNA-dependent activator of IRF3 (DAI; also referred to as DLM-1/ZBP1) was shown to participate in the activation of IRF3 downstream of dsDNA sensing (Takaoka et al., 2007). Evidence suggests that following cellular exposure to synthetic dsDNA or DNA virus infection, DAI can form a signaling complex together with TBK1 and IRF3 to initiate production of type I IFNs. However, reports also suggested that cells from DAI-deficient mice did not exhibit impaired responses to synthetic B-form dsDNA and DNA genomes derived from bacteria (Ishii et al., 2008).

IFI16 (IFN-γ-inducible protein 16), a PYHIN family member protein, was also shown to be involved in the recognition of synthetic dsDNA and DNA genomes of nuclearly replicating viruses (e.g., HSV-1, KSHV, HCMV, and EBV) (Unterholzner et al., 2010). In addition, IFI16 was also shown to participate in the DNA damage response through promoting apoptosis and senescence, suggesting that such processes may mediate inflammatory diseases (Unterholzner, 2013). Related to IFI16 is yet another putative cytosolic DNA sensor, absent in melanoma 2 (AIM2), which, through an inflammasome dependent pathway, induces production and secretion of IL-1β and IL-18 rather than type-I IFNs after DNA exposure (Burckstummer et al., 2009; Fernandes-Alnemri et al., 2009; Hornung et al., 2009). More recently, AIM2-like receptors (ALRs), consisting of 13 members possessing a Pyrin signaling domain and a DNA-binding HIN domain, have been shown to not contribute to type I IFN production downstream of DNA sensing, nor to autoimmune diseases like as AGS which are associated with responses to self-DNA (Gray et al., 2016). The establishment of human IFI16-deficient cells further illustrated competent type-I IFN responses to HCMV infection. Nevertheless, IFI16 may participate in the recognition of cytosolic DNA in a cell type and/or species-specific manner.

An additional putative DNA sensor, the DDX41 helicase, was identified through RNAi screening and shown to be involved in DNA recognition in immunocompetent cells rather than epithelial cells (Zhang et al., 2011). Shortly after its discovery, DDX41 mediated signaling was shown to go through the bacterial-derived second messenger cyclic dinucleotide (CDN) molecules cyclic di-AMP and cyclic di-GMP (Parvatiyar et al., 2012). Solved crystal structures confirmed that the binding regions for these ligands overlapped, suggesting that DDX41 can recognize multiple ligands via the DEAD domain (Omura et al., 2016). Of note, both IFI16 and DDX41 have been proposed to act upstream of STING (see below) through physical interactions. Yet, their collaborative role in modulating cytosolic sensing of DNA requires further investigation using genetically engineered mice. Additionally, several DNA damage inducible host factors like the catalytic subunit of DNA-dependent protein kinase (DNA-PKcs), and its binding co-factors Ku70/80, or Mre11 (Meiotic recombination 11) have also been implicated in cytosolic sensing through direct interactions with DNA ligands (Ferguson et al., 2012; Kondo et al., 2013). Importantly, while DNA-PKcs appear to have critical roles in DNA-mediated immune responses, DNA-PKcs deficient cells exhibit normal expression of many ISGs upon stimulation with DNA and DNA virus infection. Nonetheless, these observations reinforce the notion that DNA damage response induces type I IFN production (Brzostek-Racine et al., 2011). Use of DNA damage induced agents like 7,12-dimethylbenz-α-anthracene (DMBA) has helped shed light on the underlying events initiate the DNA damage-induced immune response via cytosolic DNA sensing pathway and implicates nucleosome leakage in eliciting cGAS/STING-dependent signal activation via recognition of self-DNA (Barber, 2015). Finally, LRRFIP1, cytosolic protein, has been shown to recognize both dsDNA and dsRNA derived from pathogens, and may control the production of type-I IFNs through the transcriptional co-activator β-catenin rather than IRF3—potentially acting as an amplifier of cytosolic NAS (Yang et al., 2010). Collectively, evidence indicates that while the long list of candidate cytosolic DNA sensors may have redundant functions, they vary in ligand specificity as well as cell type and tissue distribution.

3. DNA SENSING VIA THE cGAS/STING PATHWAY

Between 2008 and 2009, prior to the identification of several candidate DNA sensors described the above, Barber and colleagues reported that STING (Stimulator of IFN genes, also referred as MITA, MPYS, or ERIS, and encoded by TMEM173), an endoplasmic reticulum (ER) localized protein consisting of multiple transmembrane regions, acts as an essential molecule for cytosolic sensing of DNA (Barber, 2014; Ishikawa and Barber, 2008; Ishikawa et al., 2009). Several DNA species appear to trigger STING-dependent signaling via TBK1/IRF3 and IKK/NF-κB axis in a length dependent manner and STING has also been attributed with a role in responses to plasmid-based vaccines and induction of long-term immunity. And though evidence suggests that STING may interact directly with dsDNA, the physiological relevance of these observations remains to be explored (Abe et al., 2013). Nevertheless, upon ligand stimulation, STING translocates from the ER to the perinuclear-Golgi region, forms a signaling complex with TBK1 and leads to phosphorylation and activation of IRF3. In addition, STING signaling appears to self-regulate by inducing protein degradation through a ubiquitin-mediated proteasome pathway and termination of signal transduction. A subsequent study indicated that STING dimers bind cyclic dinucleotides (CDNs), suggesting that STING acts as the direct innate immune sensor for CDNs (Burdette et al., 2011). Other reports indicated that the type-I IFN and pro-inflammatory cytokine inducing flavonoid compounds, 5,6-dimethylxanthenone-4-acetic acid (DMXAA, also known as Vadimezan or ASA404 and initially identified as a potent tumor vasculature disrupting agent in mice), and 10-carboxymethyl-9-acridanone (CMA) may trigger STING dependent signal activation through direct interactions (Prantner et al., 2012; Roberts et al., 2007). However, it was also known that these compounds displayed such STING activating properties in mice but not human cells. Consistent with the unfortunate failure of these drugs in human clinical trials, these observations may be explained by structural differences in the DMXAA-binding sites across the human and mouse proteins (Gao et al., 2013).

While the identification of STING undoubtedly contributed to our understanding of the underlying molecular events that control cytosolic DNA-mediated immune responses, it was universally accepted that a sensor upstream of STING remained to be discovered. Indeed, the breakthrough came in 2013 when Chen and colleagues uncovered a role for cyclic GMP-AMP (cGAMP), and the catalytic enzyme (cyclic GMP-AMP synthase (cGAS), encoded by MB21D1 and C6orf150) in response to several cytosolic DNA ligands (Sun et al., 2013; Wu et al., 2013). cGAMP is a CDNs that consist of varying phosphodiester linkages (a 2′5′-phosphodiester linkage and a canonical 3′5′-phosphodiester linkage; the cGAMP isomer is known as 2′3′- cGAMP) and cGAS undergoes conformational rearrangement following direct binding to DNA which leads to the synthesis of cGAMP from cellular stores of ATP and GTP. cGAMP produced by cGAS upon DNA stimulation in turn acts as a cognate STING ligand to activate signaling and downstream production of type I IFN (Barber, 2014)—it is interesting to note that similarly to other PRRs (TLR, RLR and the inflammasome pathway) cGAS activation of the STING pathway does not require direct interaction the sensor and adaptor that mediates signal transduction. Importantly, the cGAS/STING pathway plays a crucial role in the induction of autoimmune and inflammatory diseases (see Section 4), highlighting the need to properly control the DNA sensing pathway for the maintenance of cellular homeostasis and immune response. Taken together, while the functional and physiological relevance of several candidate DNA sensors (e.g., DDX41, IFI16, and DNA-PKcs) needs to be fully investigated, cGAS and STING have emerged as bona fide players in cytosolic sensing of DNA (Fig. 1).

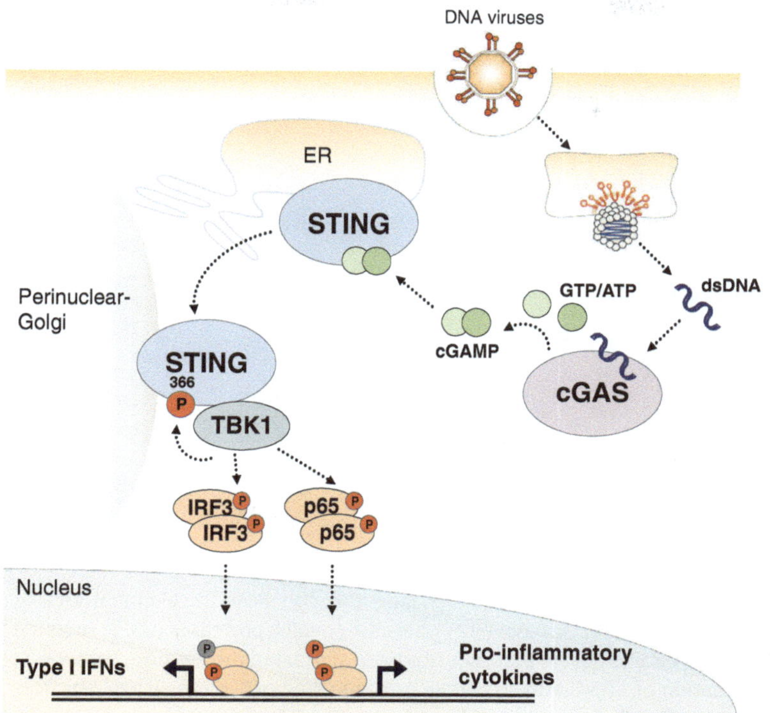

Fig. 1 Cytosolic sensing of DNA via cGAS/STING. Upon viral infection, cGAS recognizes viral dsDNA and utilizes GTP/ATP to catalyze cGAMP and trigger activation of STING. In addition to foreign DNA, cGAS promotes STING-dependent signal activation in response to nucleosome and micronuclear DNA that results from DNA damage. Following cGAMP binding, STING translocates from the ER to perinuclear-Golgi, and forms a signaling complex with TBK1 (phosphorylation of STING at S-366 occurs after translocation) leading to IRF3-mediated type I IFN production as well as NF-κB (p65)-mediated signaling. *Abbreviations: cGAS*, cyclic GMP-AMP Synthase; *STING*, Stimulator of IFN Genes; *cGAMP*, cyclic GMP-AMP; *dsDNA*, double-stranded DNA; *ER*, Endoplasmic Reticulum; *TBK1*, TANK-Binding Kinase 1; *IRF3*, Interferon Regulatory Factor 3; *P*, Phosphorylation.

4. REGULATION OF cGAS/STING DEPENDENT SIGNALING

Intracellular Post-Translational Modifications (PTM), including phosphorylation, ubiquitination, and ubiquitin-like modifications like sumoylation and ISGylation, glutamylation, acetylation or methylation, are critical in controlling cellular responses. These modulators of pathway activity participate in signal transduction by tuning enzymatic states, subcellular localization,

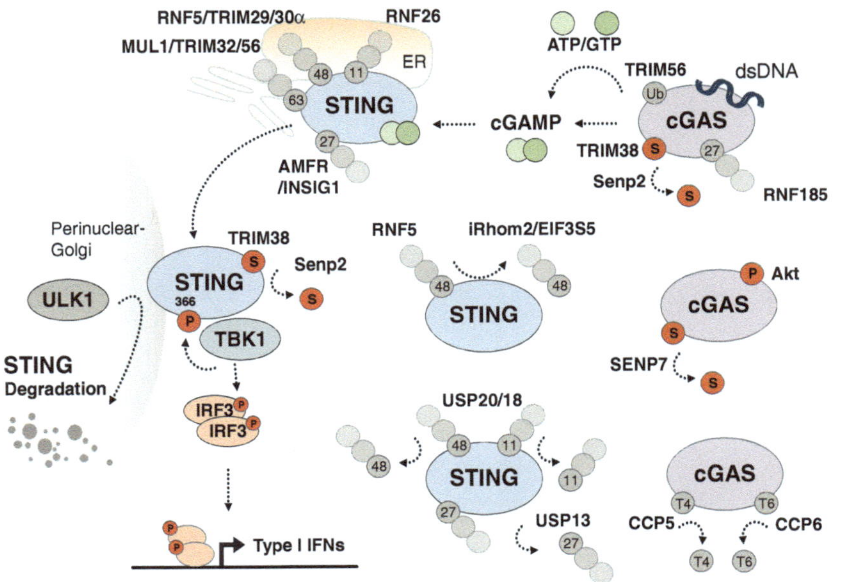

Fig. 2 Regulation of cGAS/STING-signaling through Post-Translational Modifications (PTMs). Ubiquitin E3 ligases as well as ubiquitin specific proteases (USPs) regulate STING function and localization. In addition, ubiquitin-mediated proteasome pathway can lead to degradation of STING and termination of signaling activation—ULK1 controls STING degradation. Detailed description and identity of targeted residues in cGAS and STING are described in the text. *Ub*, Ubiquitin; *S*, SUMOylation; *P*, Phosphorylation; *11*, K11-linked Ub; *27*, K27-linked Ub; *33*, K33-linked Ub; *48*, K48-linked Ub; *63*, K63-linked Ub; *T4*, TTLL (tubulin tyrosine ligase-like enzymes) 4; *T6*, TTLL6; *ULK1*, UNC-51-like kinase.

protein stability and degradation, as well as protein-protein interactions. Not surprisingly, recent evidence implicates several PTMs in the cGAS/STING pathway (summarized in Fig. 2). TRIM56 and TRIM32, both E3 ubiquitin ligases, can positively regulate STING-dependent signal activation through conjugation of K63-linked poly-ubiquitination of STING (TRIM56 at position K150, and TRIM32 at K20, K224, and K236 (Tsuchida et al., 2010; Zhang et al., 2012). Similarly, K27-linked poly-ubiquitination of STING at four Lysine residues (K137, K150, K224, and K236), mediated by E3 ubiquitin ligase complexes of AMFR (Autocrine motility factor receptor)-GP78/INSIG1 (Insulin-induced gene 1), also leads to positive regulation of STING function (Wang et al., 2014).

More recently, MUL1 (Mitochondrial E3 ubiquitin protein ligase 1) was also shown to be responsible for K63-linked poly-ubiquitination of STING at K224, leading to specific enhancement of IRF3-dependent signaling but not NF-κB (Ni et al., 2017). Conversely, two distinct ubiquitin ligases, RNF5 (Ring finger protein 5) and TRIM30α (Tripartite motif containing 30α), lead to termination of signal activation through K48-linked poly-ubiquitination of STING at K150 and K275 which target it to subsequent degradation (Wang et al., 2015a; Zhong et al., 2009). Epstein-Barr virus (EBV), a DNA virus, coopts this regulatory property of the DNA sensing pathway to evade host innate immune response. EBV induces expression of TRIM29 which in turn induces K48-linked poly-ubiquitination of STING, and attenuation of STING-mediated antiviral response (Xing et al., 2017).

Proper regulation of signal transduction requires the poly-ubiquitin conjugation system to be reversible. Indeed, inactive rhomboid protein 2 (iRhom2) contributes to stabilize STING protein by recruiting the de-ubiquitinating enzyme EIF3S5 (Eukaryotic translation initiation factor 3 subunit 5) and de-conjugating RNF5-mediated K48-linked poly-ubiquitination (Luo et al., 2016). In addition, it appears that iRhom2 may regulate the translocation of STING in response to DNA through recruitment of ER translocon-associated protein TRAPβ, which was previously identified as a STING-interacting protein (Ishikawa and Barber, 2008). Conversely, the ubiquitin specific protease 20 (USP20) can counteract K33-linked poly-ubiquitination and RNF5-mediated K48-linked poly-ubiquitination of STING, thereby facilitating STING-mediated signaling (Zhang et al., 2016). Similarly, RNF26 catalyzes K11-linked poly-ubiquitination of STING at the same conjugating site as RNF5 (K150), thereby preventing RNF5-mediated poly-ubiquitination of STING, which positively regulates STING function (Qin et al., 2014). Another ubiquitin specific protease, USP18, also participates in this process, suggesting that collaboration between two distinct USPs may catalyze de-ubiquitination of STING for signal activation. Moreover, USP13 has also been shown to act as a de-conjugating enzyme of K27-linked and K33-linked poly-ubiquitination of STING, though the role of K33-linked poly-ubiquitination of STING remains to be defined (Sun et al., 2017). Nevertheless, de-ubiquitinated STING fails to form signaling complexes with TBK1, thereby suppressing signal transduction, supporting a repressive role for USP13.

While multiple ubiquitin ligases have been implicated in directly regulating STING protein, much less is known about their modulatory role on cGAS function. A recent report illustrated that the ubiquitin ligase RNF185 acts as a positive regulator of cGAS via K27-linked poly-ubiquitination at both K173 and K384 and may contribute to the enzymatic activity of cGAS and promoting synthesis of cGAMP (Wang et al., 2017a). It was also reported that RNF185 expression is elevated in PBMCs of SLE patients, suggesting that it could play a role in auto-inflammatory disorders. Though an initial report described a role for TRIM56 in targeting STING for ubiquitination, it may also conjugate mono-ubiquitin to cGAS at K335 (Seo et al., 2018). Thus, negative regulation of cGAS function by a ubiquitin ligase remains to be identified. In short, while the role of multiple lysine (K)-residues in STING protein has been implicated in regulating its activity, it remains unclear how ubiquitination of STING and cGAS is regulated to orchestrate productive and appropriate immune responses. Additionally, the biological significance and machinery regulating non-degradative poly-ubiquitin linkage (like K27 and K11) of STING remains to be fully resolved.

Recent studies have also implicated SUMOylation in modulating signaling through the cGAS/STING axis. The first report indicated that TRIM38, an E3 ubiquitin ligase, may promote stabilization of cGAS and STING by targeting them for SUMOylation (Hu et al., 2016). In contrast, Senp2, a SUMO-specific protease, was shown to mark these proteins for degradation via proteasomal and chaperone-mediated autophagy pathways. Though cGAS K217 and K464 (which are conserved between mouse and human), and STING K338 (corresponding to K337 in murine STING), were shown to regulate this process, a conflicting report suggested that SUMOylation at K335, K372, and K382 residues of cGAS suppressed functions including DNA-binding, conformational rearrangement, and enzymatic activity—and that SENP7 may reverse this suppression through catalyzing de-SUMOylation of cGAS (Cui et al., 2017).

Other PTMs, including glutamylation, mediated by tubulin tyrosine ligase-like enzymes (TTLLs), have also been shown to be involved in modulating cGAS-mediated immune responses (Xia et al., 2016). Poly-glutamylation and mono-glutamylation of cGAS, mediated by TTLL6 and TTLL4, respectively, negatively regulate cGAS-mediated DNA binding and enzymatic activity. Conversely, the intracellular carboxypeptidases CCP5 and CCP6 counteract mono-glutamylation and poly-glutamylation of cGAS, suggesting that cGAS function is tightly regulated through cellular glutamylation and de-glutamylation. In addition to glutamylation, recent work also implicates palmitoylation, the covalent attachment of fatty acids (like palmitic acid) to cysteine (C) residues of a substrate proteins, in modulating STING (specifically, at positions C88 and C91) mediated signaling and type I IFN production downstream of DNA sensing (Mukai et al., 2016).

Perhaps most recognized and best studied PTM is the phosphorylation of substrate proteins, and STING is no exception. Recent studies have shown that STING possesses a number of serine (S) residues in the C-terminal region that can act as potential phosphorylation sites, though the primary phosphorylation site is Serine-366 (S366) (Konno et al., 2013; Tanaka and Chen, 2012). While TBK1, whose primary target is IRF3, was suggested to be involved in the phosphorylation of STING, in vitro experiments hinted otherwise (Konno et al., 2013). In addition, STING-phosphorylation and downstream degradation appear normal in TBK1-deficient mouse embryonic fibroblasts stimulated with STING agonists, suggesting that if TBK1 does indeed target STING, its role is redundant (Abe and Barber, 2014). Subsequent studies reported that the autophagy-related serine/threonine protein kinases ULK1 and ULK2 as well as the ribosomal protein S6 kinase 1 (S6K1) are involved the phosphorylation of STING (Konno et al., 2013; Wang et al., 2016). While multiple candidate STING kinases are being explored, only one has been implicated in directly controlling cGAS function, and its identity came out of studies on HSV-host interactions. Induced by HSV-1, Akt, also known as protein kinase B (PKB), phosphorylates cGAS at S305 (S291 in murine cGAS), thereby suppressing the synthesis of cGAMP and attenuating cGAS/STING-mediated antiviral responses (Seo et al., 2015). Thus, HSV infection has coopted cellular machinery to negatively regulate cGAS function and evade immune responses. Taken together, while the PTMs that control cGAS/STING function remain to be fully defined, understanding the underlying network that controls these processes may contribute to the establishment of potent therapeutic approaches for tuning cGAS/STING-mediated signaling (see Section 4).

5. AUTO-INFLAMMATORY RESPONSES MEDIATED BY cGAS/STING

While cytosolic sensing of nucleic acids and induction of innate immune responses is critical for eliminating invading pathogens, inappropriate responses to DNA from necrotic or apoptotic cells can result in the development of autoimmune diseases like SLE, AGS (characterized by high levels of anti-nuclear antibodies (ANA) and high levels of circulating proinflammatory cytokines). To limit such aberrant inflammatory responses, vertebrates have evolved mechanisms to tightly control the availability of self DNA and regulate inflammatory responses through cGAS/STING dependent signaling (Fig. 3). For example, mice lacking DNase II, which acts in

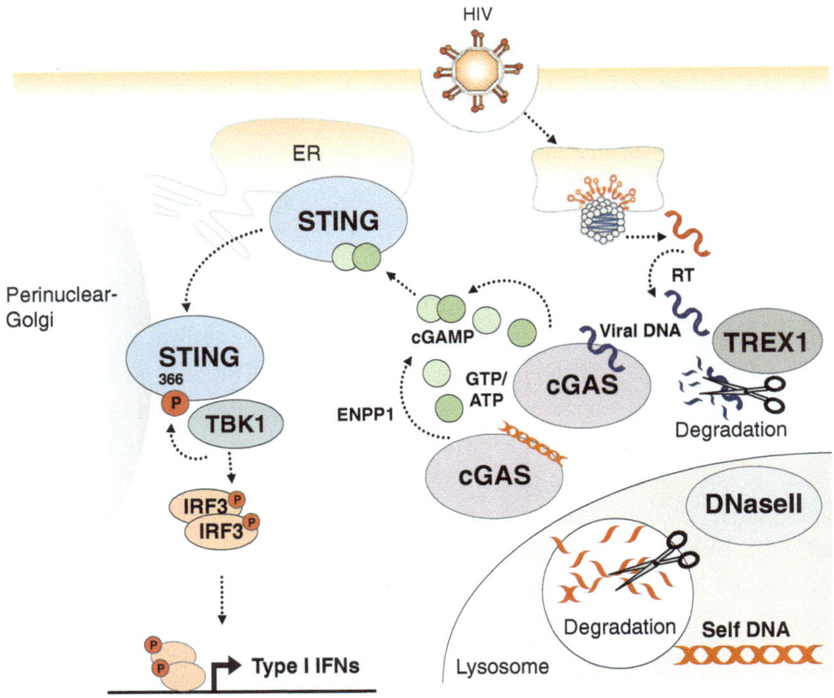

Fig. 3 Intracellular nucleases that modulate cGAS/STING signaling responses: Recognition of self and non-self DNA can augment production of pro-inflammatory genes in virtually all nucleated cells. To prevent an aberrant inflammatory, the three-prime repair exonuclease 1 (TREX1) targets self and non-self DNA for degradation in the cytoplasm, while DNase II plays a similar role in lysosomes. Hydrolysis of cGAMP by ENPP1 also prevents inappropriate activation of STING. *Abbreviations: RT,* Reverse transcription; *HIV-1,* Human Immunodeficiency virus-1.

macrophages to degrade chromosomal DNA derived from apoptotic cells, exhibit embryonic lethality due to anemia induced by the accumulation of undigested DNA (Kawane et al., 2001). Rescue of these mice through deletion of the type I IFN receptor, IFNAR highlights the pathogenic potential of IFN, though aged mice develop TNFα-mediated severe polyarthritis (Kawane et al., 2006). Additionally, mice lacking the three-prime repair exonuclease 1, TREX1 (encoded by DNase III), which degrades nicked dsDNA as well as single-stranded DNA, exhibit a significantly shortened lifespan due to constitutive activation of pro-inflammatory genes across multiple organs (Yang et al., 2007). Similarly, using genetically engineered mice, DNase I has also been implicated in the development of SLE (Napirei et al., 2000), though its role in nucleic acid sensing remains to be elucidated.

Importantly, the dramatic phenotypes observed in both DNase II and TREX1 deficient mice can be reversed through abrogation of cGAS or STING gene expression, suggesting that the cGAS/STING axis can initiate auto-inflammatory diseases (Ahn et al., 2012; Gall et al., 2012; Gao et al., 2015; Gray et al., 2015). Adding credibility to these studies, mutations in TREX1 have also been implicated in AGS and SLE patients. Furthermore, a recent report suggests that mutations in STING (specifically, N154S, V155M, and V147L) may cause vascular and pulmonary syndrome (VAPS), an auto-inflammatory disease characterized by abnormal inflammation across multiple tissues, including skin, vasculature, and lung (Liu et al., 2014)—in vitro interrogation of these mutations revealed that they result in a gain-of-function phenotype and likely contribute to constitutive production type I IFN (Jeremiah et al., 2014). Furthermore, mutations in RNaseH2, which degrades RNA/DNA hybrids, have also been implicated in the development of AGS (Crow et al., 2006), and mutant mice possessing mutated human allele (G37S) die perinatally (Pokatayev et al., 2016). Collectively, these studies and observations highlight the importance of tightly regulating DNA sensing and point to intracellular DNases like DNase II and TREX1 critical gatekeepers in controlling cGAS/STING access to self-DNA.

6. VIRAL STRATEGIES TO MODULATE cGAS/STING FUNCTION

Strong selection pressure, coupled with high mutation rates and short generation times, has led to intricate strategies employed by viruses to counter host immune surveillance and establishment of infection. While the subject is more comprehensively covered by others, here we focus on a few examples from recent reports that highlight the roles of PTMs and protein stability in regulating cytosolic sensing of DNA (Fig. 4). For example, the hepatitis B virus (HBV) polymerase suppresses cytosolic sensing of DNA by interfering with K63-linked poly-ubiquitination of STING (Liu et al., 2015). Though the clinical significance of this observation remains to be resolved—given that hepatocytes lack functional STING dependent signaling (Thomsen et al., 2016)—the data highlight the role of poly-ubiquitination in regulating STING function. Indeed, other viruses have evolved machinery to modulate K63-linked poly-ubiquitination of STING. The human coronavirus (HCoV)-NL63, severe acute respiratory syndrome (SARS) CoV, porcine epidemic diarrhea virus (PEDV) papain-like protease,

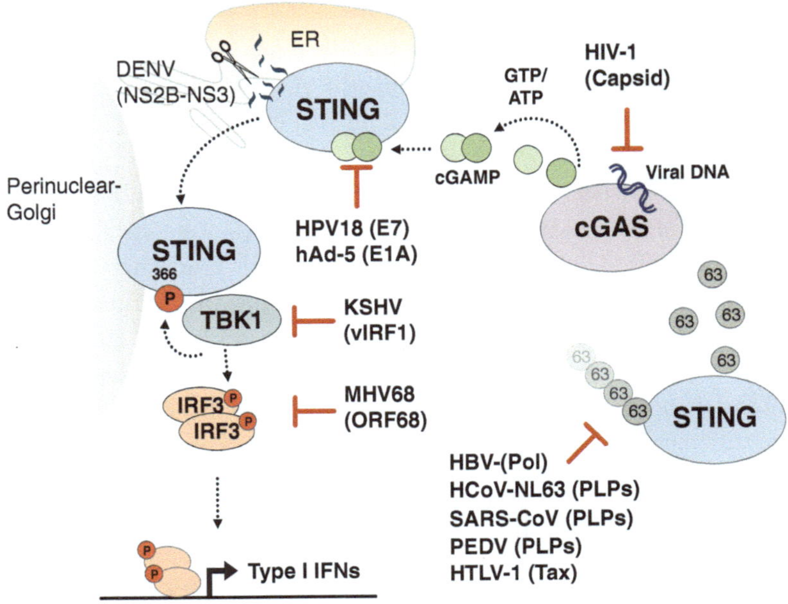

Fig. 4 Viral strategies to evade cytosolic DNA sensing: To escape cytosolic sensing, HIV-1 capsid protein suppresses cGAS-mediated sensing of DNA by recruiting cyclophilin-A (Cyp-A). Other viral proteins, including HTLV-1 Tax protein, HBV polymerase (HBV-Pol), papain-like proteases (PLPs) derived from human coronavirus (HCoV)-NL63, severe acute respiratory syndrome (SARS), and porcine epidemic diarrhea virus (PEDV), interfere K63-linked poly-ubiquitination of STING. E7 and E1A of human papillomavirus 18 (HPV18) and human adenovirus type-5 (hAd5) suppress DNA sensing through direct association with STING. Kaposi sarcoma-associated herpesvirus (KHSV) encodes a viral interferon regulatory factor 1 (vIRF1) to suppress IRF3-mediated production of IFN. The NS2B/NS3 protease of dengue virus (DENV) cleaves human STING at the N-terminal of TM3 region. Finally, murine gammaherpesvirus 68 (MHV68) encodes a de-ubiquitination (DUB) enzyme that contributes to the establishment of latent infection. *Abbreviation*: *HTLV-1*, Human T Lymphotropic Virus type-1; *HIV-1*, Human Immunodeficiency Virus-1; *HBV*, Hepatitis B Virus; *ORF*, Open Reading Frame.

hepatitis C virus non-structural protein 4B (NS4B), as well as the Tax protein of human T lymphotropic virus type-1 (HTLV-1), all suppress STING through either direct physical interactions or regulation of the poly-ubiquitination process (Chen et al., 2014; Ding et al., 2013; Nitta et al., 2013; Sun et al., 2012; Wang et al., 2017b; Yang et al., 2014).

Other examples include the human papillomavirus 18 (HPV18) derived E7 protein and human adenovirus type-5 (hAd5) derived E1A protein, both of which suppress cytosolic sensing of DNA through direct interactions with STING (Lau et al., 2015). An LxCxE motif encoded in both proteins, and conserved among many DNA tumor viruses, mediates disruption of STING-dependent signaling, suggesting a possible functional link between oncogenesis and antagonization of STING function. Another oncogenic DNA virus, Kaposi sarcoma-associated herpesvirus (KHSV) encodes a viral interferon regulatory factor 1 (vIRF1) gene capable of preventing DNA sensing via direct association with STING (Ma et al., 2015). Additionally, though details remain to be resolved, the murine gammaherpesvirus 68 (MHV68) encoded de-ubiquitination (DUB) enzyme ORF68 was shown to antagonize cytosolic sensing of DNA (Sun et al., 2015). In addition to targeting STING (described above), the HCV employs the NS3/4A protease to target the mitochondrial adaptor IPS-1 for cleavage, thereby short-circuiting the signaling cascade (Chan and Gack, 2016). In a rather exquisite example of host restriction, dengue virus (DENV), a mosquito-borne flavivirus which infects hundreds of millions of people annually, encodes the NS2B/NS3 protease which cleaves human but not murine STING (Aguirre et al., 2012). Importantly, as TBK1 mediated signaling is not suppressed by NS2B/NS3 in murine cells, mice produce higher levels of type I IFNs and effectively restrict DENV replication. Though far fewer examples exist of viral targeting of cGAS, recent reports suggest that capsid proteins of HIV-1 and HIV-2 may dampen cGAS function by recruiting cellular CPSF6 and cyclophilin-A (Lahaye et al., 2013; Rasaiyaah et al., 2013). What is clear is that understanding virus-mediated immune evasion strategies can provide critical insights into regulatory functions that control cellular responses to DNA in the cytosol.

7. MODULATION OF INNATE IMMUNE RESPONSE BY NLRs

The nucleotide-binding domain leucine-rich repeat-containing receptor (NLR) family contains five subfamilies that are subclassified according to the composition of domains in their N-termini—the NLRC family contains a caspase activation and recruitment domain (CARD), and the NLRP family contains a pyrin domain (PYD). NLRs are known for their role in triggering caspases to cleave pro-IL-1β and IL-18 into mature proinflammatory cytokines in response to various pathogen associated molecular patterns (PAMPs) and danger associated molecular patterns (DAMPs) (Saxena and Yeretssian, 2014; Zhong et al., 2013). However, recent advances have revealed that some members of the NLR family may also regulate tissue homeostasis and modulate innate immune signaling pathways (Kufer and Sansonetti, 2011; Saxena and Yeretssian, 2014; Zhong et al., 2013).

Several NLRPs (including NLRP2, 4, 5, 7, 9, and 14) are expressed primarily in mammalian oocytes, and NLRP5, NLRP7, and NLRP14 have been shown to have important roles in reproduction and development (McDaniel and Wu, 2009; Murdoch et al., 2006; Tong et al., 2000; Westerveld et al., 2006; Zhang et al., 2008). Additionally, NLRC3 and NLRC5, as well as several NLRPs, have been demonstrated to function as negative regulators of PRRs (Table 1). For example, NLRP2 suppresses

Table 1 Summary of Inhibitory Functions Mediated by Inflammasome Family Members

NLRP subfamily	Function	Reference
NLRP2	Negative regulator of TLR/TNFR pathway	Bruey et al. (2004), Fontalba et al. (2007)
NLRP4	Negative regulator of RIG-I/IPS-1 pathway	Cui et al. (2012)
NLRP6/12	Negative regulator of NF-κB (The role of anti-viral in intestinal)	Zaki et al. (2011), Allen et al. (2012), Lupfer and Kanneganti (2013), Chen et al. (2014), Anand et al. (2012), Wang et al. (2015b)
NLRP11	Negative regulator of TLR pathway Negative regulator of RIG-I/IPS-1 pathway	Wu et al. (2017), Qin et al. (2017b), Ellwanger et al. (2018)
NLRP14	Negative regulator of cGAS/STING pathway Negative regulator of RIG-I/IPS-1 pathway	Abe et al. (2017)
NLR		
NLRX1	Negative regulator of RIG-I/IPS-1 pathway Negative regulator of cGAS/STING pathway	Moore et al. (2008), Guo et al. (2016), Qin et al. (2017a), Ma et al. (2017)
NLRC3	Negative regulator of cGAS/STING pathway	Zhang et al. (2014), Tocker et al. (2017)
NLRC5	Negative regulator of TLR pathway	Cui et al. (2010)

both TLR and TNFα-mediated NF-κB activation at the level of IKK complex formation (Bruey et al., 2004; Fontalba et al., 2007). Similarly, NLRP4 inhibits RIG-I signaling by recruiting DTX4, an E3 ubiquitin ligase which marks TBK1 for degradation (Cui et al., 2012). NLRP6 and NLRP12 suppress NF-κB signaling and are involved in the maintenance of intestinal homeostasis and tumorigenesis (Allen et al., 2012; Anand et al., 2012; Chen, 2014; Lupfer and Kanneganti, 2013; Zaki et al., 2011)—more recent reports also point to a role for NLRP6 in anti-viral immunity in mouse intestines (Wang et al., 2015b). NLRX1, the only mitochondrially localized member of this family, modulates sensing of dsRNA and virus-induced ROS production (Guo et al., 2016; Ma et al., 2017; Moore et al., 2008; Qin et al., 2017a; Tattoli et al., 2008). In collaboration with IQGAP1, NLRC3 suppresses DNA sensing lymphoid cells, myeloid cells, and epithelial cells through interactions with STING (Tocker et al., 2017; Zhang et al., 2014). In addition, NLRC5 can suppress TLR signaling by modulating IKK activation (Cui et al., 2010), though NLRC5 deficiency does not affect this signaling pathway (Kumar et al., 2011b). Finally, the primate specific NLRP11 suppresses TLR-mediated activation of NF-κB by targeting TRAF6 for degradation in myeloid cells and B-cells (Ellwanger et al., 2018; Qin et al., 2017b; Wu et al., 2017). Together, these emerging properties call into question whether the established role for this family in activating proinflammatory responses may in fact represent an exception rather than a rule.

8. NLRP14: A NEW RHEOSTAT OF CYTOSOLIC NUCLEIC ACID SENSING

As highlighted above, inappropriate activation of the NAS pathway can result in highly destructive immune responses. Yet, molecular dampeners, structural determinants and topological barriers described above as well as novel checks and balances yet to be discovered tightly regulate the STING/cGAS axis and defend against such circumstances. One instance when distinguishing between self and non-self DNA is particularly important is fertilization, when sperm cell derived DNA can be found in oocyte cytoplasm. Yet, while STING/cGAS and rest of the NAS pathway (which are all expressed in oocytes) are afforded access to this foreign DNA, a nucleic acid sensing response is not triggered. In a recent publication, Abe and colleagues reasoned that germ cells must therefore possess a robust mechanism to negatively regulate cytosolic nucleic acid sensing

(Abe et al., 2017). Through a combination of data-mining and experimental approaches, the group identified 20 candidate genes that may serve as putative regulators of NAS in oocytes. Among these, four (TRIM42, TGIF2LX, C4orf42, ZPLD1) significantly suppressed cGAS-mediated signaling, and one, NLRP14 (NACHT, LRR and PYD Domains-Containing 14; a component of the inflammasome family of proteins), shut off NAS virtually all together. What followed were a series of observations that delineate a role for NLRP14 as a rheostat for nucleic acid sensing, highlight the importance of controlling innate immune responses to foreign and endogenous ligands, and suggest that tight regulation of these processes is critical in maintaining proper immunologic homeostasis in germline. Specifically, the authors demonstrate that NLRP14 associates with both STING and MAVS and prevents downstream signal transduction (Fig. 5). In turn, these adaptor

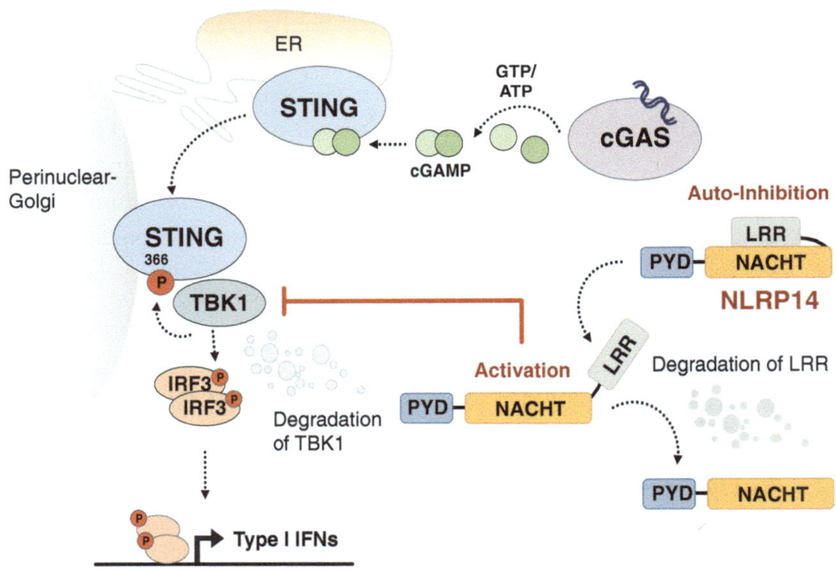

Fig. 5 NLRP14-mediated inhibition of cytosolic nucleic acid sensing. Upon activation of STING dependent signaling, conformational rearrangement from inhibitory-form to an active-form of NLRP14 is induced. Targeting of TBK1 for K48-linked poly-ubiquitination is then initiated and results in proteasome dependent degradation of TBK1 and termination of STING dependent as well as RIG-I dependent signaling. Similarly, NLRP14 is targeted for degradation to avoid persistent immunosuppression. *Abbreviations: NLRP14*, Nucleotide-binding oligomerization domain, Leucine rich Repeat and Pyrin domain containing protein 14; *PYD*, Pyrin domain; *NACHT*, NAIP (neuronal apoptosis inhibitory protein), CIITA (MHC class II transcription activator), HET-E (incompatibility locus protein from Podospora anserine) and TP1 (telomerase-associated protein), *LRR*, Leucine-rich repeat; *Ub*, Ubiquitin; *P*, Phosphorylation.

molecules induce the proteasomal degradation of NLRP14, a feedback loop that may be critical in preventing persistent immunosuppression and proper induction of innate immune responses under appropriate conditions (for example, post fertilization of oocytes). Importantly, several human diseases, including Crohn's disease, Celiac disease, Blau syndrome, rheumatoid arthritis, Type-1 and -2 diabetes, SLE, Cryopyrin-associated periodic syndromes (CAPS), inflammatory bowel diseases, colitis as well as colon cancer, have been shown to be associated with mutations in NLRPs (Saxena and Yeretssian, 2014; Zhong et al., 2013). Other NLRPs (like NLRP2, NLRP5, and NLRP7) have been associated with miscarriage and infertility (Docherty et al., 2015; Huang et al., 2013). In line with these observations ectopic expression of a K108X NLRP14 allele (rs76274604; coding for a nonsense mutation that introduces an early STOP codon in NLRP14) results in reduced suppression of TBK1-mediated signaling (Abe et al., 2017). With an allele frequency of 1.7% in the human population, and a minor allele frequency of 3% in East Asian and Ad Mixed American populations, infertility associated with homozygosity of this gene may affect 3 in 10,000 individuals. As an immunological rheostat, NLRP14 safeguards against inappropriate cytosolic responses to nucleic acids and the acquisition of such a function may have been a prerequisite to sexual reproduction. Indeed, homologs of the inflammasome family of proteins have been found widely across life—arguing in favor of their involvement in immunity and broader cellular processes, as is observed with other expanded gene families.

9. CONCLUDING REMARKS

The past decade has seen rapid advancement in our understanding of how cells sense cytosolic DNA, including the identification of receptors and molecular machinery that participate in this process as well as recognition of multiple diseases associated with this pathway. Recent findings highlight the importance of controlling innate immune responses to DNA and suggest that tight regulation of these processes is critical in maintaining proper immunologic homeostasis across multiple organs. Thus, understanding the underlying network of regulators that control signaling through the cGAS/STING axis will undoubtedly contribute to the development of potent therapeutic agents against auto-immune and inflammatory diseases, as well as certain cancers.

ACKNOWLEDGMENTS

We apologize to authors whose work could not be cited due to space constraints and narrow focus of the review. S.D.S. acknowledges funding from NIH R01 (GM109018-05) and NIH U54 (CA121852-07).

REFERENCES

Abe, T., Barber, G.N., 2014. Cytosolic-DNA-mediated, STING-dependent proinflammatory gene induction necessitates canonical NF-kappaB activation through TBK1. J. Virol. 88 (10), 5328–5341.

Abe, T., et al., 2013. STING recognition of cytoplasmic DNA instigates cellular defense. Mol. Cell 50 (1), 5–15.

Abe, T., et al., 2017. Germ-cell-specific inflammasome component NLRP14 negatively regulates cytosolic nucleic acid sensing to promote fertilization. Immunity 46 (4), 621–634.

Aguirre, S., et al., 2012. DENV inhibits type I IFN production in infected cells by cleaving human STING. PLoS Pathog. 8 (10), e1002934.

Ahn, J., et al., 2012. STING manifests self DNA-dependent inflammatory disease. Proc. Natl. Acad. Sci. U.S.A. 109 (47), 19386–19391.

Allen, I.C., et al., 2012. NLRP12 suppresses colon inflammation and tumorigenesis through the negative regulation of noncanonical NF-kappaB signaling. Immunity 36 (5), 742–754.

Anand, P.K., et al., 2012. NLRP6 negatively regulates innate immunity and host defence against bacterial pathogens. Nature 488 (7411), 389–393.

Barber, G.N., 2014. STING-dependent cytosolic DNA sensing pathways. Trends Immunol. 35 (2), 88–93.

Barber, G.N., 2015. STING: infection, inflammation and cancer. Nat. Rev. Immunol. 15 (12), 760–770.

Barber, G.N., 2017. The birds, the bees, and innate immunity. Immunity 46 (4), 521–522.

Broz, P., Monack, D.M., 2013. Newly described pattern recognition receptors team up against intracellular pathogens. Nat. Rev. Immunol. 13 (8), 551–565.

Bruey, J.M., et al., 2004. PAN1/NALP2/PYPAF2, an inducible inflammatory mediator that regulates NF-kappaB and caspase-1 activation in macrophages. J. Biol. Chem. 279 (50), 51897–51907.

Brzostek-Racine, S., et al., 2011. The DNA damage response induces IFN. J. Immunol. 187 (10), 5336–5345.

Burckstummer, T., et al., 2009. An orthogonal proteomic-genomic screen identifies AIM2 as a cytoplasmic DNA sensor for the inflammasome. Nat. Immunol. 10 (3), 266–272.

Burdette, D.L., et al., 2011. STING is a direct innate immune sensor of cyclic di-GMP. Nature 478 (7370), 515–518.

Chan, Y.K., Gack, M.U., 2016. Viral evasion of intracellular DNA and RNA sensing. Nat. Rev. Microbiol. 14 (6), 360–373.

Chen, G.Y., 2014. Role of Nlrp6 and Nlrp12 in the maintenance of intestinal homeostasis. Eur. J. Immunol. 44 (2), 321–327.

Chen, X., et al., 2014. SARS coronavirus papain-like protease inhibits the type I interferon signaling pathway through interaction with the STING-TRAF3-TBK1 complex. Protein Cell 5 (5), 369–381.

Crow, Y.J., et al., 2006. Mutations in genes encoding ribonuclease H2 subunits cause Aicardi-Goutières syndrome and mimic congenital viral brain infection. Nat. Genet. 38 (8), 910–916.

Crowl, J.T., et al., 2017. Intracellular nucleic acid detection in autoimmunity. Annu. Rev. Immunol. 35, 313–336.

Cui, J., et al., 2010. NLRC5 negatively regulates the NF-kappaB and type I interferon signaling pathways. Cell 141 (3), 483–496.

Cui, J., et al., 2012. NLRP4 negatively regulates type I interferon signaling by targeting the kinase TBK1 for degradation via the ubiquitin ligase DTX4. Nat. Immunol. 13 (4), 387–395.

Cui, Y., et al., 2017. SENP7 potentiates cGAS activation by relieving SUMO-mediated inhibition of cytosolic DNA sensing. PLoS Pathog. 13 (1), e1006156.

de Oliveira Mann, C.C., Kranzusch, P.J., 2017. cGAS conducts micronuclei DNA surveillance. Trends Cell Biol. 27 (10), 697–698.

Ding, Q., et al., 2013. Hepatitis C virus NS4B blocks the interaction of STING and TBK1 to evade host innate immunity. J. Hepatol. 59 (1), 52–58.

Docherty, L.E., et al., 2015. Mutations in NLRP5 are associated with reproductive wastage and multilocus imprinting disorders in humans. Nat. Commun. 6, 8086.

Ellwanger, K., et al., 2018. The NLR family pyrin domain-containing 11 protein contributes to the regulation of inflammatory signaling. J. Biol. Chem. 293 (8), 2701–2710.

Ferguson, B.J., et al., 2012. DNA-PK is a DNA sensor for IRF-3-dependent innate immunity. elife 1, e00047.

Fernandes-Alnemri, T., et al., 2009. AIM2 activates the inflammasome and cell death in response to cytoplasmic DNA. Nature 458 (7237), 509–513.

Fontalba, A., Gutierrez, O., Fernandez-Luna, J.L., 2007. NLRP2, an inhibitor of the NF-kappaB pathway, is transcriptionally activated by NF-kappaB and exhibits a nonfunctional allelic variant. J. Immunol. 179 (12), 8519–8524.

Gall, A., et al., 2012. Autoimmunity initiates in nonhematopoietic cells and progresses via lymphocytes in an interferon-dependent autoimmune disease. Immunity 36 (1), 120–131.

Gao, P., et al., 2013. Structure-function analysis of STING activation by c[G(2′,5′)pA(3′,5′)p] and targeting by antiviral DMXAA. Cell 154 (4), 748–762.

Gao, D., et al., 2015. Activation of cyclic GMP-AMP synthase by self-DNA causes autoimmune diseases. Proc. Natl. Acad. Sci. U.S.A. 112 (42), E5699–E5705.

Gluck, S., et al., 2017. Innate immune sensing of cytosolic chromatin fragments through cGAS promotes senescence. Nat. Cell Biol. 19 (9), 1061–1070.

Gray, E.E., et al., 2015. Cutting edge: cGAS is required for lethal autoimmune disease in the Trex1-deficient mouse model of Aicardi-Goutières syndrome. J. Immunol. 195 (5), 1939–1943.

Gray, E.E., et al., 2016. The AIM2-like receptors are dispensable for the interferon response to intracellular DNA. Immunity 45 (2), 255–266.

Guo, H., et al., 2016. NLRX1 sequesters STING to negatively regulate the interferon response, thereby facilitating the replication of HIV-1 and DNA viruses. Cell Host Microbe 19 (4), 515–528.

Harding, S.M., et al., 2017. Mitotic progression following DNA damage enables pattern recognition within micronuclei. Nature 548 (7668), 466–470.

Hornung, V., et al., 2009. AIM2 recognizes cytosolic dsDNA and forms a caspase-1-activating inflammasome with ASC. Nature 458 (7237), 514–518.

Hu, M.M., et al., 2016. Sumoylation promotes the stability of the DNA sensor cGAS and the adaptor STING to regulate the kinetics of response to DNA virus. Immunity 45 (3), 555–569.

Huang, J.Y., et al., 2013. A genetic association study of NLRP2 and NLRP7 genes in idiopathic recurrent miscarriage. Hum. Reprod. 28 (4), 1127–1134.

Ishii, K.J., et al., 2006. A toll-like receptor-independent antiviral response induced by double-stranded B-form DNA. Nat. Immunol. 7 (1), 40–48.

Ishii, K.J., et al., 2008. TANK-binding kinase-1 delineates innate and adaptive immune responses to DNA vaccines. Nature 451 (7179), 725–729.

Ishikawa, H., Barber, G.N., 2008. STING is an endoplasmic reticulum adaptor that facilitates innate immune signalling. Nature 455 (7213), 674–678.

Ishikawa, H., Ma, Z., Barber, G.N., 2009. STING regulates intracellular DNA-mediated, type I interferon-dependent innate immunity. Nature 461 (7265), 788–792.

Jeremiah, N., et al., 2014. Inherited STING-activating mutation underlies a familial inflammatory syndrome with lupus-like manifestations. J. Clin. Invest. 124 (12), 5516–5520.

Kawane, K., et al., 2001. Requirement of DNase II for definitive erythropoiesis in the mouse fetal liver. Science 292 (5521), 1546–1549.

Kawane, K., et al., 2006. Chronic polyarthritis caused by mammalian DNA that escapes from degradation in macrophages. Nature 443 (7114), 998–1002.

Kieser, K.J., Kagan, J.C., 2017. Multi-receptor detection of individual bacterial products by the innate immune system. Nat. Rev. Immunol. 17 (6), 376–390.

Kondo, T., et al., 2013. DNA damage sensor MRE11 recognizes cytosolic double-stranded DNA and induces type I interferon by regulating STING trafficking. Proc. Natl. Acad. Sci. U.S.A. 110 (8), 2969–2974.

Konno, H., Konno, K., Barber, G.N., 2013. Cyclic dinucleotides trigger ULK1 (ATG1) phosphorylation of STING to prevent sustained innate immune signaling. Cell 155 (3), 688–698.

Kufer, T.A., Sansonetti, P.J., 2011. NLR functions beyond pathogen recognition. Nat. Immunol. 12 (2), 121–128.

Kumar, H., Kawai, T., Akira, S., 2011a. Pathogen recognition by the innate immune system. Int. Rev. Immunol. 30 (1), 16–34.

Kumar, H., et al., 2011b. NLRC5 deficiency does not influence cytokine induction by virus and bacteria infections. J. Immunol. 186 (2), 994–1000.

Lahaye, X., et al., 2013. The capsids of HIV-1 and HIV-2 determine immune detection of the viral cDNA by the innate sensor cGAS in dendritic cells. Immunity 39 (6), 1132–1142.

Lau, L., et al., 2015. DNA tumor virus oncogenes antagonize the cGAS-STING DNA-sensing pathway. Science 350 (6260), 568–571.

Liu, Y., et al., 2014. Activated STING in a vascular and pulmonary syndrome. N. Engl. J. Med. 371 (6), 507–518.

Liu, Y., et al., 2015. Hepatitis B virus polymerase disrupts K63-linked ubiquitination of STING to block innate cytosolic DNA-sensing pathways. J. Virol. 89 (4), 2287–2300.

Luo, W.W., et al., 2016. iRhom2 is essential for innate immunity to DNA viruses by mediating trafficking and stability of the adaptor STING. Nat. Immunol. 17 (9), 1057–1066.

Lupfer, C., Kanneganti, T.D., 2013. Unsolved mysteries in NLR biology. Front. Immunol. 4, 285.

Ma, Z., et al., 2015. Modulation of the cGAS-STING DNA sensing pathway by gammaherpesviruses. Proc. Natl. Acad. Sci. U.S.A. 112 (31), E4306–E4315.

Ma, Z., et al., 2017. NLRX1 negatively modulates type I IFN to facilitate KSHV reactivation from latency. PLoS Pathog. 13 (5), e1006350.

Mackenzie, K.J., et al., 2017. cGAS surveillance of micronuclei links genome instability to innate immunity. Nature 548 (7668), 461–465.

McArthur, K., et al., 2018. BAK/BAX macropores facilitate mitochondrial herniation and mtDNA efflux during apoptosis. Science 359 (6378). https://doi.org/10.1126/science.aao6047.

McDaniel, P., Wu, X., 2009. Identification of oocyte-selective NLRP genes in rhesus macaque monkeys (Macaca mulatta). Mol. Reprod. Dev. 76 (2), 151–159.

Moore, C.B., et al., 2008. NLRX1 is a regulator of mitochondrial antiviral immunity. Nature 451 (7178), 573–577.

Mukai, K., et al., 2016. Activation of STING requires palmitoylation at the Golgi. Nat. Commun. 7, 11932.

Murdoch, S., et al., 2006. Mutations in NALP7 cause recurrent hydatidiform moles and reproductive wastage in humans. Nat. Genet. 38 (3), 300–302.
Napirei, M., et al., 2000. Features of systemic lupus erythematosus in Dnase1-deficient mice. Nat. Genet. 25 (2), 177–181.
Ni, G., Konno, H., Barber, G.N., 2017. Ubiquitination of STING at lysine 224 controls IRF3 activation. Sci. Immunol. 2 (11). pii: eaah7119.
Nitta, S., et al., 2013. Hepatitis C virus NS4B protein targets STING and abrogates RIG-I-mediated type I interferon-dependent innate immunity. Hepatology 57 (1), 46–58.
Omura, H., et al., 2016. Structural and functional analysis of DDX41: a bispecific immune receptor for DNA and cyclic dinucleotide. Sci. Rep. 6, 34756.
Parvatiyar, K., et al., 2012. The helicase DDX41 recognizes the bacterial secondary messengers cyclic di-GMP and cyclic di-AMP to activate a type I interferon immune response. Nat. Immunol. 13 (12), 1155–1161.
Pokatayev, V., et al., 2016. RNase H2 catalytic core Aicardi-Goutières syndrome-related mutant invokes cGAS-STING innate immune-sensing pathway in mice. J. Exp. Med. 213 (3), 329–336.
Prantner, D., et al., 2012. 5,6-Dimethylxanthenone-4-acetic acid (DMXAA) activates stimulator of interferon gene (STING)-dependent innate immune pathways and is regulated by mitochondrial membrane potential. J. Biol. Chem. 287 (47), 39776–39788.
Qin, Y., et al., 2014. RNF26 temporally regulates virus-triggered type I interferon induction by two distinct mechanisms. PLoS Pathog. 10 (9), e1004358.
Qin, Y., et al., 2017a. NLRX1 mediates MAVS degradation to attenuate hepatitis C virus-induced innate immune response through PCBP2. J. Virol. 91 (23), e01264-17.
Qin, Y., et al., 2017b. NLRP11 disrupts MAVS signalosome to inhibit type I interferon signaling and virus-induced apoptosis. EMBO Rep. 18 (12), 2160–2171.
Rasaiyaah, J., et al., 2013. HIV-1 evades innate immune recognition through specific cofactor recruitment. Nature 503 (7476), 402–405.
Roberts, Z.J., et al., 2007. The chemotherapeutic agent DMXAA potently and specifically activates the TBK1-IRF-3 signaling axis. J. Exp. Med. 204 (7), 1559–1569.
Rongvaux, A., et al., 2014. Apoptotic caspases prevent the induction of type I interferons by mitochondrial DNA. Cell 159 (7), 1563–1577.
Saxena, M., Yeretssian, G., 2014. NOD-like receptors: master regulators of inflammation and cancer. Front. Immunol. 5, 327.
Seo, G.J., et al., 2015. Akt kinase-mediated checkpoint of cGAS DNA sensing pathway. Cell Rep. 13 (2), 440–449.
Seo, G.J., et al., 2018. TRIM56-mediated monoubiquitination of cGAS for cytosolic DNA sensing. Nat. Commun. 9 (1), 613.
Stetson, D.B., Medzhitov, R., 2006. Recognition of cytosolic DNA activates an IRF3-dependent innate immune response. Immunity 24 (1), 93–103.
Sun, L., et al., 2012. Coronavirus papain-like proteases negatively regulate antiviral innate immune response through disruption of STING-mediated signaling. PLoS One 7 (2), e30802.
Sun, L., et al., 2013. Cyclic GMP-AMP synthase is a cytosolic DNA sensor that activates the type I interferon pathway. Science 339 (6121), 786–791.
Sun, C., et al., 2015. Evasion of innate cytosolic DNA sensing by a gammaherpesvirus facilitates establishment of latent infection. J. Immunol. 194 (4), 1819–1831.
Sun, H., et al., 2017. USP13 negatively regulates antiviral responses by deubiquitinating STING. Nat. Commun. 8, 15534.
Takaoka, A., et al., 2007. DAI (DLM-1/ZBP1) is a cytosolic DNA sensor and an activator of innate immune response. Nature 448 (7152), 501–505.
Tanaka, Y., Chen, Z.J., 2012. STING specifies IRF3 phosphorylation by TBK1 in the cytosolic DNA signaling pathway. Sci. Signal. 5 (214), ra20.

Tattoli, I., et al., 2008. NLRX1 is a mitochondrial NOD-like receptor that amplifies NF-kappaB and JNK pathways by inducing reactive oxygen species production. EMBO Rep. 9 (3), 293–300.

Thomsen, M.K., et al., 2016. Lack of immunological DNA sensing in hepatocytes facilitates hepatitis B virus infection. Hepatology 64 (3), 746–759.

Tocker, A.M., et al., 2017. The scaffolding protein IQGAP1 interacts with NLRC3 and inhibits type I IFN production. J. Immunol. 199 (8), 2896–2909.

Tong, Z.B., et al., 2000. Mater, a maternal effect gene required for early embryonic development in mice. Nat. Genet. 26 (3), 267–268.

Tsuchida, T., et al., 2010. The ubiquitin ligase TRIM56 regulates innate immune responses to intracellular double-stranded DNA. Immunity 33 (5), 765–776.

Unterholzner, L., 2013. The interferon response to intracellular DNA: why so many receptors? Immunobiology 218 (11), 1312–1321.

Unterholzner, L., et al., 2010. IFI16 is an innate immune sensor for intracellular DNA. Nat. Immunol. 11 (11), 997–1004.

Wang, Q., et al., 2014. The E3 ubiquitin ligase AMFR and INSIG1 bridge the activation of TBK1 kinase by modifying the adaptor STING. Immunity 41 (6), 919–933.

Wang, Y., et al., 2015a. TRIM30alpha is a negative-feedback regulator of the intracellular DNA and DNA virus-triggered response by targeting STING. PLoS Pathog. 11 (6), e1005012.

Wang, P., et al., 2015b. Nlrp6 regulates intestinal antiviral innate immunity. Science 350 (6262), 826–830.

Wang, F., et al., 2016. S6K-STING interaction regulates cytosolic DNA-mediated activation of the transcription factor IRF3. Nat. Immunol. 17 (5), 514–522.

Wang, Q., et al., 2017a. The E3 ubiquitin ligase RNF185 facilitates the cGAS-mediated innate immune response. PLoS Pathog. 13 (3), e1006264.

Wang, J., et al., 2017b. HTLV-1 tax impairs K63-linked ubiquitination of STING to evade host innate immunity. Virus Res. 232, 13–21.

Westerveld, G.H., et al., 2006. Mutations in the testis-specific NALP14 gene in men suffering from spermatogenic failure. Hum. Reprod. 21 (12), 3178–3184.

White, M.J., et al., 2014. Apoptotic caspases suppress mtDNA-induced STING-mediated type I IFN production. Cell 159 (7), 1549–1562.

Wu, J., Chen, Z.J., 2014. Innate immune sensing and signaling of cytosolic nucleic acids. Annu. Rev. Immunol. 32, 461–488.

Wu, J., et al., 2013. Cyclic GMP-AMP is an endogenous second messenger in innate immune signaling by cytosolic DNA. Science 339 (6121), 826–830.

Wu, C., et al., 2017. NLRP11 attenuates toll-like receptor signalling by targeting TRAF6 for degradation via the ubiquitin ligase RNF19A. Nat. Commun. 8 (1), 1977.

Xia, P., et al., 2016. Glutamylation of the DNA sensor cGAS regulates its binding and synthase activity in antiviral immunity. Nat. Immunol. 17 (4), 369–378.

Xing, J., et al., 2017. TRIM29 promotes DNA virus infections by inhibiting innate immune response. Nat. Commun. 8 (1), 945.

Yang, Y.G., Lindahl, T., Barnes, D.E., 2007. Trex1 exonuclease degrades ssDNA to prevent chronic checkpoint activation and autoimmune disease. Cell 131 (5), 873–886.

Yang, P., et al., 2010. The cytosolic nucleic acid sensor LRRFIP1 mediates the production of type I interferon via a beta-catenin-dependent pathway. Nat. Immunol. 11 (6), 487–494.

Yang, X., et al., 2014. Proteolytic processing, deubiquitinase and interferon antagonist activities of Middle East respiratory syndrome coronavirus papain-like protease. J. Gen. Virol. 95 (Pt. 3), 614–626.

Yang, H., et al., 2017. cGAS is essential for cellular senescence. Proc. Natl. Acad. Sci. U.S.A. 114 (23), E4612–E4620.

Zaki, M.H., et al., 2011. The NOD-like receptor NLRP12 attenuates colon inflammation and tumorigenesis. Cancer Cell 20 (5), 649–660.
Zhang, P., et al., 2008. Expression analysis of the NLRP gene family suggests a role in human preimplantation development. PLoS One 3 (7), e2755.
Zhang, Z., et al., 2011. The helicase DDX41 senses intracellular DNA mediated by the adaptor STING in dendritic cells. Nat. Immunol. 12 (10), 959–965.
Zhang, J., et al., 2012. TRIM32 protein modulates type I interferon induction and cellular antiviral response by targeting MITA/STING protein for K63-linked ubiquitination. J. Biol. Chem. 287 (34), 28646–28655.
Zhang, L., et al., 2014. NLRC3, a member of the NLR family of proteins, is a negative regulator of innate immune signaling induced by the DNA sensor STING. Immunity 40 (3), 329–341.
Zhang, M., et al., 2016. USP18 recruits USP20 to promote innate antiviral response through deubiquitinating STING/MITA. Cell Res. 26 (12), 1302–1319.
Zhong, B., et al., 2009. The ubiquitin ligase RNF5 regulates antiviral responses by mediating degradation of the adaptor protein MITA. Immunity 30 (3), 397–407.
Zhong, Y., Kinio, A., Saleh, M., 2013. Functions of NOD-like receptors in human diseases. Front. Immunol. 4, 333.

CHAPTER FOUR

Nucleic Acid Sensing Perturbation: How Aberrant Recognition of Self-Nucleic Acids May Contribute to Autoimmune and Autoinflammatory Diseases

Valentina Bordignon[1], Ilaria Cavallo, Giovanna D'Agosto, Elisabetta Trento, Martina Pontone, Elva Abril, Enea Gino Di Domenico, Fabrizio Ensoli

Clinical Pathology and Microbiology, San Gallicano Dermatological Institute IRCCS, Rome, Italy
[1]Corresponding author: e-mail address: valentina.bordignon@ifo.gov.it

Contents

Abstract

Bacteria and mammalian cells have developed sophisticated sensing mechanisms to detect and eliminate foreign genetic material or to restrict its expression and replication. Progress has been made in the understanding of these mechanisms, which keep foreign or unwanted nucleic acids in check. The complex of mechanisms involved in RNA and DNA sensing is part of a system which is now appreciated as "immune sensing

International Review of Cell and Molecular Biology, Volume 344
ISSN 1937-6448
https://doi.org/10.1016/bs.ircmb.2018.09.001

of nucleic acids" or better "nucleic acid immunity." Nucleic acids, which are critical components for inheriting genetic information in all species, including pathogens, are key structures recognized by the innate immune system. However, while nucleic acid recognition is required for host defense against pathogens, there is a potential risk of self-nucleic acids recognition. In fact, besides its essential contribution to antiviral or microbial defense and restriction of endogenous retro elements, deregulation of nucleic acid immunity can also lead to human diseases due to erroneous detection and response to self-nucleic acids, causing sterile inflammation and autoimmunity.

In this review we will discuss the roles of nucleic acid receptors in guarding against pathogen invasion, and how the microbial environment could interfere or influence immune sensing in discriminating between self and non-self and how this may contribute to autoimmunity or inflammatory diseases.

1. INTRODUCTION

To preserve genome integrity, organisms have established sophisticated mechanisms to detect and eliminate foreign genetic material or to restrict its function and replication. In particular, mammals have evolved complex innate and adaptive immune sensing mechanisms for nucleic acids recognition.

The innate immune system detects the presence of foreign organisms and initiates a coordinated response to eliminate infectious threats. Among the microbial products recognized by innate immune effectors, efficient recognition of nucleic acids (DNA and RNA) is critical, as suggested by the existence of several families of receptors specific to these ligands (Campbell et al., 1999). However, the detection of nucleic acids also presents a risk for self-recognition. In turn, self-activation in response to host nucleic acids may lead to autoimmune diseases. Inappropriate detection of self-molecules is prevented through subcellular compartmentalization of receptors, degradation of self-nucleic acids by endogenous nucleases, and specialization of innate receptors, which specifically detect conserved microbial features absent from the host (Coutinho et al., 2015).

Humans constantly receive physical, chemical and biological cues from the environment. In this review, we will focus primarily on how this sophisticated immune sensory system discriminates self from non-self-nucleic acids and on the connection between autoimmunity and the microbial environment, which includes both pathogenic microbes and commensal microbiota.

2. PRINCIPLES OF NUCLEIC ACID SENSING

In vertebrates, the principle of self- versus non-self-nucleic acid sensing is based on different important criteria, in fact the detection of foreign nucleic acids is based on their availability (concentration, degradation, shielding), localization (extracellular, endolysosome, cytosol) and structure (sequence, conformation, modification). The simplest mechanism to avoid immune recognition of self-DNA that may accumulate in the circulation, in the endolysosomes and in the cytosol of phagocytic cells, is based on the activity of three nucleases that rapidly degrade most self-nucleic acids before they can be sensed by nucleic acid receptors (Lan et al., 2014; Napirei et al., 2000; Schlee and Hartmann, 2016).

Specifically, the three fundamental check-points in nucleic acid sensing include: the localization of nucleic acid ligands (i.e., outside the cell membrane, in the endolysosomal compartment or in the cytosol); the structure of nucleic acid ligands, as characterized by sequence motifs, conformation (i.e., base-paired versus unpaired); and chemical modification. In many cases, a combination of all these aspects contributes to the effective recognition of potentially dangerous foreign nucleic acids thus leading to the induction of an appropriate innate immune response (Ablasser et al., 2013).

The immune and molecular circuits involved in the nucleic acid sensing are represented in Fig. 1.

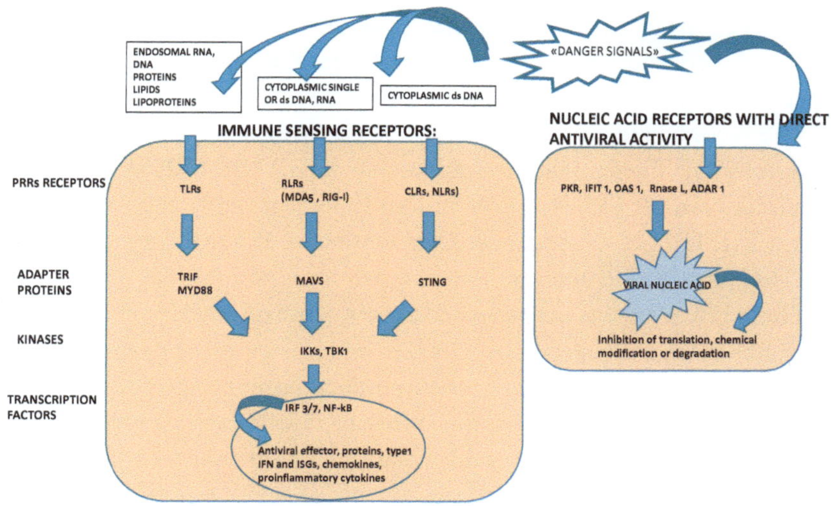

Fig. 1 Immune and molecular circuits involved in the nucleic acid sensing.

3. THE NUCLEIC ACID RECEPTORS

Invasion of microbes to the host induces a defense response that is initiated by the pattern recognition receptors (PRRs) (Oliveira, 2014). PRRs are divided into four types based on structural homology: Toll-like receptors (TLRs), RIG-I-like receptors (RLRs), C-type lectin receptors (CLRs), and NOD-like receptors (NLRs) (Kawai and Akira, 2011; Takeuchi and Akira, 2010).

Moreover, the nucleic acid receptors can be divided into two main categories: immune sensing receptors and nucleic acid receptors with direct antiviral activity.

The first category of nucleic acid receptors includes PRRs belonging to the TLRs family, specifically TLR3, TLR7, TLR8 and TLR9, to the RLR family of RNA sensors (such as retinoic acid inducible gene I (RIG-I or DDX58) (Sato et al., 2015; Schlee, 2013) and melanoma differentiation associated gene 5 (MDA5 or IFIH1) (Glas et al., 2013), the DNA sensors absent in melanoma 2 (AIM2) and cyclic GMP–AMP synthetase (cGAS). The PRRs directly or indirectly induce transcription factors, including nuclear factor-κB (NF-κB) (Alexopoulou et al., 2001) and IFN-regulatory factor 3 (IRF3) (Chattopadhyay et al., 2010), that upregulate the expression of antiviral effector proteins, chemokines and cytokines (Kumagai et al., 2007) such as interleukin-1 (IL-1), IL-6, IL-12 and tumor necrosis factor alpha (TNFα), which initiate and sustain inflammation as well as homing and activation of adaptive immune cells (T and B lymphocytes). Each PRR induces production of type I interferons (IFNs) and IFN-stimulated genes (ISGs) (Roberts et al., 1998).

The second category comprises nucleic acid receptors with direct antiviral activity. These include double-stranded RNA (dsRNA)-activated protein kinase R (PKR), IFN-induced protein with tetratricopeptide repeats 1 (IFIT1), 2′-5′-oligoadenylate synthetase 1 (OAS1) and ribonuclease L (RNase L), and adenosine deaminase acting on RNA 1 (ADAR1).

These DNA sensors are proposed as new types of nucleic acid receptors, in fact like PRRs, they recognize the biochemical features of foreign nucleic acids but their major function is not to induce immune responses via transcription factors and cytokines, but rather to directly act on viral RNA by either inhibiting translation or by chemical modification or degradation of the target nucleic acid.

4. INNATE IMMUNE DETECTION OF MICROBIAL NUCLEIC ACIDS

Once a microbe has overcome the physical and chemical barriers of epithelia, it will be immediately recognized by cells of the innate immune system, such as tissue-residing macrophages and dendritic cells (DCs), via cell associated PRRs. PRRs recognize distinct, conserved microbial structures such as proteins, lipids, lipoproteins, glycans and nucleic acids of bacteria, viruses, parasites and fungi, referred to as pathogen-associated molecular patterns (PAMPs) (Janeway, 1989). Upon PAMP recognition, most PRRs activate a signaling cascade leading to the activation of a defense system that includes various immune genes especially coding for inflammatory cytokines, chemokines and type I interferons (IFNs), in order to establish an appropriate immune response. As PAMPs usually differ from host molecules, they are thought to be well-suited for the initial discrimination of infectious non-self (pathogens and microbial environment) from non-infectious self (Blander and Sander, 2012).

5. NUCLEIC ACID SENSING BY TLRS OF THE PRR FAMILY IN DISTINCT CELLULAR COMPARTMENTS

Extensive work over the past 15 years has led to the discovery and characterization of various PRR families present in distinct cellular compartments. However, very recently there has been an acceleration of research and discoveries related to how, and why, PRRs sense pathogen-derived nucleic acids (Gürtler and Bowie, 2013). The innate immune system's ability to detect RNA and DNA via endosomal and cytosolic PRRs is an essential mechanism to mount protective immune responses especially to viruses, as they do not have many other features suitable for detection.

TLRs are generally expressed in immune cells including monocytes, macrophages, DCs, and B cells as well as non-immune cells such as keratinocytes or epithelial cells (Novak et al., 2010).

TLRs involved in nucleic acid recognition are also differentially expressed in specific subsets of DCs and their activation induces maturation. In this way DCs acquire the capacity to capture and present antigen on major histocompatibility complex (MHC) to activate CD4+ or CD8+ T cells. Activation of CD4+ helper T cells by specific antigens induces the production of antibody to improve defense against the pathogen (Kurts et al., 2010). CD8+ T cells are stimulated through a process known as cross-presentation by antigens that are presented on the MHC class I, which is important for elimination of virally infected cells and is facilitated by TLR3, TLR7, and TLR9 (Schulz et al., 2005).

According to their cellular location and to their initial activation sites in the cells, TLRs are divided into two subgroups (Kawasaki et al., 2011). One group, including TLR1, TLR2, TLR4, TLR5, TLR6, and TLR11 mainly distributes on the cell surface and sense proteins, lipids, and lipoprotein unique to bacteria. The other group includes TLR3, TLR7, TLR8, and TLR9, which sense nucleic acids and reside in the intracellular vesicles such as endoplasmic reticulum (ER), lysosomes, and endosomes (Beutler, 2009; Blasius and Beutler, 2010).

5.1 Nucleic Acid Recognition by Cellular TLRs

TLRs have conserved homology regions, containing leucine-rich repeats (LRRs) in the extracellular space linked by the transmembrane region to the intracellular Toll/IL-1 receptor (TIR) domain. Specific patterns from microbial pathogens are recognized by LRRs, which are predicted to create a dimer for ligand binding. The TLRs are characterized by a domain for ligand recognition, whereas the TIR domain is highly homologous, as demonstrated by structural analysis (Nyman et al., 2008). Activation of the TLRs by ligand binding conducts a signal from its TIR domain to the adapter protein that contains the TIR domain. The adapter protein acts mainly via two signaling pathways: the TIR domain, containing the adapter inducing IFNβ (TRIF)-dependent pathway, and the myeloid differentiation primary response protein (MyD88)-dependent pathway. TLR3 depends on TRIF, whereas TLR7, TLR8, and TLR9 depend on MyD88 (Akira et al., 2006). Both adapter proteins culminate in the activation of transcription factors, such as the nuclear factor kappa-light-chain-enhancer of activated B cells (NF-κB), IFN-regulatory factor (IRF) 3 and IRF7 for transcription of proinflammatory cytokines, and type I IFNs.

The physiological restricted localization of TLRs is necessary for the proper functioning of the receptor. In fact, TLR9, but not TLR7, requires acidification at the endolysosome (Lund et al., 2003).

TLR9 was originally discovered to recognize unmethylated 2-deoxyribo (cytidine-phosphate-guanosine) (CpG-motif containing DNA) DNA present in bacterial DNA, in the endolysosomal compartment, and

DNA–RNA hybrids that have not been degraded by nucleases I and II (Hemmi et al., 2000). TLR9 is initially localized in the ER as a precursor (Sepulveda et al., 2009) and changes its conformation when it binds to CpG DNA (Latz et al., 2007).

TLR9 triggers type I IFN production in DCs and polyclonal activation of B cells via the myeloid differentiation primary response protein 88 (MYD88) and IRF7 signaling pathway. In contrast to the endolysosomal compartment, DNA is usually absent in the cytosol, and the mere presence of cytosolic DNA, indicative of non-self or "dangerous" DNA, promotes inflammasome activation and the induction of type I IFN by the PYHIN (pyrin and HIN domain-containing protein) family member AIM2 (Hornung et al., 2009; Rathinam et al., 2010). The AIM2-related PYHIN domain-containing IFNγ-inducible protein 16 (IFI16) has also been implicated in inflammasome activation in response to short dsDNA (Kerur et al., 2011) while the DNA-damage sensor RAD50 was reported to activate the CARD9–BCL-10 pathway leading to pro-IL-1β mRNA induction after stimulation with long dsDNA (such as poly(dAdT), poly(dGdC) and calf thymus DNA) (Roth et al., 2014).

5.2 Nucleic Acid Recognition by Other DNA Sensing Receptors

In contrast to adaptive immunity which uses specific immune receptors for each antigen, the innate immune system utilizes PRRs as a unique set of molecules that have been selected to recognize molecular patterns derived from pathogens and damaged cells (Gregersen, 2007), which are thought to represent a threat to host's homeostasis ("danger" signals). As such, PRRs are strategically located in specific compartments of the eukaryotic cells. In addition to the TLRs described above, which recognize nucleic acids on the cell membranes or on endolysosomal compartments but not in the cytosol, we distinguish other DNA sensing receptors: the nucleotide binding and oligomerization domain (NOD) receptors (NLRs), which monitor the cytosolic compartment closely interacting with TLR signaling pathways; the retinoid acid inducible gene (RIG)-I-like receptors or melanoma differentiation associated gene 5 (MDA5) that sense single- or double-stranded RNA or DNA in the cytoplasm (RLRs) (Kato et al., 2006; Takeuchi and Akira, 2008; Yoneyama et al., 2004). Recently, the mitochondrial adaptor protein stimulator of IFN genes (STING) was reported as a direct sensor of cyclic diguanylate monophosphate (cyclic GMP–AMP) (Burdette et al., 2011). In the cytosol, cyclic GMP–AMP synthetase (cGAS), IFNγ-inducible protein 16 (IFI16) and absent in melanoma 2 (AIM2) detect DNA that has not been degraded by nucleases such as 3′ repair exonuclease 1 (TREX1) and cooperates with STING as a frontline of host defense initiating adaptive immune response (Palm and Medzhitov, 2009). It is broadly accepted that STING is essential for the activation of the IRF3-dependent pathway (Ishikawa et al., 2009; Zhong et al., 2008) and was found to sense both cyclic dinucleotides secreted by intracellular bacteria (Burdette et al., 2011; Sauer et al., 2011) and the endogenous cyclic GMP–AMP, which contains one 2′-5′-phosphodiester linkage and a canonical 3′-5′ linkage (c[G(2′-5′) pA(3′-5′)p]).

Many cytosolic DNA sensors upstream of the STING pathway have been proposed. The idea being that direct engagement of cytosolic DNA by such a sensor somehow mobilizes STING to engage TBK1 and ultimately cause type I IFN induction. Such pathways would also be subject to negative regulation in order to avoid constant IFN induction by endogenous DNA (Gürtler and Bowie, 2013).

6. CAN SEQUENCE PATTERNS INFLUENCE RECOGNITION?

As demonstrated with recent computational approaches (Vabret et al., 2017), the innate DNA sensors seem to detect and discriminate foreign sequences from self-motifs by a sequence-specific sensing. Microbial genomes display specific nucleic acid patterns at a different frequency compared to the human genome. Typical microbial features recognized by the innate immune system include distinct products of post-transcriptional processing, such as nucleotide chemical modifications and specific 5′- or 3′-moieties. These variations, such as frequencies of nucleotide words, are sufficient to assign biological sequences to taxonomic spaces, at least broadly (Coutinho et al., 2015).

Thus, sequence-specific sensing, including coding and non-coding motifs, should be based on the specific sequence properties to distinguish "self" from foreign genomes.

Functional analysis of the cGAS–STING–IRF3 pathway and the AIM2 inflammasome pathway, however, revealed length-dependent but not sequence-dependent requirements for DNA recognition, which is consistent with structural data showing that the DNA phosphate backbone binds to the HIN domain of AIM2 and the carboxyl terminus of cGAS (Jin et al., 2012). Further evidence showed that stem-loop ssDNA structures, which are characteristic of retroviruses and retroviral elements such as ssDNA reverse transcripts from HIV1, potently activates cGAS, although the stems in such stem-loop structures are shorter than 40 bp (Herzner et al., 2015). TLR9 and cGAS can also recognize DNA–RNA. The hybridization of a 30mer GU-repetitive RNA strand with a 30mer AC-repetitive DNA strand yields a potent TLR9 stimulus, yet abrogates the TLR7 stimulating activity of the GU-repetitive RNA sequence (Rigby et al., 2014). Of note, the AC-repetitive DNA strand that lacks CpG has no TLR9 activity as a single strand, underlining the RNA–DNA hybrid as a real recognition motif. Based on structural models, the mechanism through which RNA–DNA hybrids and dsDNA bind to cGAS appears similar (Mankan et al., 2014).

7. REGULATION OF DNA SENSING

To prevent erroneous sensing, PRRs are strategically located on cell membranes, in the cytosolic and in the endosomal compartments of the eukaryotic cells (Kontaki and Boumpas, 2010). In addition to sequence preference, all nucleic acid sensing TLRs are localized in the endosome. This specific localization appears clearly aimed at protecting the cells, by avoiding TLRs access to self-nucleic acid (Barton et al., 2006). However, besides playing an important role in protection from pathogens, the innate immune signaling has the potential for triggering autoimmune diseases, since the mechanism of self-discrimination can fail.

Indeed, nucleic acid sensing TLRs do not have strong mechanisms for preventing self-nucleic acid recognition. However, structural differences, such as the high levels of unmethylated CpG motifs in viral DNA for TLR9 and clusters of U or GU-rich sequences in viral RNA for TLR7, have been considered key factors in the discrimination between self and non-self nucleic acids.

Any incorrectly located, naked dsDNA of microbial or host origin is likely recognized as an "alarm" or "danger" signal by the innate immune system. While ensuring host defense against many pathogens, especially intracellular bacteria and viruses, this may have detrimental consequences in cases where accumulated host DNA are recognized, leading to the development of autoimmune diseases. To prevent this, deoxyribonucleases (DNases) are located at extra- and intracellular locations to immediately degrade endogenous DNA released from apoptotic and necrotic cells. DNase I is found in the extracellular space and in serum. Mutations of this enzyme have been linked with systemic lupus erythematosus (SLE)-like symptoms (Napirei et al., 2000; Yasutomo et al., 2001). Cytosolic ER-associated DNase III, termed TREX1, digests reverse-transcribed ssDNA intermediates from endogenous retroelements. Mutations in TREX1 are associated with Aicardi-Goutières syndrome (AGS), chilblain lupus and SLE (Stetson et al., 2008), and STING has been implicated in the development of AGS in a mouse model (Gall et al., 2012).

Although TLR7 and TLR9 were shown to play a role in disease pathogenesis (Kawai and Akira, 2010), cytosolic DNA sensors are likely to be involved.

In addition to these mechanisms, a new controlling mechanism has been described (Brinkmann et al., 2007). A number of chaperone proteins are required for shuttling nucleic acid sensors from the ER to endosomes (Lee and Barton, 2014). Each chaperone protein is involved with its specific receptor, assisting in trafficking, folding, and final activation (Pandey et al., 2015). Trafficking of NA sensing TLR7 and 9 is dependent on a multi-transmembrane chaperon protein Unc93 homolog B1 (Unc93B1), as the loss-of-function mutation of Unc93B1 abolishes NA sensing by these TLRs, due to the lack of trafficking of these TLRs (Tabeta et al., 2006). TLR7 and TLR9 are anchored on ER membrane in their steady state, and bind to Unc93B1. When cells are stimulated, Unc93B1 transport the TLRs to endolysosome where TLRs recognize the ligands (Kim et al., 2008). This function depends on N-terminal region of Unc93B1. Binding activity between TLRs and Unc93B1 is critical for the proper balance between TLR7 and TLR9 to inhibit TLR7 hyper-response. In fact an unbalanced binding affinity leads to a TLR7 skewed response and lethal systemic inflammation (Fukui et al., 2009).

8. TLRs AND AUTOIMMUNITY

The evidence for involvement of PRRs (either in the case they work as inducers of cell signaling or exclusively as opsonins) in autoimmunity provides important insights in our understanding the pathogenesis and classification of autoimmune diseases. In some diseases, enhanced PRR signaling is key to disease pathogenesis, since PRR overexpression appears to be a critical factor in the development of systemic autoimmunity. Other diseases involve PRRs as triggers and/or modifiers of autoimmunity. However, the difference between triggers (which are required) and modifiers (which can affect the course of the disease by enhancement or exacerbation) is not easy to establish.

Evidence indicates that PAMPs are shared between pathogenic and non pathogenic microorganisms and that self eukaryotic molecules as nucleic acids can be recognized by signaling PRRs (Haas et al., 2008). However, in physiological conditions this apparent promiscuity is solved at the organismal level by either physical (Gewirtz et al., 2001) or functional segregation (Lee et al., 2006) of non pathogenic microbes and PRRs, or removal of nucleic acids from the extracellular spaces (Napirei et al., 2000) or the cytosol (Stetson et al., 2008), or rapid removal of dead cells. In pathophysiological conditions these protective mechanisms can be broken (Gaipl et al., 2007).

Several DNA-binding proteins are supposed to participate in triggering autoimmune or autoinflammatory processes (Lande et al., 2007). In support to this notion, the antimicrobial peptide LL37, the expression of which is induced by microbial activation of PRRs (Schauber et al., 2007), can activate type I IFNs production by delivering DNA complexes to DCs (Gilliet and Lande, 2008). In this scenario, LL37, acting through a pathogen-like induced molecular pathway, has been implicated in mediating TLR9 and TLR7 responses to self-nucleic acids in psoriasis, a common chronic inflammatory disease of the skin and joints (Ganguly et al., 2009). In systemic autoimmune diseases such as systemic lupus erythematosus (SLE), scleroderma, and Sjögren's syndrome, antibodies against self-DNA, complexed with DNA from dead cells, are internalized by cell surface receptors FcεRII on B cells and DCs (Marshak-Rothstein, 2006). The antibody complexes activate TLRs or cytosolic DNA sensors to produce an excess of type I IFNs, which precipitates the development and progression of systemic autoimmune diseases (Lau et al., 2005; Tian et al., 2007).

In experimental models, TLR ligands are commonly used as adjuvants to generate organ-specific autoimmune diseases such as arthritis and encephalitis in mice. Moreover, mice with deficiency of negative regulators of TLR signals, such as SHP1 (Croker et al., 2008), A20 (Turer et al., 2008), TANK (Kawagoe et al., 2009), and Zc3h12a (Matsushita et al., 2009), spontaneously develop autoimmune diseases by aberrant production of inflammatory cytokines and type I IFNs.

Involvement of TLRs in autoimmune diseases such as SLE has been shown by experimental model systems (Lau et al., 2005; Leadbetter et al., 2002).

SLE is a complex autoimmune disease with a wide spectrum of clinical and immunological features characterized by production of autoantibodies and immune complexes, which is attributed to loss of self-tolerance

(Bai et al., 2018; Kuhn and Landmann, 2014). Cutaneous and articular manifestation are the most common clinical signs of SLE (Ribero et al., 2017).

In SLE evidence points to the involvement of TLR-7 and TLR-9 (Kontaki and Boumpas, 2010; Papadimitraki et al., 2006). In mice, production of antinuclear antibodies depends on endosomal TLRs that bind dsDNA (TLR-9) or ssRNA (TLR7) (Urbonaviciute et al., 2008). This may account for the high-frequency and dominance of antinuclear antibody response in lupus (Kono et al., 2009).

To stimulate adaptive immunity by self-DNA-antibody, it is necessary for the antibody complexes to be internalized, to allow the activation of TLR7 or TLR9. Aberrant expression of type I IFNs is correlated with disease severity in SLE (Marshak-Rothstein, 2006). Several processes for internalization of DNA–antibody complexes have been proposed for the activation of TLRs. One major activation process involved in disease development in lpr/lpr mice relay on their internalization through FccRIII or FccRIIa on DCs, which are the main source of type I IFNs (Means et al., 2005). The Fc region of DNA–antibody complexes is recognized by FccR and internalized DNA is recognized by TLR9 in the endosomes. Blocking peptides specific to the FccR did not stimulate pDCs by immune complex from SLE patients (Bave et al., 2003). An alternative process is based on the internalization through the BCR-mediated endocytosis pathway. B cells expressing an antigen receptor specific for self-IgG internalize DNA–IgG2a–chromatin immune complexes by synergistic engagement of the antigen receptor (Leadbetter et al., 2002).

9. COMMENSALS: A POTENT PLAYER IN IMMUNITY

The microbiome is defined as the totality of microorganisms and their genes inhabiting a unique gut environment. Human microbiota consist of trillions of commensal microorganisms, 10 times more cells than the number of cells constituting the human body. In fact, the total number of human genes is outnumbered by several orders by those of gut bacteria. Commensal microbes are present all over the body including skin, nasopharynx, buccal cavity, genital tract and digestive tract (Ley et al., 2006). Gut is colonized by bacteria soon after the birth, deriving from the surfaces contacted during or after delivery (Dominguez-Bello et al., 2010). Humans have a diverse consortium of bacteria which includes both commensals and pathogens and it is still poorly understood how the immune system can elicit immunity against pathogens but not to commensals. Emerging evidence suggests that hosts are tolerant but not ignorant (Sathyabama et al., 2014). Studies suggest that signals derived from gut microbiota are critical for the development of the immune system (Duerkop et al., 2009; Vieira et al., 2014). Studies on germ-free (GF) animal models, reared in a sterile environment and thus never been exposed to any microorganisms, shed some light on the importance of the microbiota in shaping both innate and adaptive immunity (Smith et al., 2007). On the other hand, the perturbation of microbiota, either by antibiotic treatment or microbiota reconstitution, provided key evidence for the role of microbiota in immune homeostasis (Wu and Wu, 2012).

Several reports point at a pivotal role exerted by the gut microbiota in regulating the development of APCs. DCs of Peyer's patches produce high levels of interleukin-10 (IL-10), compared with splenic DCs activated under similar conditions (Iwasaki and Kelsall, 1999). Similar to DCs, gut macrophages are located in close proximity to the intestinal microbiota, and they develop a unique phenotype, so-called inflammation anergy, referring to the non inflammatory profile of intestinal macrophages when they encounter microbial stimuli in homeostatic conditions (Smythies et al., 2005). For example, intestinal macrophages do not produce proinflammatory cytokines in response to microbial stimuli such as TLR ligands, a set of microbe associated molecular patterns (Smythies et al., 2010).

Recent findings show that gut microbiota has a much greater influence on the systemic immune system than previously anticipated. It has been shown that microbiota can concur in controlling T-cell differentiation in the lamina propria (LP), where the IL-17 expressing CD4+ T cell subset (Th17 cells) is especially abundant (Ivanov et al., 2008). Th17 cells produce many proinflammatory cytokines, like IL-17, IL-21 and IL-22, and are thought to play a pro-pathogenic role in autoimmune diseases like arthritis and experimental autoimmune encephalomyelitis (EAE) (Bettelli et al., 2007). The presence of Th17 cells in the small and large intestine is dependent on gut microbiota, since their number in the LP of GF mice is drastically reduced, whereas the number of Foxp3+ regulatory T cells (Treg cells) is increased (Atarashi et al., 2008).

In an effort to determine whether the gut microbiota in general or just a specific subset of bacteria were responsible for the Th17 cell development, Ivanov et al. treated mice with antibiotics selectively killing different subsets of bacteria. They identified vancomycin-sensitive bacteria as the specific subset able to induce Th17-cell differentiation in the LP of the small intestine (Ivanov et al., 2008). Later, the same group identified segmented filamentous bacteria (SFB) to be potent inducers of LP Th17 cells (Ivanov et al., 2009).

Other studies revealed an association of specific bacterial species with the development of particular T-cell subtypes. For example, *Bacteroides fragilis* was shown to induce the development of a systemic Th1 response through its polysaccharide A (PSA) molecules (Mazmanian et al., 2005).

10. DO MICROBES HAVE A ROLE IN AUTOIMMUNITY?

What are the indications that the microbial environment may affect autoimmunity?

As the gut microbiota has such profound effects on both the innate and adaptive immune system, it is not surprising that some members of the gut microbiota have been linked to autoimmune diseases (Wu and Wu, 2012).

Studies of autoimmune diseases in identical twins, emigrants from areas with high or low incidence of a particular autoimmune disease to areas with either low or high incidence, respectively, or in genetically related populations living in dissimilar conditions, all strongly suggest that the environment play a part on disease onset (Bach, 2005). For example, studies on identical twins showed a limited concordance of autoimmune diseases such as type 1 diabetes (T1D) (Zipris, 2009). Assuming that environmental factors play a role, the 50% discordance in T1D in identical twins can be explained in two mutually exclusive ways: an infection promoted T1D in the unlucky twin, or something, probably microbes (Pandey et al., 2015), prevented disease development in the unaffected twin. It seems, however, that both scenarios are possible.

Even in animal models, it seems that other factors, in addition to genetic predisposition, may influence autoimmunity (Maldonado et al., 1999). For example, in male mice of the B10.BR strain, which acquire a spontaneous autoimmune ankylosing enthesopathy, the disease does not develop in germ-free conditions (Rehakova et al., 2000). The gnotobiotic approach, which involves colonization with a defined microbiota, revealed some specificity in the expression of autoimmune ankylosing enthesopathy, since it was driven by a group of commensal anaerobic bacteria, but not by *Lactobacillus* species (Sinkorova et al., 2008). Thus, use of germ-free and gnotobiotic mice has revealed that tolerance to self-antigens can be altered by microbes and that, at least in some diseases, microbial triggers are required to promote the pathogenesis in genetically predisposed animals. This is true for both polygenic and monogenic models of autoimmunity.

It is conceivable that microbes may be involved in autoimmunity in different ways. The presence of a particular microorganism may be required for disease. Alternatively, it may trigger autoimmunity in genetically susceptible individuals, even when the persistence of microorganisms is not required, or it may be able to accelerate the onset or progression of the disease. Conversely, microbes might also act in preventing autoimmunity (Chervonsky, 2009, 2010; Gangaplara et al., 2013; Sathyabama et al., 2014).

The first aspect is exemplified by rheumatic fever, in which antibodies against streptococcal antigens cross-react with a myocardial antigen (Malkiel et al., 2000). The concept of cross-reactivity or molecular mimicry was introduced (Oldstone, 1998) to explain autoimmunity induced by similarity or identity between MHC-bound peptides derived from pathogens and MHC-bound peptides derived from self.

Pathogenic mechanisms based on molecular mimicry have been suggested in the Guillain–Barré syndrome, where amino acid similarities between the gangliosides of the nervous system and the LPS of *Campylobacter jejuni* have been described (Houliston et al., 2011).

Molecular mimicry may be involved in some, but clearly not all, cases of autoimmunity (Benoist and Mathis, 2001).

A genetic predisposition must also play a part in molecular mimicry-associated diseases, since mimic peptides derived from pathogens show restricted binding to particular MHC alleles. Lastly, mimicry or not, the issue of microbial influence on autoimmunity is directly linked to the complex circuitries that regulate the interactions between the innate and adaptive immune systems (Janeway, 1989; Medzhitov, 2009; Miller et al., 1997; Munz et al., 2009), specifically considering the role of PRRs in the provision of signals required for the activation of adaptive immune responses (Kawai and Akira, 2009; Palm and Medzhitov, 2009).

11. GUT MICROBIOTA AND AUTOIMMUNITY: THE FUTURE

The influence of commensals on health and disease through the regulation of immune function has emerged as an area of scientific and clinical importance. It is now evident that the gut microbiota has a profound effect on the host immune system and can affect autoimmune-related diseases both within and outside the gut. Even though there is a little knowledge of the human gut microbiota, the recent advancements in "next-generation" sequencing have led to a revolution in developing a culture-independent and thorough method to characterize gut microbial communities.

The ensuing knowledge on the relationship between commensal communities and the host immune system and their influence in autoimmune disorders highlights the need for a better understanding of the pathways that initiates and maintains the host-commensal relationship. Digging out mechanism employed by gut microbes in preventing or breaking T and B cell tolerance against host components represents an exciting area of research, which promise to have a big impact on new generation therapies.

12. METHODS

The present review focuses on the description of immune and molecular circuitries involved in the process of nucleic acid sensing and how they act in response to pathogen- or damage-associated signals as well as their putative roles in the pathogenesis of autoimmunity.

REFERENCES

Ablasser, A., Hertrich, C., Wassermann, R., et al., 2013. Nucleic acid driven sterile inflammation. Clin. Immunol. 147, 207–215.

Akira, S., Uematsu, S., Takeuchi, O., 2006. Pathogen recognition and innate immunity. Cell 124, 783–801.

Alexopoulou, L., Holt, A.C., Medzhitov, R., et al., 2001. Recognition of double-stranded RNA and activation of NF-κB by toll-like receptor 3. Nature 413, 732–738.

Atarashi, K., Nishimura, J., Shima, T., et al., 2008. ATP drives lamina propria T(H)17 cell differentiation. Nature 455, 808–812.

Bach, J.F., 2005. Infections and autoimmune diseases. J. Autoimmun. 25, 74–80.

Bai, Y., Tong, Y., Liu, Y., Hu, H., 2018. Self-dsDNA in the pathogenesis of systemic lupus erythematosus. Clin. Exp. Immunol. 191, 1–10.

Barton, G.M., Kagan, J.C., Medzhitov, R., 2006. Intracellular localization of toll-like receptor 9 prevents recognition of self DNA but facilitates access to viral DNA. Nat. Immunol. 7, 49–56.

Bave, U., Magnusson, M., Eloranta, M.L., et al., 2003. Fc gamma RIIa is expressed on natural IFN-alpha-producing cells (plasmacytoid dendritic cells) and is required for the IFN-alpha production induced by apoptotic cells combined with lupus IgG. J. Immunol. 171, 3296–3302.

Benoist, C., Mathis, D., 2001. Autoimmunity provoked by infection: how good is the case for T cell epitope mimicry? Nat. Immunol. 2, 797–801.

Bettelli, E., Oukka, M., Kuchroo, V.K., 2007. T(H)-17 cells in the circle of immunity and autoimmunity. Nat. Immunol. 8, 345–350.

Beutler, B., 2009. Microbe sensing, positive feedback loops, and the pathogenesis of inflammatory diseases. Immunol. Rev. 227, 248–263.

Blander, J.M., Sander, L.E., 2012. Beyond pattern recognition: five immune checkpoints for scaling the microbial threat. Nat. Rev. Immunol. 12, 215–225.

Blasius, A.L., Beutler, B., 2010. Intracellular toll-like receptors. Immunity 32, 305–315.

Brinkmann, M.M., Spooner, E., Hoebe, K., et al., 2007. The interaction between the ER membrane protein UNC93B and TLR3, 7, and 9 is crucial for TLR signaling. J. Cell. Biol. 177, 265–275.
Burdette, D.L., Monroe, K.M., Sotelo-Troha, K., et al., 2011. STING is a direct innate immune sensor of cyclic di-GMP. Nature 478, 515–518.
Campbell, A., Mrázek, J., Karlin, S., 1999. Genome signature comparisons among prokaryote, plasmid, and mitochondrial DNA. Proc. Natl. Acad. Sci. U.S.A. 96, 9184–9189.
Chattopadhyay, S., Marques, J.T., Yamashita, M., 2010. Viral apoptosis is induced by IRF-3-mediated activation of Bax. EMBO J. 29, 1762–1773.
Chervonsky, A., 2009. Innate receptors and microbes in induction of autoimmunity. Curr. Opin. Immunol. 21, 641–647.
Chervonsky, A.V., 2010. Influence of microbial environment on autoimmunity. Nat. Immunol. 11, 28–35.
Coutinho, T.J., Franco, G.R., Lobo, F.P., 2015. Homology-independent metrics for comparative genomics. Comput. Struct. Biotechnol. J. 13, 352–357.
Croker, B.A., Lawson, B.R., Rutschmann, S., et al., 2008. Inflammation and autoimmunity caused by a SHP1 mutation depend on IL-1, MyD88, and a microbial trigger. Proc. Natl. Acad. Sci. U.S.A. 105, 15028–15033.
Dominguez-Bello, M.G., Costello, E.K., Contreras, M., et al., 2010. Delivery mode shapes the acquisition and structure of the initial microbiota across multiple body habitats in newborns. Proc. Natl. Acad. Sci. U.S.A. 107, 11971–11975.
Duerkop, B.A., Vaishnava, S., Hooper, L.V., 2009. Immune responses to the microbiota at the intestinal mucosal surface. Immunity 31, 368–376.
Fukui, R., Saitoh, S., Matsumoto, F., et al., 2009. Unc93B1 biases toll-like receptor responses to nucleic acid in dendritic cells toward DNA- but against RNA-sensing. J. Exp. Med. 206, 1339–1350.
Gaipl, U.S., Munoz, L.E., Grossmayer, G., et al., 2007. Clearance deficiency and systemic lupus erythematosus (SLE). J. Autoimmun. 28, 114–121.
Gall, A., Treuting, P., Elkon, K.B., et al., 2012. Autoimmunity initiates in nonhematopoietic cells and progresses via lymphocytes in an interferon-dependent autoimmune disease. Immunity 36, 120–131.
Gangaplara, A., Massilamany, C., Steffen, D., et al., 2013. Mimicry epitope from Ehrlichia canis for interphotoreceptor retinoid-binding protein 201-216 prevents autoimmune uveoretinitis by acting as altered peptide ligand. J. Neuroimmunol. 263, 98–107.
Ganguly, D., Chamilos, G., Lande, R., et al., 2009. Self-RNA-antimicrobial peptide complexes activate human dendritic cells through TLR7 and TLR8. J. Exp. Med. 206, 1983–1994.
Gewirtz, A.T., Navas, T.A., Lyons, S., et al., 2001. Cutting edge: bacterial flagellin activates basolaterally expressed TLR5 to induce epithelial proinflammatory gene expression. J. Immunol. 167, 1882–1885.
Gilliet, M., Lande, R., 2008. Antimicrobial peptides and self-DNA in autoimmune skin inflammation. Curr. Opin. Immunol. 20, 401–407.
Glas, M., Coch, C., Trageser, D., 2013. Targeting the cytosolic innate immune receptors RIG-I and MDA5 effectively counteracts cancer cell heterogeneity in glioblastoma. Stem Cells 31, 1064–1074.
Gregersen, P.K., 2007. Modern genetics, ancient defenses, and potential therapies. N. Engl. J. Med. 356, 1263e6.
Gürtler, C., Bowie, A.G., 2013. Innate immune detection of microbial nucleic acids. Trends Microbiol. 21, 413–420.
Haas, T., Metzger, J., Schmitz, F., 2008. The DNA sugar backbone 2′ deoxyribose determines toll-like receptor 9 activation. Immunity 28, 315–323.

Hemmi, H., Takeuchi, O., Kawai, T., et al., 2000. A toll-like receptor recognizes bacterial DNA. Nature 408, 740–745.

Herzner, A.M., Hagmann, C.A., Goldeck, M., et al., 2015. Sequence-specific activation of the DNA sensor cGAS by Y-form DNA structures as found in primary HIV-1 cDNA. Nat. Immunol. 16, 1025–1033.

Hornung, V., Ablasser, A., Charrel-Dennis, M., et al., 2009. AIM2 recognizes cytosolic dsDNA and forms a caspase-1-activating inflammasome with ASC. Nature 458, 514–518.

Houliston, R.S., Vinogradov, E., Dzieciatkowska, M., et al., 2011. Lipooligosaccharide of Campylobacter jejuni: similarity with multiple types of mammalian glycans beyond gangliosides. J. Biol. Chem. 286, 12361–12370.

Ishikawa, H., Ma, Z., Barber, G.N., 2009. STING regulates intracellular DNA-mediated, type I interferon-dependent innate immunity. Nature 461, 788–792.

Ivanov, I.I., Frutos Rde, L., Manel, N., et al., 2008. Specific microbiota direct the differentiation of IL-17-producing T-helper cells in the mucosa of the small intestine. Cell Host Microbe 4, 337–349.

Ivanov, I.I., Atarashi, K., Manel, N., et al., 2009. Induction of intestinal Th17 cells by segmented filamentous bacteria. Cell 139, 485–498.

Iwasaki, A., Kelsall, B.L., 1999. Freshly isolated Peyer's patch, but not spleen, dendritic cells produce interleukin 10 and induce the differentiation of T helper type 2 cells. J. Exp. Med. 190, 229–239.

Janeway Jr., C.A., 1989. Approaching the asymptote? Evolution and revolution in immunology. Cold Spring Harb. Symp. Quant. Biol. 54, 1–13.

Jin, T., Perry, A., Jiang, J., et al., 2012. Structures of the HIN domain:DNA complexes reveal ligand binding and activation mechanisms of the AIM2 inflammasome and IFI16 receptor. Immunity 36, 561–571.

Kato, H., Takeuchi, O., Sato, S., et al., 2006. Differential roles of MDA5 and RIG-I helicases in the recognition of RNA viruses. Nature 441, 101–105.

Kawagoe, T., Takeuchi, O., Takabatake, Y., et al., 2009. TANK is a negative regulator of Toll-like receptor signaling and is critical for the prevention of autoimmune nephritis. Nat. Immunol. 10, 965–972.

Kawai, T., Akira, S., 2009. The roles of TLRs, RLRs and NLRs in pathogen recognition. Int. Immunol. 21, 317–337.

Kawai, T., Akira, S., 2010. The role of pattern-recognition receptors in innate immunity: update on toll-like receptors. Nat. Immunol. 11, 373–384.

Kawai, T., Akira, S., 2011. Toll-like receptors and their crosstalk with other innate receptors in infection and immunity. Immunity 34, 637–650.

Kawasaki, T., Kawai, T., Akira, S., 2011. Recognition of nucleic acids by pattern-recognition receptors and its relevance in autoimmunity. Immunol. Rev. 243, 61–73.

Kerur, N., Veettil, M.V., Sharma-Walia, N., et al., 2011. IFI16 acts as a nuclear pathogen sensor to induce the inflammasome in response to Kaposi sarcoma-associated herpesvirus infection. Cell Host Microbe 9, 363–375.

Kim, Y.M., Brinkmann, M.M., Paquet, M.E., et al., 2008. UNC93B1 delivers nucleotide-sensing toll-like receptors to endolysosomes. Nature 452, 234–238.

Kono, D.H., Haraldsson, M.K., Lawson, B.R., et al., 2009. Endosomal TLR signaling is required for anti-nucleic acid and rheumatoid factor autoantibodies in lupus. Proc. Natl. Acad. Sci. U.S.A. 106, 12061e6.

Kontaki, E., Boumpas, D.T., 2010. Innate immunity in systemic lupus erythematosus: sensing endogenous nucleic acids. J. Autoimmun. 35, 206–211.

Kuhn, A., Landmann, A., 2014. The classification and diagnosis of cutaneous lupus erythematosus. J. Autoimmun. 49, 14–19.

Kumagai, Y., Takeuchi, O., Kato, H., 2007. Alveolar macrophages are the primary interferon-alpha producer in pulmonary infection with RNA viruses. Immunity 27, 240–252.

Kurts, C., Robinson, B.W., Knolle, P.A., 2010. Crosspriming in health and disease. Nat. Rev. Immunol. 10, 403–414.

Lan, Y.Y., Londono, D., Bouley, R., 2014. DNase2a deficiency uncovers lysosomal clearance of damaged nuclear DNA via autophagy. Cell Rep. 9, 180–192.

Lande, R., Gregorio, J., Facchinetti, V., et al., 2007. Plasmacytoid dendritic cells sense self-DNA coupled with antimicrobial peptide. Nature 449, 564–569.

Latz, E., Verma, A., Visintin, A., et al., 2007. Ligand-induced conformational changes allosterically activate toll-like receptor 9. Nat. Immunol. 8, 772–779.

Lau, C.M., Broughton, C., Tabor, A.S., et al., 2005. RNA-associated autoantigens activate B cells by combined B cell antigen receptor/toll-like receptor 7 engagement. J. Exp. Med. 202, 1171e7.

Leadbetter, E.A., Rifkin, I.R., Hohlbaum, A.M., et al., 2002. Chromatin-IgG complexes activate B cells by dual engagement of IgM and toll-like receptors. Nature 416, 603–607.

Lee, B.L., Barton, G.M., 2014. Trafficking of endosomal toll-like receptors. Trends Cell Biol. 24, 360–369.

Lee, J., Mo, J.H., Katakura, K., 2006. Maintenance of colonic homeostasis by distinctive apical TLR9 signalling in intestinal epithelial cells. Nat. Cell Biol. 8, 1327–1336.

Ley, R.E., Turnbaugh, P.J., Klein, S., et al., 2006. Microbial ecology: human gut microbes associated with obesity. Nature 444, 1022–1023.

Lund, J., Sato, A., Akira, S., et al., 2003. Toll-like receptor 9-mediated recognition of Herpes simplex virus-2 by plasmacytoid dendritic cells. J. Exp. Med. 198, 513–520.

Maldonado, M.A., Kakkanaiah, V., MacDonald, G.C., et al., 1999. The role of environmental antigens in the spontaneous development of autoimmunity in MRL-lpr mice. J. Immunol. 162, 6322–6330.

Malkiel, S., Liao, L., Cunningham, M.W., et al., 2000. T-cell-dependent antibody response to the dominant epitope of streptococcal polysaccharide, N-acetylglucosamine, is cross-reactive with cardiac myosin. Infect. Immun. 68, 5803–5808.

Mankan, A.K., Schmidt, T., Chauhan, D., et al., 2014. Cytosolic RNA:DNA hybrids activate the cGAS-STING axis. EMBO J. 33, 2937–2946.

Marshak-Rothstein, A., 2006. Toll-like receptors in systemic autoimmune disease. Nat. Rev. Immunol. 6, 823–835.

Matsushita, K., Takeuchi, O., Standley, D.M., et al., 2009. Zc3h12a is an RNase essential for controlling immune responses by regulating mRNA decay. Nature 458, 1185–1190.

Mazmanian, S.K., Liu, C.H., Tzianabos, A.O., et al., 2005. An immunomodulatory molecule of symbiotic bacteria directs maturation of the host immune system. Cell 122, 107–118.

Means, T.K., Latz, E., Hayashi, F., et al., 2005. Human lupus autoantibody-DNA complexes activate DCs through cooperation of CD32 and TLR9. J. Clin. Invest. 115, 407–417.

Medzhitov, R., 2009. Approaching the asymptote: 20 years later. Immunity 30, 766–775.

Miller, S.D., Vanderlugt, C.L., Begolka, W.S., et al., 1997. Persistent infection with Theiler's virus leads to CNS autoimmunity via epitope spreading. Nat. Med. 3, 1133–1136.

Munz, C., Lunemann, J.D., Getts, M.T., et al., 2009. Antiviral immune responses: triggers of or triggered by autoimmunity? Nat. Rev. Immunol. 9, 246–258.

Napirei, M., Karsunky, H., Zevnik, B., 2000. Features of systemic lupus erythematosus in DNase1-deficient mice. Nat. Genet. 25, 177–181.

Novak, N., Koch, S., Allam, J.P., et al., 2010. Dendritic cells: bridging innate and adaptive immunity in atopic dermatitis. J. Allergy Clin. Immunol. 125, 50–59.

Nyman, T., Stenmark, P., Flodin, S., et al., 2008. The crystal structure of the human toll-like receptor 10 cytoplasmic domain reveals a putative signalling dimer. J. Biol. Chem. 283, 11861–11865.

Oldstone, M.B., 1998. Molecular mimicry and immune-mediated diseases. FASEB J. 12, 1255–1265.

Oliveira, S.C., 2014. Innate immune sensing of nucleic acids from pathogens. Microbes Infect. 16, 977–978.

Palm, N.W., Medzhitov, R., 2009. Pattern recognition receptors and control of adaptive immunity. Immunol. Rev. 227, 221–233.

Pandey, S., Kawai, T., Akira, S., 2015. Microbial sensing by toll-like receptors and intracellular nucleic acid sensors. Cold Spring Harb. Perspect. Biol. 7, a016246.

Papadimitraki, E.D., Choulaki, C., Koutala, E., et al., 2006. Expansion of toll-like receptor 9-expressing B cells in active systemic lupus erythematosus: implications for the induction and maintenance of the autoimmune process. Arthritis Rheum. 54, 3601e11.

Rathinam, V.A., Jiang, Z., Waggoner, S.N., et al., 2010. The AIM2 inflammasome is essential for host defense against cytosolic bacteria and DNA viruses. Nat. Immunol. 11, 395–402.

Rehakova, Z., Capková, J., Stěpánková, R., et al., 2000. Germ-free mice do not develop ankylosing enthesopathy, a spontaneous joint disease. Hum. Immunol. 61, 555–558.

Ribero, S., Sciascia, S., Borradori, L., et al., 2017. The cutaneous spectrum of lupus erythematosus. Clin. Rev. Allergy Immunol. 53, 291–305.

Rigby, R.E., Webb, L.M., Mackenzie, K.J., et al., 2014. RNA:DNA hybrids are a novel molecular pattern sensed by TLR9. EMBO J. 33, 542–558.

Roberts, R.M., Liu, L., Guo, Q., Leaman, D., et al., 1998. The evolution of the type I interferons. J. Interferon Cytokine Res. 18, 805–816.

Roth, S., Rottach, A., Lotz-Havla, A.S., et al., 2014. Rad50-CARD9 interactions link cytosolic DNA sensing to IL-1β production. Nat. Immunol. 15, 538–545.

Sathyabama, S., Khan, N., Agrewala, J.N., 2014. Friendly pathogens: prevent or provoke autoimmunity. Crit. Rev. Microbiol. 40, 273–280.

Sato, S., Li, K., Kameyama, T., 2015. The RNA sensor RIG-I dually functions as an innate sensor and direct antiviral factor for hepatitis B virus. Immunity 42, 123–132.

Sauer, J.D., Sotelo-Troha, K., von Moltke, J., et al., 2011. The *N*-ethyl-*N*-nitrosourea-induced goldenticket mouse mutant reveals an essential function of Sting in the *in vivo* interferon response to *Listeria monocytogenes* and cyclic dinucleotides. Infect. Immun. 79, 688–694.

Schauber, J., Dorschner, R.A., Coda, A.B., et al., 2007. Injury enhances TLR2 function and antimicrobial peptide expression through a vitamin D-dependent mechanism. J. Clin. Invest. 117, 803–811.

Schlee, M., 2013. Master sensors of pathogenic RNA-RIG-I like receptors. Immunobiology 218, 1322–1335.

Schlee, M., Hartmann, G., 2016. Discriminating self from non-self in nucleic acid sensing. Nat. Rev. Immunol. 16, 566–580.

Schulz, O., Diebold, S.S., Chen, M., et al., 2005. Toll-like receptor 3 promotes cross-priming to virus-infected cells. Nature 433, 887–892.

Sepulveda, F.E., Maschalidi, S., Colisson, R., et al., 2009. Critical role for asparagine endopeptidase in endocytic toll-like receptor signaling in dendritic cells. Immunity 31, 737–748.

Sinkorova, Z., Capkova, J., Niederlova, J., et al., 2008. Commensal intestinal bacterial strains trigger ankylosing enthesopathy of the ankle in inbred B10.BR (H-2(k)) male mice. Hum. Immunol. 69, 845–850.

Smith, K., McCoy, K.D., Macpherson, A.J., 2007. Use of axenic animals in studying the adaptation of mammals to their commensal intestinal microbiota. Semin. Immunol. 19, 59–69.
Smythies, L.E., Sellers, M., Clements, R.H., et al., 2005. Human intestinal macrophages display profound inflammatory anergy despite avid phagocytic and bacteriocidal activity. J. Clin. Invest. 115, 66–75.
Smythies, L.E., Shen, R., Bimczok, D., et al., 2010. Inflammation anergy in human intestinal macrophages is due to Smad-induced IkappaBalpha expression and NF-kappaB inactivation. J. Biol. Chem. 285, 19593–19604.
Stetson, D.B., Ko, J.S., Heidmann, T., et al., 2008. Trex1 prevents cell-intrinsic initiation of autoimmunity. Cell 134, 587–598.
Tabeta, K., Hoebe, K., Janssen, E.M., et al., 2006. The Unc93b1 mutation 3d disrupts exogenous antigen presentation and signaling via Toll-like receptors 3, 7 and 9. Nat. Immunol. 7, 156–164.
Takeuchi, O., Akira, S., 2008. MDA5/RIG-I and virus recognition. Curr. Opin. Immunol. 20, 17–22.
Takeuchi, O., Akira, S., 2010. Pattern recognition receptors and inflammation. Cell 140, 805–820.
Tian, J., Avalos, A.M., Mao, S.Y., et al., 2007. Toll-like receptor 9-dependent activation by DNA-containing immune complexes is mediated by HMGB1 and RAGE. Nat. Immunol. 8, 487e96.
Turer, E.E., Tavares, R.M., Mortier, E., et al., 2008. Homeostatic MyD88-dependent signals cause lethal inflammation in the absence of A20. J. Exp. Med. 205, 451–464.
Urbonaviciute, V., Furnrohr, B.G., Meister, S., et al., 2008. Induction of inflammatory and immune responses by HMGB1-nucleosome complexes: implications for the pathogenesis of SLE. J. Exp. Med. 205, 3007e18.
Vabret, N., Bhardwaj, N., Greenbaum, B.D., 2017. Sequence-specific sensing of nucleic acids. Trends Immunol. 38, 53–65.
Vieira, S.M., Pagovich, O.E., Kriegel, M.A., 2014. Diet, microbiota and autoimmune diseases. Lupus 23, 518–526.
Wu, H.J., Wu, E., 2012. The role of gut microbiota in immune homeostasis and autoimmunity. Gut Microbes 3, 4–14.
Yasutomo, K., Horiuchi, T., Kagami, S., 2001. Mutation of *DNA SE1* in people with systemic lupus erythematosus. Nat. Genet. 28, 313–314.
Yoneyama, M., Kikuchi, M., Natsukawa, T., et al., 2004. The RNA helicase RIG-I has an essential function in double-stranded RNA-induced innate antiviral responses. Nat. Immunol. 5, 730–737.
Zhong, B., Yang, Y., Li, S., et al., 2008. The adaptor protein MITA links virus-sensing receptors to IRF3 transcription factor activation. Immunity 29, 538–550.
Zipris, D., 2009. Epidemiology of type 1 diabetes and what animal models teach us about the role of viruses in disease mechanisms. Clin. Immunol. 131, 11–23.

CHAPTER FIVE

Common Differences: The Ability of Inflammasomes to Distinguish Between Self and Pathogen Nucleic Acids During Infection

Christopher R. Lupfer[a,*], Meagan D. Rippee-Brooks[a], Paras K. Anand[b,*]

[a]Department of Biology, Missouri State University, Springfield, MO, United States
[b]Infectious Diseases and Immunity, Imperial College London, London, United Kingdom
*Corresponding authors: e-mail address: christopherlupfer@missouristate.edu; paras.anand@imperial.ac.uk

Contents

Abstract

The innate immune system detects the presence of pathogens based on detection of non-self. In other words, most pathogens possess intrinsic differences that can distinguish them from host cells. For example, bacteria and fungi have cell walls comprised

International Review of Cell and Molecular Biology, Volume 344
ISSN 1937-6448
https://doi.org/10.1016/bs.ircmb.2018.10.001

of peptidoglycan and carbohydrates (like mannans), respectively. Germline encoded pattern recognition receptors (PRRs) of the Toll-like receptor (TLR) and C-type lectin receptor (CLR) family have the ability to detect such unique pathogen associated features. However, some TLRs and members of the RIG-I-like receptor (RLR), NOD-like receptor (NLR), or AIM2-like receptor (ALR) family can sense pathogen invasion based on pathogen nucleic acids. Nucleic acids are not unique to pathogens, thus raising the question of how such PRRs evolved to detect pathogens but not self. In this chapter, we will examine the PRRs that sense pathogen nucleic acids and subsequently activate the inflammasome signaling pathway. We will examine the selective mechanisms by which these receptors distinguish pathogens from "self" and discuss the importance of such pathways in disease development in animal models and human patients.

1. INTRODUCTION

The immune system maintains and restores homeostasis of the host in response to physical injury, pathogen invasion, or metabolic dysfunction. The first line of defense is the innate immune system, which consists of leukocytes, epithelial cells, and endothelial cells possessing the ability to recognize pathogens and damage to self-tissues. Through strategically placed cells present in almost every tissue in the host, the innate immune system responds almost immediately to disturbances in homeostasis by detecting the presence of pathogen associated molecular patterns (PAMPs) or damage associated molecular patterns (DAMPs). PAMPs consist of conserved pathogen molecules including bacterial and fungal cell wall components or DNA and RNA from pathogens that are different from the host (CpG DNA from bacteria). In addition, DNA and RNA present in cellular or extracellular compartments where such molecules are not found under homeostasis (DNA in the cytoplasm) can further trigger an immune response (Burckstummer et al., 2009; Fernandes-Alnemri et al., 2009). DAMPs consist of cellular molecules released during cell damage and include ATP or nuclear proteins like HMGB1 released from damaged cells into the extracellular environment (Barlan et al., 2011b). Thus, the innate immune system can detect a wide range of pathogens or pathological conditions with a limited set of pattern recognition receptors (PRRs) that possess the ability to detect conserved PAMPs or DAMPs instead of responding to specific pathogen features individually (i.e., toxins or receptor proteins from pathogens).

Several classes of PRRs are involved in PAMP and DAMP recognition; however, this chapter will focus on a specific group of PRRs with the ability to recognize nucleic acids and activate an immune signaling complex known as the inflammasome. The inflammasome is a multimeric protein structure consisting of an activating PRR, the adaptor protein ASC and proteases caspase-1 and caspase-11. Of note, not all inflammasomes require all caspases, thus the inclusion or exclusion of caspases provides some specificity to the immune response to different stimuli (Cunha et al., 2017; Pierini et al., 2012). The functions of the inflammasome are twofold. Active caspase-1, and in some cases caspase-8, cleaves cytokine precursors pro-IL-1β and pro-IL-18 into their active forms (Karki et al., 2015). Active caspase-1 and caspase-11 (caspase-4 and -5 in humans) can also cleave the autoinhibitory C-terminal fragment of the protein gasdermin D, which then forms pores in the cell membrane resulting in an inflammatory form of cell death termed pyroptosis (He et al., 2015; Kayagaki et al., 2015; Shi et al., 2015). Although the overall function of the inflammasome is cytokine production and cell death, numerous PRRs can activate the inflammasome in response to a myriad of stimuli. Some members of the NOD-like receptor family of PRRs can activate the inflammasome. These include NLRP1, NLRP3, NLRP7, NLRP9b, and NLRC4. AIM2 and IFI16, as well as the protein pyrin, also activate the inflammasome. In addition, several NLR apoptosis inhibitory proteins (NAIP1, NAIP2, NAIP5, NAIP6) recognize ligands from bacterial secretion systems and activate NLRC4 (Kofoed and Vance, 2011; Rayamajhi et al., 2013; Yang et al., 2013a; Zhao et al., 2011). Of the PRRs listed above, NLRP3, AIM2, NLRP9b, and IFI16 are reported to sense nucleic acids and activate the inflammasome. Of importance are the mechanisms by which these PRRs can distinguish between self and pathogen DNA or RNA, as there are very few structural differences between host and pathogen nucleic acids. The specificity of these PRRs for distinct pathogens or pathological conditions and their regulation are discussed in this chapter.

2. OVERVIEW OF THE AIM2 INFLAMMASOME

There are two pathways associated with the detection of cytoplasmic DNA: the interferon-stimulatory DNA (ISD) pathway consisting of cGAS and STING and the caspase-1 inflammasome pathway (Gray et al., 2016). Here, the inflammasome pathway will be discussed in more detail. The AIM2 inflammasome, short for "absent in melanoma 2," is essential in the recognition and defense against self and pathogen associated double-stranded DNA (dsDNA) (Burckstummer et al., 2009; Fernandes-Alnemri et al., 2009). Therefore, this defense mechanism can be simply referred to as a cytosolic dsDNA sensor. DNA sensing by AIM2 is initiated through the recognition of DNA and results in the secretion of mature interleukin (IL)-1β and IL-18, followed by pyroptosis (Gray et al., 2016; Strittmatter et al., 2016). Current research suggests AIM2 participates in the inflammasome by direct binding of pathogen-derived DNA (Strittmatter et al., 2016). After DNA binding, AIM2 forms multimers, allowing it to recruit the adaptor protein ASC and subsequently ASC recruits caspase-1 into the inflammasome complex (Fig. 1). AIM2 appears to lack sequence specificity due to the fact that it interacts with the sugar-phosphate backbone and does not interact with bases in the major or minor groves (Jin et al., 2012). Thus, AIM2 has the ability to detect any and all DNA in the cytoplasm that is at least 80 bp long and is "naked" or not masked by the pathogen in some way

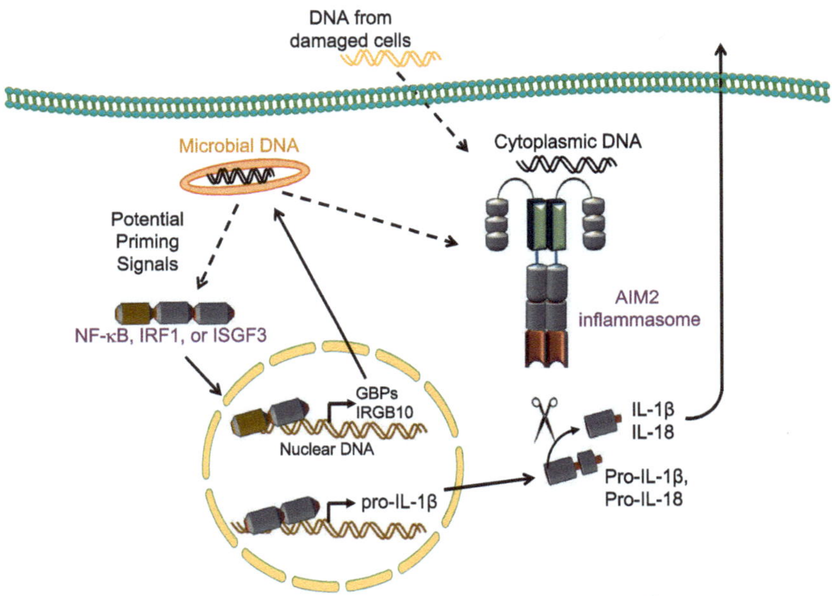

Fig. 1 AIM2 inflammasome activation mechanisms. Absent in melanoma 2 (AIM2) is present in an inactive conformation in the cytoplasm of cells. Upon binding to cytoplasmic DNA in a sequence independent manner, AIM2 adopts an open, active conformation that allows for binding to the adaptor protein ASC, which binds to caspase-1. This complete inflammasome allows for activation of caspase-1 and proteolytic processing and activation of the cytokines IL-1β and IL-18. The DNA from some pathogens must be released from inside the pathogen into the cytoplasm. Guanylate-binding proteins (GBPs) and IRGB10 are induced by the type-I interferon response. GBPs and IRGB10 then bind to pathogens and disrupt the pathogen membranes to facilitate release of pathogen DNA into the cytoplasm where it can activate AIM2.

(i.e., DNA binding proteins or capsids/membranes) (Jin et al., 2012). As a result, AIM2 is able to respond to pathogen DNA from viruses, bacteria, fungi, and parasites if such DNA is naked in the cytoplasm. AIM2 is also capable of responding to transfected human DNA and may play a role in some forms of autoimmunity such as systemic lupus erythematosus (SLE) (Zhang et al., 2013). However, we will only discuss the role for AIM2 in pathogen detection in this chapter. In addition, we will discuss the host and pathogen encoded proteins involved in activating or suppressing AIM2 inflammasome activation.

2.1 AIM2 During Viral Infection

To combat viral pathogens, the innate immune system must communicate the sensing of nucleic acids within the cytoplasm to effectively induce antiviral immune responses. The sensing of nucleic acids during viral infection is particularly important as viruses are diverse in capsid and other protein structures such that they are poor PAMPs. Instead, viruses have conserved nucleic acid structures that serve as PAMPs, either due to unique features of their nucleic acids or due to the unique location of these nucleic acids. For example, healthy cells should not have DNA in the cytoplasm. Thus, any pathogen DNA in the cytoplasm indicates a disease state. AIM2 was initially reported to activate the inflammasome in response to cytoplasmic DNA from vaccinia virus (VV) or murine cytomegalovirus (MCMV) (Fernandes-Alnemri et al., 2010; Hornung et al., 2009; Rathinam et al., 2010). More recently, the secondary messenger, 2′3′-cGAMP, which is typically associated with type-I interferon induction, was also showed to play a role in the activation of the AIM2 and NLRP3 inflammasomes during MCMV infection (Swanson et al., 2017). Alternatively, pox viruses possess pyrin only proteins with the ability to inhibit inflammasome activation (Dorfleutner et al., 2007). Recently, human encoded pyrin only protein 3 (POP3) was reported to inhibit the AIM2 inflammasome during VV or MCMV infection and expression of human POP3 in mice, which naturally lack POP3, impaired the immune response to MCMV (Khare et al., 2014). Thus, human cells can regulate the activation of the AIM2 inflammasome via induced expression of POP3, but the importance of this in humans remains to be seen.

Human cytomegalovirus (HCMV) is an enveloped linear double-stranded DNA virus, and member of the *Herpesviridae* family and *Betaherpesvirinae* subfamily. The AIM2 inflammasome is important for control of virus replication in human THP1 monocytes (Huang et al., 2017a). Furthermore, HCMV encoded pUL83 protein inhibits the AIM2 inflammasome, presumably to facilitate virus infection (Huang et al., 2017b). This appeared to be the result of pUL83 interaction with AIM2 that resulted in decreased protein levels of AIM2, pro-caspase-1, and pro-IL1β (Huang et al., 2017b).

Herpes simplex virus 1 (HSV-1) is an enveloped linear double-stranded DNA virus and member of the *Alphaherpesvirinae* subfamily. Mucocutaneous portions of the skin, mainly keratinocytes, are the target of HSV-1 infections, whereas HSV1 becomes latent in nerves of the trigeminal ganglion (Theil et al., 2003). Strittmatter et al. investigated the role of (HSV-1) within keratinocytes (Strittmatter et al., 2016). Here, they found that AIM2 played the key role in the assembly of the inflammasome response (Strittmatter et al., 2016). Their results indicate that IL-1β and IL-18 originating from keratinocytes form a feedback loop that is important for the control of HSV-1 skin infections. However, the epithelial cells must be primed with interferon-γ in order to secrete active forms of pro-inflammatory cytokines during HSV-1 infections (Strittmatter et al., 2016). Their research also suggests that keratinocytes use AIM2 as their sensor, but other cell types, such as primary human monocytes (PHM) or THP1 cells, use NLRP3 as the key sensor and that IFI16 is the key sensor in fibroblasts. Therefore, the PRR responsible for inflammasome activation depends on which cellular model is used in HSV-1 infection (Strittmatter et al., 2016). There are reports of keratinocytes lacking the ability to secrete IL-1β upon HSV-1 infection or that the levels are too low to detect (Milora et al., 2014). This could indicate that HSV-1 is a weak inflammasome activator or suppresses IL-1β; as *Herpesviridae* family members possess many inhibitors of the immune response, suppression of IL-1β production seems likely (Paludan et al., 2011). Strittmatter et al. confirmed that HSV-1 suppresses IL-1β production specifically at the posttranscriptional level. Furthermore, HSV-1 inhibits inflammasome activation. However, priming keratinocytes with IFN-γ stimulated caspase-1 activation and secretion of IL-1β and IL-18 (Strittmatter et al., 2016). Finally, HSV-1 VP22 protein was recently reported to inhibit AIM2 inflammasome assembly (Maruzuru et al., 2018). In addition to HSV-1 inhibition of the AIM2 inflammasome, the cellular protein TRIM11 also reduces AIM2 inflammasome activation via autophagic degradation of AIM2 during HSV-1 infection (Liu et al., 2016a). Therefore, host and virus mediated inhibition of AIM2 are involved in regulating inflammasome activation and treatment with IFN-γ to enhance priming of the AIM2 inflammasome may have therapeutic benefit.

Epstein-Barr virus (EBV), also known as human herpesvirus 4 (HHV-4) or mononucleosis, is also an enveloped double-stranded DNA virus of the *Gammaherpesvirinae* subfamily. Little is known about the activation of the inflammasome in response to primary EBV infections. Epithelial and B cells are the two common cell types infected by EBV. Recently, Torii et al. discovered that EBV infects human monocytes, specifically THP-1 cells and PHM and an inflammasome response develops. Specifically, caspase-dependent IL-1β production occurs within these cells but not within B- or T cell lines (Torii et al., 2017). However, Ansari et al. provided evidence that inflammasome activation does occur in B lymphocytes and epithelial cells, but only during latency (Ansari et al., 2013). To identify the sensor molecule responsible for EBV mediated inflammasome activation in monocytes, mRNA and protein levels of AIM2 and NLRP3 were examined. Increased levels of AIM2 were observed in EBV-infected THP-1 cells and PHM, whereas levels of NLRP3 did not show remarkable change. Knockdown of AIM2 through small interfering RNA (siRNA) also attenuated caspase-1 activation (Torii et al., 2017). Therefore, the AIM2 inflammasome was involved in the inflammatory response of monocytes to EBV. Further data by Torii et al. showed that serum IL-18 levels were high in patients with acute EBV, suggesting that monocytes and macrophages may play a role in the primary infection or reactivation of EBV, or that some other cell type responds to EBV in vivo (Torii et al., 2017). One question that remains is how EBV infects monocytes. Monocytes lack CD21, the primary receptor for EBV, but infection could occur through CD35 and HLA-DR in CD21-deficent cells (Ogembo et al., 2013). Interestingly, it appears that infection of monocytes with EBV is not productive, as EBV BZLF1 gene expression was elevated in EBV-infected THP-1 cells, but EBV DNA replication was not observed (Torii et al., 2017). One possible explanation is that AIM2 inflammasome activation in monocytes results in pyroptosis from successful caspase-1 activation (Torii et al., 2017).

Hepatitis B virus (HBV) is a member of the *Hepadnavirdae* family and is another enveloped, double-stranded DNA virus. Worldwide, HBV poses an imperative health risk. An associated issue to contracting the disease is kidney inflammation, namely glomerulonephritis. This specific condition, known as HBV-GN, is associated with increased expression of AIM2, caspase-1, and IL-1β (Zhen et al., 2014). Through an in vitro model with primary human glomerular mesangial (HGM) cells transfected with HBV DNA, enhanced levels of caspase-1, IL-1β, and IL-18 were observed. Utilization of siRNA knockdown of AIM2 decreased inflammasome activation

(Zhen et al., 2014). However, it remains to be seen if natural infection also results in AIM2 inflammasome activation, instead of transfected viral DNA. Immunohistochemistry analysis of human clinical samples provides further evidence to the positive correlation of AIM2, caspase-1, IL-1β, and IL-18, as these were all present in tissue affected by HBV-GN (Zhen et al., 2014). Another study showed through immunohistochemistry of liver biopsies that caspase-1, IL-1β, and IL-18 were all elevated in HBV-infected patients (Han et al., 2015). Thus, a clear association between the AIM2 inflammasome and HBV infection exists, but further research is needed to decipher the details.

Finally, adenoviruses (AdVs) are non-enveloped linear double-stranded DNA viruses often used for viral transduction or for pseudotyping as vaccines. Interestingly, AdVs do not activate the AIM2 inflammasome unless the cells are first primed through the type-I IFN pathway (Liu et al., 2017). Instead, AdVs activate the NLRP3 inflammasome (Barlan et al., 2011a, b; Muruve et al., 2008). Alternatively, the adenovirus-based canarypox virus vector was able to activate the AIM2 inflammasome due to simultaneous type-I IFN pathway activation (Liu et al., 2017). Additionally, antibody binding to AdVs following repeated AdV exposure facilitates virus uptake by dendritic cells and subsequent AIM2 inflammasome activation (Eichholz et al., 2016). Thus, AIM2 inflammasome activation to some viruses requires further priming or cytosolic delivery mechanisms.

RNA-based viruses, such as influenza A virus (IAV), are major public health risks around the world. IAV is a member of the *Orthomyxoviridae* family and has an eight RNA segmented genome. Although an RNA virus, two groups have reported a role for the AIM2 inflammasome during IAV infection in mice. As IAV has a negative sense single-stranded RNA genome, the only plausible source of DNA to activate AIM2 is host derived. Specifically, host DNA released into the lung is recognized by alveolar macrophages, resulting in AIM2 inflammasome activation (Schattgen et al., 2016; Zhang et al., 2017). These two reports have aided our understanding of inflammation by demonstrating a cross-talk between PAMPs and DAMPs through analysis of the release of host DNA into the lung microenvironments from dead cells in response to IAV infection. The endogenous DNA is likely derived from neutrophil extracellular traps (NETs) or necrotic bronchiolar epithelial cells. Levels of dsDNA in mice infected with IAV were increased by bronchiolar lavage (BAL) analysis (Schattgen et al., 2016). Treatment of IAV-infected mice with recombinant adeno-associated virus (AAV) that expressed DNase-I reduced the amount of DNA in the BAL (Schattgen et al., 2016). FACS analysis of leukocyte populations within the lungs of AAV-DNase-I treated mice, or *Aim2*$^{-/-}$ mice revealed a significant increase in the number of $CD4^+$ and $CD8^+$ T cells compared to control mice at 5-days postinfection. In these experiments, host DNA present in the lung microenvironment due to IAV infection activated AIM2, which protected mice from overt inflammation (Schattgen et al., 2016). However, Zhang et al. also reported that AIM2 can respond to host derived DNA in the lungs of IAV-infected mice, but reported that *Aim2*$^{-/-}$ mice survive IAV infection better than WT mice (Zhang et al., 2017). Improved survival in *Aim2*$^{-/-}$ mice was associated with increased monocytes and decreased neutrophils in the lungs (Zhang et al., 2017). In both reports, viral load within the lungs of WT and *Aim2*$^{-/-}$ mice was not different (Schattgen et al., 2016; Zhang et al., 2017). This suggests that AIM2 does not directly impact IAV replication. However, as each group reported different findings with regard to survival, more research is needed to understand the role for AIM2 during IAV and other RNA virus infections. As DNA in this setting is host derived, it would be a DAMP instead of a PAMPs, and future research into host DAMPs and their role during infection is needed.

Another RNA virus, enterovirus A71 (EV-A71), which is a positive sense single-stranded RNA virus, may also require AIM2 for induction of neuronal pathology. Yogarajah et al. observed that AIM2, caspase-1, and IL-1β expression were all elevated in EV-A71-infected neurons and caspase-1 activation occurred (Yogarajah et al., 2017, 2018). Furthermore, siRNA mediated knockdown of AIM2 impaired cell death and resulted in increased virus replication (Yogarajah et al., 2017). However, it remains to be seen how AIM2 is activated during EV-A71 infection, as this is an RNA virus.

2.2 AIM2 and Bacteria

Shifting to bacterial infections, the AIM2 inflammasome mediates recognition of many cytosolic bacteria. AIM2 activation by *Francisella tularensis* subspecies *novicida* (*F. novicida*) and *Listeria monocytogenes* (*L. monocytogenes*) was reported nearly a decade ago (Fernandes-Alnemri et al., 2010; Jones et al., 2010; Rathinam et al., 2010). Importantly, *F. novicida* is only sensed by the AIM2 inflammasome, whereas *L. monocytogenes* is also sensed by the NLRP3 and NLRC4 inflammasomes (Kim et al., 2010; Rathinam et al., 2010; Tsuchiya et al., 2010; Warren et al., 2010; Wu et al., 2010). The obvious connection is that both bacteria enter the cytoplasm and DNA is sensed by AIM2 to induce inflammasome activation, and that *F. novicida* lacks a flagellum to activate NLRC4. However, it is not clear how *F. novicida* avoids NLRP3 activation. One possible explanation comes from another bacterium which also activates AIM2 and NLRP3, namely *Legionella pneumophila*. During *Legionella pneumophila* infection, inflammasome activation is sequential, where AIM2 initially forms inflammasomes in response to bacterial DNA in the cytoplasm and activates caspase-11 resulting in cell membrane damage. This then allows for NLRP3 inflammasome activation (Cunha et al., 2017). Finally, in the absence of caspase-1, AIM2 and ASC activate caspase-8 mediated apoptosis to inhibit bacterial replication (Pierini et al., 2012). These findings demonstrate that differential responses of the immune system to specific pathogens can be mediated by inflammasome activation of different caspases and further research in this area is needed to better understand these mechanisms.

In order for AIM2 to detect *F. novicida*, *L. monocytogenes*, and presumably other bacteria in the cytoplasm, bacterial lysis is required. Some genetically modified strains of *F. novicida* and *L. monocytogenes* can lyse more easily, resulting in improved AIM2 activation (Peng et al., 2011; Sauer et al., 2010). Even WT bacteria lyse periodically, resulting in low levels of AIM2 inflammasome activation (Sauer et al., 2010). This begs the question of whether the AIM2 response is a passive or active, host-driven, immune mechanism. Numerous reports indicate that type-I interferon signaling is required for robust AIM2 inflammasome activation during *F. novicida* (Belhocine and Monack, 2012; Fernandes-Alnemri et al., 2010; Jones et al., 2010) and even *Streptococcus pneumoniae* infection (Fang et al., 2014). This would suggest some priming or cell activation event is required for complete AIM2 inflammasome activation. Several studies conclusively demonstrate that type-I interferon signaling upregulates guanylate-binding proteins (GBPs). Specifically, GBP2 and GBP5 are key AIM2 activators during *F. novicida* infection (Man et al., 2015; Meunier et al., 2015; Wallet et al., 2017). When *F. novicida* becomes cytosolic, promotion of bacteriolysis occurs. This is achieved through GBP mediated localization of another protein, IRGB10, to the bacterial membrane (Man et al., 2015; Meunier et al., 2015). There, IRGB10 inserts into bacterial membranes and destabilizes them, releasing bacterial DNA and other ligands into the cytoplasm (Man et al., 2016). A similar requirement for GBPs for AIM2 inflammasome activation during infection with *Brucella abortus* and *Legionella pneumophila* was also reported (Costa Franco et al., 2018; Liu et al., 2018). However, endogenous levels of GBPs were sufficient for bacterial ligand release in *Legionella pneumophila* infected cells (Liu et al., 2018). GBPs are also important for early AIM2 activation during infection with *Chlamydia muridarum* and *Chlamydia trachomatis*, but decrease in importance over time, suggesting other mechanisms also play a role in AIM2 activation with *Chlamydia* (Finethy et al., 2015). Finally, it must be noted that AIM2 activation via transfection of synthetic DNA or during DNA viral infections is independent of this GBP pathway (Man et al., 2015) and that GBPs possess other, inflammasome independent, mechanisms for inhibiting bacterial growth (Wallet et al., 2017).

Several *Mycobacterium* species also activate the AIM2 inflammasome, including *Mycobacterium tuberculosis*, *M. smegmatis* and *M. bovis* (Saiga et al., 2012; Shah et al., 2013; Yang et al., 2013b). Interestingly, virulent strains of *M. tuberculosis* failed to activate the AIM2 inflammasome despite the presence of bacterial DNA in the cytoplasm (Shah et al., 2013). Accordingly, the ESX-1 protein of virulent *M. tuberculosis* inhibited type-I interferon mediated induction of AIM2 and potentially directly inhibits AIM2 inflammasome activation as well (Shah et al., 2013). On the other hand, AIM2 may competitively inhibit type-I interferon and autophagy responses during infection with *M. bovis* by binding to bacterial DNA and preventing detection of bacterial DNA by other cytoplasmic DNA sensors like STING (Liu et al., 2016b).

In conclusion, much research has been done to understand the role for the AIM2 inflammasome during bacterial infection and a myriad of other bacterial species are able to activate the AIM2 inflammasome. These include *Porphyromonas gingivalis* (Park et al., 2014), *Staphylococcus aureus* infection of the central nervous system (Hanamsagar et al., 2014), and *Brucella abortus* (Gomes et al., 2013; Miraglia et al., 2016). However, as the recent discovery of GBPs indicates, there is still much we do not understand about this inflammasome.

2.3 AIM2 and Fungi

Fungal pathogens are extremely interesting in their virulence patterns. Most fungal pathogens do not release their genetic material within the cytosol during infection (Niemiec et al., 2017). Therefore, the number of fungal pathogens activating the AIM2 inflammasome is scarce to our knowledge. However, *Aspergillus fumigatus* is gaining attention in recent years due to its involvement in infecting as an opportunistic pathogen. This fungus is saprophytic and ubiquitous, making contact with immunocompromised individuals inevitable. There is limited research pertaining to the involvement of

A. fumigatus and its affects with the cytosolic dsDNA sensor, AIM2. However, AIM2 and NLRP3 activation form a dual cytoplasmic surveillance system that orchestrates responses against *A. fumigatus* infection (Karki et al., 2015). Both AIM2 and NLRP3 engage the inflammasome to trigger innate immune responses against *A. fumigatus* infections. Mice lacking AIM2 and NLRP3 are observed to be hyper susceptible to invasive aspergillosis (Karki et al., 2015). This is due to a failure to confine *Aspergillus* hyphae in the lung and leads to widespread hyphal dissemination to lung blood vessels. Both of these inflammasome receptors activate caspase-1 and induce processing of IL-1β and IL-18 by recruiting caspase-8 to the same inflammasome complex. From there, IL-1β, IL-18, caspase-1, caspase-8, and FADD are crucial for preventing *Aspergillus*-induced mortality (Karki et al., 2015). Together, these results unveiled a dual cytoplasmic sensing mechanism governed by AIM2 and NLRP3 to orchestrate robust host responses to *A. fumigatus*. AIM2 and NLRP3 play redundant roles in driving robust inflammasome activation to mediate the secretion of IL-1β and IL-18 in response to *Aspergillus* infection (Karki et al., 2015). Interestingly, *Candida albicans* activates NLRP3 exclusively (Wellington et al., 2014). This is possibly due to the failure of this microorganism to release its DNA into the host cytosol to activate AIM2, or this specific fungus contains a virulence factor that mediates its evasion from this surveillance mechanism. Overall, the extracellular release of *A. fumigatus* DNA due to germination of this fungus may activate the AIM2 inflammasome (Karki et al., 2015).

2.4 AIM2 and Parasites

Similar to AIM2 and fungi, there exists limited data and knowledge about the correspondence of parasites and the AIM2 inflammasome. *Plasmodium falciparum* is a causative agent of malaria. As malaria continues to be one of the most devastating global health concerns in human history, understanding the immune response to this infection is critical for improving treatment options. Kalantari et al. described the possible role for AIM2 during *Plasmodium falciparum* infection (Kalantari et al., 2014). They discovered that the hemoglobin by-product hemozoin can carry parasite DNA into the cytoplasm and induce both NLRP3 and AIM2 inflammasome activation (Kalantari et al., 2014). In general, a pro-inflammatory cytokine storm results when parasitized red blood cells rupture, and IL-1β has been linked to this condition. Because of its highly synergistic nature, IL-1β can modify the expression and actions of other modulators of the immune system (Kalantari et al., 2014).

Overall, AIM2 inflammasome activation is essential for host defense and clearance of intracellular associated pathogens including bacteria, viruses, fungi, and parasites. Furthermore, a variety of pathogen and host encoded modulators or AIM2 activation can affect pathogen growth and/or inflammation and tissue damage. Therefore, the importance of AIM2 inflammasome activation deserves to be investigated further.

3. NLRP3: THE OMNISCIENT INFLAMMASOME

To say that NLRP3 responds to "numerous" stimuli would be understatement. The NLRP3 inflammasome reacts to "stuff," varying from reactive oxygen species, potassium efflux, bacterial pore-forming toxins, and non-infectious crystals (Place and Kanneganti, 2018). These stimuli are the usual basis for the activation of the NLRP3 inflammasome pathway. However, as NLRP3 has a reported role in virtually every infectious or non-infectious disease, we must limit our discussion here to a selected number of publications that indicate NLRP3 may possess the ability to directly, or through adaptor proteins, indirectly detect microbial or self-nucleic acids. We will not discuss the myriad of papers involving nucleic acid mediated priming of the NLRP3 inflammasome.

3.1 NLRP3 Detection of Pathogen Nucleic Acids

The idea that the NLRP3 inflammasome could respond to pathogen nucleic acids was first reported by Kanneganti et al. (2006a, b) (Fig. 2). Infection of macrophages with enterohemorrhagic *Escherichia coli* by another group also activated the NLRP3 inflammasome through the presence of RNA:DNA hybrids in the cytoplasm of infected cells that localized to NLRP3 inflammasome (Kailasan Vanaja et al., 2014). Furthermore, various RNA species (mRNA, tRNA, rRNA) were found to activate the NLRP3 inflammasome (Sha et al., 2014). However, no direct mechanism for NLRP3 mediated nucleic acid sensing was evident in any of these reports. Another report by Gupta et al. showed that RNA from group B Streptococcus (GBS) could activate NLRP3, and bacterial RNA could be immunoprecipitated with NLRP3 (Gupta et al., 2014). However, this does not rule out the potential for an intermediate RNA-binding protein associated with NLRP3. The missing pieces to this puzzle may have been discovered with DHX33 and DDX19A, both members of the DExD/H-box helicase family. Both proteins can form a complex with NLRP3 in human or porcine monocyte or macrophages, respectively, in response to cytoplasmic viral and bacterial RNA (Fig. 2). These interactions result

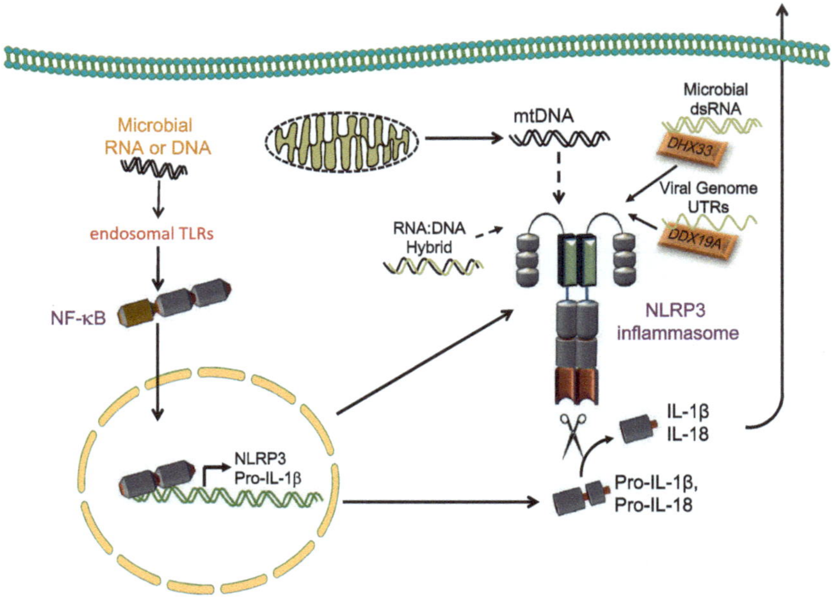

Fig. 2 NLRP3 response to cytoplasmic nucleic acids. NLRP3 is a sensor of host cell damage via DAMPs including reactive oxygen species, potassium ion efflux, and more. However, several papers demonstrate that NLRP3 responds to certain nucleic acids in the cytoplasm of cells too. In general, this appears to be mediated by specific adaptor proteins like DHX33 and DDX19A, which contain helicase domains capable of binding to nucleic acids. These adaptors then activate NLRP3. There are some species of nucleic acids such as RNA:DNA hybrids and mitochondrial DNA (mtDNA) that also activate NLRP3, but the adaptor or mechanisms involved remain to be determined.

in inflammasome activation (Li et al., 2015; Mitoma et al., 2013). Specificity of DHX33 for pathogen RNA was based on the presence of double-stranded RNA in the cytoplasm (Mitoma et al., 2013). Alternatively DDX19A recognized the 5′ and 3′ untranslated regions (UTRs) of porcine reproductive and respiration syndrome virus (PRRSV) (Li et al., 2015). Furthermore, DHX33 mediated activation of NLRP3 is regulated by TRIM33 through ubiquitination of DHX33 (Weng et al., 2014), thus allowing control of this inflammatory pathway. Lastly, the 2′,5′-oligoadenylate (2–5A) synthetase (OAS)/RNase L pathway can process viral RNA and generate ligands for DHX33 to facilitate NLRP3 inflammasome activation (Chakrabarti et al., 2015).

3.2 NLRP3 Detection of Self-Nucleic Acids During Infection

Detection of self-DNA by AIM2 during pathogenic infection was previously discussed for infection with IAV above. However, mitochondrial

DNA (mtDNA) is another potential source of host derived nucleic acid with the ability to activate the inflammasome (Fig. 2). Intriguingly, mtDNA does not appear to activate the AIM2 inflammasome, unless it is transfected into cells (Shimada et al., 2012). Instead, cellular damage during infection with *Chlamydia pneumoniae* or treatment with LPS + ATP resulted in the release of mtDNA into the cytoplasm where it could be immunoprecipitated with NLRP3 and lead to inflammasome activation (Shimada et al., 2012). Accumulation of damaged mitochondria due to impaired autophagy further allows for mtDNA mediated NLRP3 inflammasome activation (Nakahira et al., 2011). More recently, new mtDNA synthesis mediated by CMPK2 was shown to be required for NLRP3 activation (Zhong et al., 2018). However, it remains to be determined if this is a direct NLRP3 interaction, or if some intermediate protein facilitates this interaction. Overall, the studies discussed here suggest that self-mtDNA deserves further examination as a DAMP induced by pathogen invasion with the potential for NLRP3 inflammasome activation.

4. THE IFI16 INFLAMMASOME

Inflammasomes are mainly assembled by members of the NLR family, which upon activation by apt stimuli result in the cytoplasmic organization of the supramolecular complex with critical roles in protection against cytosolic pathogens. Whether inflammasomes can assemble in the nucleus and impart protection against nuclear-replicating pathogens was not known until the role of the DNA sensor interferon gamma-inducible protein 16 (IFI16) was elucidated (Kerur et al., 2011).

IFI16 is predominantly a nuclear protein and belongs to the interferon (IFN)-inducible p200 (also known as PYHIN) family. In mice, the p200 gene family includes p202a, p202b, p203, p204, MNDAL, and AIM2, while in humans, the gene family is represented by IFI16, MNDA, AIM2, and IFIX (Zhao et al., 2015). The p200 family members consist of at least one partially conserved repeat of 200 amino acid residues (the HIN-200 domain) toward their C-terminus and all of them (except p202a and p202b in mice) have a PYRIN domain (PYD) at their N-terminus allowing homotypic interactions with other PYD-containing proteins (Ludlow et al., 2005). On the other hand, the HIN-200 domain binds DNA and the IFI16 protein, in particular, contains two repeats of this domain separated by a spacer rich in serine-threonine-proline. Human IFI16 is regarded as a murine p204 homolog based on similar functions and closely related domain structure, however, further functional evidence supporting this is required

(Cridland et al., 2012). The role and mechanistic details of IFI16 in type-I interferon production have been established recently. In response to viral genomes or transfected DNA, IFI16 directly associates with DNA motifs through the HIN-200 domains leading to the oligomerization of IFI16 and activation of downstream STING, causing a conformational change in the STING dimer. STING subsequently associates with TANK binding kinase 1 (TBK1) leading to phosphorylation of STING and recruitment and phosphorylation of interferon regulatory factor 3 (IRF3) (Dobbs et al., 2015; Liu et al., 2015). Phosphorylated IRF3 then enters the nucleus triggering the type-I interferon gene program (Fig. 3). Previous studies have also

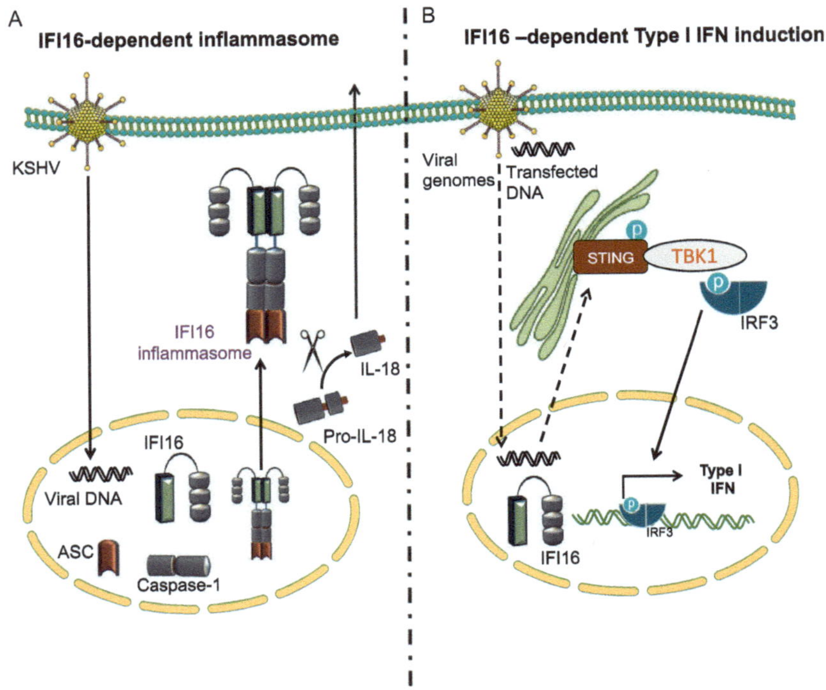

Fig. 3 IFI16 assembles inflammasome in the nucleus and regulates type-I interferon signaling. (A) Recognition of KSHV genomes by IFI16 leads to nuclear IFI16 inflammasome assembly. Once the inflammasome is formed, it translocates to the nucleus where it cleaves pro-IL-18 to its' biologically active form. (B) In response to viral genomes or transfected DNA, IFI16 directly associates with DNA motifs through the HIN-200 domains leading to the activation of ER-resident STING. STING subsequently associates with TBK1 leading to the phosphorylation of STING and recruitment and phosphorylation of IRF3. Phosphorylated IRF3 enters the nucleus to trigger the type-I interferon gene expression.

implicated IFI16 in cell cycle regulation, differentiation, and apoptosis, and these are discussed in detail elsewhere (Aglipay et al., 2003; Brunette et al., 2012; Luan et al., 2008; Xin et al., 2003; Zhang et al., 2008; Zhao et al., 2015).

4.1 IFI16 Assembles an Inflammasome in the Nucleus

The potential clue to investigate the role of IFI16 in inflammasome formation came from the presence of the PYD and HIN-200 domains in IFI16 (Xiao, 2015). Similar to IFI16, the DNA sensor AIM2 also contains both PYD and HIN-200 domains and assembles an inflammasome in response to dsDNA. The structural and functional similarities between AIM2, IFI16, and other DNA receptors prompted conception of a new family of innate DNA sensors called AIM2-like receptors (ALRs). Consistent with its localization, IFI16 was observed to trigger inflammasome assembly in response to nuclear replicating Kaposi sarcoma-associated herpesvirus (KSHV) infection of primary HMVEC-d endothelial cells (Kerur et al., 2011) (Fig. 3). The involvement of other DNA receptors was ruled out, either because they are not expressed in HMVEC-d cells (e.g., MNDA) or because they are predominantly localized in the cytosol (e.g., AIM2) and are thus unable to respond to KSHV infection. Contrary to monocytes and macrophages where caspase-1 is predominantly cytosolic, HMVEC-d cells expressed ASC and pro-caspase-1 predominantly in the nuclei of uninfected cells, which upon KSHV infection assembled the nuclear IFI16 inflammasome (Kerur et al., 2011). Of note, the complex is also observed in the cytosol. However, there is evidence that the IFI16 inflammasome first assembles in the nucleus and subsequently redistributes to the cytosol. In agreement, the active p20 band of caspase-1 was observed in nuclear fractions of infected cells as early as 2 h p.i., although it was also detectable in the cytoplasm at the earliest time examined (Kerur et al., 2011). Moreover, IFI16 directly colocalized with KSHV genomes in the nuclei of infected endothelial cells. Similarly, immunofluorescent studies demonstrated ASC and caspase-1 aggregates in the perinuclear area at later times p.i. revealing the dynamic nature of the IFI16 assembled inflammasome between subcellular compartments (Kerur et al., 2011). As IFI16 is also observed in the cytoplasm (Unterholzner et al., 2010), it is unclear what precludes inflammasome formation in the cytoplasm, but this may be dependent on the spatio-temporal expression and localization of IFI16 in the cell type involved, proximity to viral DNA, and/or any necessary IFI16 posttranslational modifications. Nonetheless, the relocation of IFI16 to the cytoplasm provides a fail-safe mechanism to avert overt caspase activity in the nucleus. Moreover, these findings suggest that IFI16 complements other plasma membrane and cytosolic pathogen sensors in the nucleus signifying how the host cells have reinforced each cellular compartment to avoid pathogen invasion.

The physiological significance of the IFI16 inflammasome in humans demands further investigation. In a clinical study conducted with mucosal biopsies obtained from two different cohorts of patients with inflammatory bowel disease (IBD), the mRNA expression of both AIM2 and IFI16 was found significantly upregulated in the inflamed colon of active patients as compared to noninflamed colon of controls and inactive patients with IBD (Vanhove et al., 2015). The transcriptional data correlated well with immunoblotting studies exhibiting increased protein expression of AIM2 and IFI16 in the colonic mucosa of active patients (Vanhove et al., 2015). These findings were independently verified by a study which confirmed elevated expression by immunohistochemistry of IFI16 in intestinal samples from both ulcerative colitis and Crohn's disease patients (Caneparo et al., 2016). Moreover, the latter study revealed elevated IFI16 expression limited to the intestinal epithelial cells of IBD patients. Although data were not included, IFI16 staining was conspicuously absent in intestinal epithelial cells from other non-IBD inflammatory conditions (Caneparo et al., 2016). Increased caspase-1 and the inflammasome downstream effector HMGB1 have also been observed in the epithelial compartment of the colonic mucosa from patients with active disease (Vanhove et al., 2015). Additional studies in genetic models of IFI16 are required to directly implicate the role of IFI16 in disease pathogenesis. As inflammasomes are critically involved in the pathogenesis of IBD, improved understanding in this area could have important implications in developing more personalized therapies for IBD patients.

4.2 IFI16 Mediated Pyroptosis Depletes $CD4^+$ T Cells During HIV Infection

Activation of the inflammasome is almost always followed by an inflammatory form of cell death, pyroptosis. Pyroptosis of the infected cells results in the destruction of the pathogen niche; however, this is also accompanied by tissue damage because of the nature of this form of cell death (Kayagaki et al., 2015; Lupfer and Anand, 2016). Depletion of quiescent "bystander" $CD4^+$ T cells is the principal driver of AIDS during HIV infection. The bystander

$CD4^+$ T cells are non-permissive to HIV infection and undergo abortive infection characterized by the cytosolic accumulation of incomplete HIV reverse transcripts (Doitsh et al., 2010). IFI16 recognizes these reverse transcripts leading to caspase-1 activation and pyroptosis followed by LDH secretion (Monroe et al., 2014). In agreement, shRNA-based knockdown of IFI16 rescued $CD4^+$ T cells cocultured with T cells infected with a HIV-GFP reporter virus from cell death. By contrast, knockdown of AIM2 and STING did not have any appreciable effect on LDH release and thus pyroptosis. Of note, cell death could be rescued in control cells upon addition of a non-nucleoside reverse transcriptase inhibitor, efavirenz, thereby supporting the notion that the $CD4^+$ T cell depletion results from abortive infection and not viral integration (Doitsh et al., 2010; Monroe et al., 2014). Moreover, IFI16 knockdown not only decreased active caspase-1 levels (determined by FLICA) but also IFN-β induction, suggesting an upstream role of IFI16 in triggering both these pathways in $CD4^+$ T cells. Altogether, these findings suggest that IFI16 induces caspase-1 mediated pyroptosis in abortively infected $CD4^+$ T cells thereby highlighting the detrimental consequences of inflammasome activity during AIDS progression. The precise localization of IFI16 and whether the assembled inflammasome redistributes between subcellular compartments during HIV infection is not known.

4.3 Acetylation of IFI16 Regulates Cytoplasmic and Nuclear Shuttling

Despite being predominantly nuclear at steady-state levels, IFI16 can unexpectedly sense pathogenic DNA in both the nucleus and the cytoplasm suggesting either that a fraction of IFI16 is always cytosolic or that IFI16 is distributed from the nucleus to the cytosol by a posttranslational modification. The latter scenario is supported by two independent studies proposing that the acetylation of IFI16 is critical for nuclear and cytoplasmic shuttling (Ansari et al., 2015; Li et al., 2012). However, precisely where IFI16 is acetylated is less clear. Li et al. revealed the presence of an evolutionarily conserved multipartite nuclear localization signal (NLS) in IFI16 and proposed that the acetylation of IFI16 NLS in macrophages restricts the protein to the cytoplasm by inhibiting nuclear import (Li et al., 2012). Thus, treatment with the broad-spectrum histone deacetylase (HDAC) inhibitor trichostatin A triggered the accumulation of cytoplasmic IFI16. Subsequent experiments suggested the involvement of acetyltransferase p300 in regulating IFI16 localization. In agreement, overexpressing p300 displayed a striking cytoplasmic accumulation of IFI16-GFP. A study by Ansari and colleagues confirmed the above findings but revealed that IFI16 modification by acetyltransferase p300 occurred in the nucleus immediately following the recognition of herpesvirus genomes (Ansari et al., 2015). Acetylation of IFI16 preceded IFI16-ASC interaction in the nucleus before the assembled IFI16 inflammasome translocated to the cytoplasm resulting in caspase-1 activation and IFN-β production. In agreement, IFI16 acetylation remained unaffected by either ASC or STING knockdown. Furthermore, IFI16 remained unmodified and contained within the nucleus upon cytoplasm restricted vaccinia virus infection, highlighting both that acetylation occurred in the nucleus and that the modification is necessary for IFI16 translocation to the cytoplasm. However, acetylation was not found to be required for DNA sensing and IFI16 physically associated with KSHV and HSV-1 genomes regardless of the presence of an acetylation inhibitor (Ansari et al., 2015). Further studies should address whether IFI16 acetylation is common to other model systems and to decipher the precise compartment where acetylation occurs. Regardless, these findings establish acetylation as a molecular toggle regulating IFI16 distribution.

4.4 Viral Evasion of the IFI16 Inflammasome

Activation of the innate immune responses including inflammasomes is essential to protect against pathogens. In turn, viruses have co-opted several mechanisms to establish a productive infection in their host cells. HSV-1 establishes life-long latency in infected cells. Similar to KSHV infection, HSV-1 infection of human skin fibroblasts activates the IFI16 inflammasome followed by relocalization of the receptor complex to the cytoplasm (Johnson et al., 2013). Additionally, in contrast to KSHV, HSV-1 infection also activates the NLRP3 inflammasome. However, IFI16 is distinctly targeted for degradation at later times postinfection in a manner dependent on ICP0, an immediate early protein of HSV-1. While the expression of NLRP3 and ASC remained unaltered, caspase-1 was also found "trapped" in actin clusters thereby blocking the overall activity of both the NLRP3 and IFI16 inflammasomes (Johnson et al., 2013). The above findings are independently confirmed by another study which demonstrated that expression of viral ICP0 protein in the nucleus causes nuclear relocalization and degradation of IFI16 thereby inhibiting the STING-IRF3 signaling pathway (Orzalli et al., 2012). However, the latter study did not observe any detectable change in IFI16 localization and the receptor remained nuclear restricted in control and HSV-1-infected human skin fibroblasts (Orzalli et al., 2012). These studies established that HSV-1 ICP0 protein targets IFI16 and caspase-1 to inhibit the inflammasome pathway. IFI16 also acts as a nuclear DNA sensor for human cytomegalovirus (HCMV) in human foreskin fibroblasts (HFFs) triggering the expression of antiviral cytokines via the STING-TBK1-IRF3 signaling pathway. However, the HCMV tegument protein pUL83 (also known as phosphoprotein pp65) interacts with the pyrin domain of IFI16 thereby blocking the oligomerization and induction of antiviral responses (Li et al., 2013). Accordingly, infection with a HCMV mutant strain unable to make pUL83-encoded pp65 enhanced IFN-β production as compared to HFFs infected with the wild-type strain (Biolatti et al., 2018). Of interest, in THP-1 derived macrophages, an interaction between recombinant pUL83 and AIM2 has been demonstrated. This has been suggested to reduce the activation of the AIM2 inflammasome upon poly(dA:dT) transfection in a recombinant inflammasome system assembled in HEK293T cells (Huang et al., 2017b). These latter findings need confirmation but would have wider implications if the tegument protein indeed affects inflammasome activation.

4.5 IFI16 Differentiates Self From Non-Self dsDNA

While the nuclear localization of DNA sensor IFI16 is advantageous in initiating rapid responses to nuclear pathogens, the receptor must also faithfully differentiate between "self" and "non-self" DNA. Crystallographic and binding studies on AIM2 and isolated HIN200 domains of IFI16 suggested that IFI16 assumes an autoinhibited conformation in which the DNA binding HIN200 domain is blocked by the PYD domain which is only displaced upon sensing foreign DNA (Jin et al., 2012). Subsequently, IFI16 one-dimensionally tracks long stretches of exposed foreign dsDNA to assemble into supramolecular signaling platforms (Stratmann et al., 2015). The initiation of IFI16 assembly requires at least 4 molecules, and the optimal oligomeric assembly is only reached in the presence of 10 IFI16 molecules with a dsDNA-binding footprint of ~15 bp for each IFI16 (Stratmann et al., 2015). The scanning of the exposed DNA is believed to increase the probability of encountering other IFI16 molecules required for IFI16 filament formation (Morrone et al., 2014). At the same time, however, the targeting of "self" DNA is averted because of the presence of host DNA into tightly packed nucleosomes which do not leave long DNA stretches unexposed, a prerequisite for IFI16 DNA tracking and ensuing filament formation (Morrone et al., 2014; Stratmann et al., 2015). Thus, IFI16 molecules cluster into distinct protein filaments only on exposed foreign DNA. Unexpectedly, the HIN200 domain of IFI16 binds DNA with weak affinity without facilitating filament formation. By contrast, the IFI16 PYD domain plays a positive role in DNA binding and drives the filament assembly (Morrone et al., 2014). Furthermore, the surface residues in the PYD domain that enables cooperative DNA binding are conserved suggesting a common strategy likely used by other ALRs (Morrone et al., 2014). These studies thus reveal how IFI16 differentiates friend from foe and highlight the property by which even a single molecule of non-self DNA can be rapidly engaged by IFI16.

5. NLRP9B: THE NEWEST NUCLEIC ACID SENSING INFLAMMASOME

As discussed above, NLRP3 can be activated in response to nucleic acids from pathogens through interaction with several helicase proteins. Similarly, NLRP9b is activated in response to short double-stranded RNA (dsRNA) fragments during intestinal infection with rotavirus, but dsRNA does not directly activate NLRP9b. Instead, dsRNA is detected by the helicase DHX9, which subsequently binds to and activates NLRP9b to activate the inflammasome (Zhu et al., 2017). Mice deficient in NLRP9b had higher viral loads, more diarrhea, and increased destruction of intestinal villi compared to WT controls (Zhu et al., 2017). Interestingly, pyroptosis mediated by NLRP9b inflammasome induced gasdermin D cleavage was responsible for control of rotavirus replication, whereas IL-18 produced by the inflammasome was dispensable (Zhu et al., 2017). Therefore, NLRP9b appears to be a true nucleic acid sensing inflammasome with the help of DHX9, but it remains to be seen how many other pathogens induce NLRP9b, or if it is specific only to rotavirus.

6. NUCLEIC ACID SENSING INFLAMMASOMES IN AUTOINFLAMMATION AND AUTOIMMUNITY

Our discussion thus far has focused on the role for the inflammasomes during infectious disease. However, the ability of some nucleic acid sensing inflammasomes to detect pathogen or mitochondrial DNA in an apparently sequence independent manner poses a risk for unintentionally detecting self-DNA. In the case of NLRP3, there are so many potential activators that determining the importance of mtDNA during in vivo disease models is experimentally difficult. Non-alcoholic fatty liver disease and non-alcoholic steatohepatitis are autoinflammatory conditions associated with obesity, glucose insensitivity, or other dysregulated metabolic conditions. Two reports now suggest that NLRP3 sensing of mtDNA is involved in these diseases and is associated with oxidized or increased presence of some fatty acids (Pan et al., 2018; Schuster et al., 2018). In a separate model, mitochondrial damage and release of mtDNA were associated with NLRP3 inflammasome activation during type-1 diabetes (Carlos et al., 2017). Thus, mtDNA mediated activation of NLRP3 can lead to autoinflammatory or autoimmune diseases.

The ability of AIM2 to distinguish friend from foe is based on the location of DNA. In general, host DNA is in the nucleus or packaged in apoptotic bodies and removed by phagocytosis. However, during acute pancreatitis, inflammation results from premature activation of digestive enzymes in the pancreas instead of the intestines. Essentially, the pancreas begins digesting itself resulting in release of nucleic acids (Shah et al., 2009). Deletion of *Aim2* was able to protect mice from tissue damage and inflammation during acute pancreatitis, suggesting that nucleic acids from damaged tissues are somehow internalized into neighboring cells, or immune cells entering the pancreas, resulting in inflammasome activation (Kang et al., 2016). In another model, $Aim2^{-/-}$ mice are protected from the damaging effects of systemic lupus erythematosus (SLE or lupus). During SLE, the immune system responds to nucleic acids or nucleosomes and produces autoantibodies to these entities. Interestingly, $Aim2^{-/-}$ mice do not develop the clinical signs of SLE, such as glomerulonephritis, but they still develop autoantibodies (Baum et al., 2015; Jakobs et al., 2015). This indicates that the presence of autoantibodies is not the only issue involved in SLE pathology, and that innate immune signaling and inflammation mediated by nucleic acid sensors such as AIM2 is essential for disease progression (Baum et al., 2015; Jakobs et al., 2015; Zhang et al., 2013). The question here is, how does self-DNA in SLE patients get into the cytoplasm and activate AIM2? One mouse model of SLE is a double knockout for $DNaseII^{-/-}$ and $Ifnar^{-/-}$ (Kawane et al., 2006). DNase II is present in phagosomes where it degrades the DNA from phagocytosed apoptotic bodies or extracellular DNA. In $DNaseII^{-/-}$ mice, DNA is not degraded and presumably has the chance to move from the phagosome into the cytoplasm (Monteith et al., 2016). Thus, the ability of AIM2 to distinguish between self and pathogen DNA relies on the ability of the host to keep self-DNA out of the cytoplasm. If DNase II cannot degrade DNA from extracellular sources, then it eventually enters the cytoplasm and activates AIM2. Similarly, the sequestration of mtDNA in the mitochondria limits NLRP3 activation, and only under chronic cellular or metabolic stress does mtDNA escape into the cytoplasm and activate NLRP3.

IFNs are widely recognized to trigger and accelerate the pathogenesis and perpetuation of autoimmune responses. Studies have demonstrated that IFN-inducing IFI16 levels are enhanced in the sera of patients affected by systemic autoimmune diseases including SLE, systemic sclerosis, rheumatoid arthritis, and Sjogren's syndrome (SS) (Alunno et al., 2015; Gugliesi et al., 2013). Particularly, IFI16 has been relatively well characterized in SS where its' levels are elevated and it serves as an autoantigen. Consequently, the levels of both IFI16 and anti-IFI16 antibodies are dominant in the sera of SS patients and are associated with severe outcomes (Alunno et al., 2015). How IFI16 becomes extracellular and adds to the initiation and progression of autoimmune diseases is not fully understood. It has been suggested that IFI16 mislocalization from its' predominantly nuclear site may be involved in initiating secretion to the extracellular milieu (Bawadekar et al., 2015). This may be achieved by tissue injury/apoptosis or inflammasome-induced pyroptosis followed by IFI16 release via exosomes. Increased expression and cytoplasmic mislocalization of IFI16 have been reported in acinar and duct epithelial cells of labial salivary glands of SS patients (Alunno et al., 2015; Antiochos et al., 2018). In these cells, IFI16 was observed to exist in filamentous structures in the cytoplasm (Antiochos et al., 2018). This filamentous nature of IFI16 required its' association with dsDNA in cultured epithelial cells. Though DNA nucleases could access IFI16-bound dsDNA and degrade it, this did not result in dissociation of IFI16 filaments suggesting the persistence of IFI16 filaments once formed (Antiochos et al., 2018). Moreover, human anti-IFI16-positive SS sera immunoprecipitated filamentous dsDNA-bound IFI16 approximately two- to threefold more strongly suggesting that autoantibodies preferred epitopes selectively exposed in filamentous IFI16 (Antiochos et al., 2018). These findings outline potential mechanisms and hint that dsDNA-induced oligomerization of IFI16 and subsequent release determines IFI16's status as an antigen in systemic autoimmune diseases.

7. CONCLUSIONS

The immune cells are armed with a limited set of PRRs to respond to diverse PAMPs and DAMPs. Inflammasome forming PPRs have gained tremendous attention in recent years because of the numerous infectious and sterile stimuli they can respond to thereby implicating them in varied diseases. Nucleic acid sensing PRRs including those that assemble inflammasomes are clearly important in maintaining immune homeostasis as they have the obligation to respond to pathogen DNA while differentiating it from hosts' own "self" DNA. Understanding how these PRRs recognize nucleic acids and maintain specificity is imperative to gain improved realization for translational purposes.

Mammalian cells display a variety of nucleic acid receptors and four of these (AIM2, NLRP3, IFI16, and NLRP9b) are known to establish inflammasomes. While the broad functions of these PRRs are generally known, we still have major gaps in our understanding of what is actually recognized, especially since few structural differences exist between pathogen and self-DNA. Clearly, this is not of relevance for NLRP3, as no nucleic acid stimuli appear to bind directly to this receptor. However, some progress has been made in deciphering the principles that AIM2 and IFI16 follow in nucleic acid sensing. Of note, both AIM2 and IFI16 lack sequence specificity. AIM2 recognizes any dsDNA of approximately 80 bp in length, so far as it is "naked" or not masked by the pathogen. In the same way, IFI16 needs long stretches of exposed DNA, which is not tightly packed as in host nucleosomes. While they faithfully respond to their cognate ligands most of the time, any aberrant activity of these PRRs is associated with autoimmune diseases including systemic lupus erythematosus and Sjögren syndrome.

In order to completely appreciate the biology of nucleic acid sensing inflammasome forming PRRs, there is a need to conduct cell type and tissue-specific studies in relevant disease models. IFI16, and the assembled inflammasome, shuttles between different compartments; therefore, there is a need to decrypt the precise functions of nucleic acid sensing PRRs in distinct compartments. Such studies should also aid in determining how pathogens may evade inflammasomes. In the evolutionary arms race, pathogens have co-opted advanced mechanisms to counter host strategies. Therefore, it is not unlikely that pathogens recruit histone proteins to their DNA to uniquely evade inflammasome recognition. Furthermore, more animal and clinical studies are needed to unravel the specific mechanisms in a physiological setting to make clinical applications possible. The improved knowledge of the higher order biochemical and functional details of how inflammasomes sense, oligomerize, and assemble on nucleic acids should help to effectively induce or restore antiviral immunity. Moreover, these insights will likely lead to new treatment avenues for autoimmune diseases. Future studies in this area will help to translate these discoveries into practice.

ACKNOWLEDGMENTS

We apologize to numerous investigators whose work could not be cited due to space limitations. C.R.L. is supported by the Department of Biology, Missouri State University. Work in the laboratory of P.K.A. is supported by funds from The Wellcome Trust (108248/Z/15/Z), The Royal Society (RG150535), the Medical Research Council (UK) through an MRC New Investigator Grant (MR/S00968X/1), and core funds (F28072) from Imperial College London.

REFERENCES

Aglipay, J.A., Lee, S.W., Okada, S., Fujiuchi, N., Ohtsuka, T., Kwak, J.C., Wang, Y., Johnstone, R.W., Deng, C., Qin, J., Ouchi, T., 2003. A member of the pyrin family, IFI16, is a novel BRCA1-associated protein involved in the p53-mediated apoptosis pathway. Oncogene 22, 8931–8938.

Alunno, A., Caneparo, V., Carubbi, F., Bistoni, O., Caterbi, S., Bartoloni, E., Giacomelli, R., Gariglio, M., Landolfo, S., Gerli, R., 2015. Interferon gamma-inducible protein 16 in primary Sjögren's syndrome: a novel player in disease pathogenesis? Arthritis Res. Ther. 17, 208.

Ansari, M.A., Singh, V.V., Dutta, S., Veettil, M.V., Dutta, D., Chikoti, L., Lu, J., Everly, D., Chandran, B., 2013. Constitutive interferon-inducible protein 16-inflammasome activation during Epstein-Barr virus latency I, II, and III in B and epithelial cells. J. Virol. 87, 8606–8623.

Ansari, M.A., Dutta, S., Veettil, M.V., Dutta, D., Iqbal, J., Kumar, B., Roy, A., Chikoti, L., Singh, V.V., Chandran, B., 2015. Herpesvirus genome recognition induced acetylation of nuclear IFI16 is essential for its cytoplasmic translocation, inflammasome and IFN-β responses. PLoS Pathog. 11, e1005019.

Antiochos, B., Matyszewski, M., Sohn, J., Casciola-Rosen, L., Rosen, A., 2018. IFI16 filament formation in salivary epithelial cells shapes the anti-IFI16 immune response in Sjogren's syndrome. JCI Insight 3, e120179.

Barlan, A.U., Danthi, P., Wiethoff, C.M., 2011a. Lysosomal localization and mechanism of membrane penetration influence nonenveloped virus activation of the NLRP3 inflammasome. Virology 412, 306–314.

Barlan, A.U., Griffin, T.M., McGuire, K.A., Wiethoff, C.M., 2011b. Adenovirus membrane penetration activates the NLRP3 inflammasome. J. Virol. 85, 146–155.

Baum, R., Sharma, S., Carpenter, S., Li, Q.Z., Busto, P., Fitzgerald, K.A., Marshak-Rothstein, A., Gravallese, E.M., 2015. Cutting edge: AIM2 and endosomal TLRs differentially regulate arthritis and autoantibody production in DNase II-deficient mice. J. Immunol. 194, 873–877.

Bawadekar, M., De Andrea, M., Gariglio, M., Landolfo, S., 2015. Mislocalization of the interferon inducible protein IFI16 by environmental insults: implications in autoimmunity. Cytokine Growth Factor Rev. 26, 213–219.

Belhocine, K., Monack, D.M., 2012. Francisella infection triggers activation of the AIM2 inflammasome in murine dendritic cells. Cell. Microbiol. 14, 71–80.

Biolatti, M., Dell'Oste, V., Pautasso, S., Gugliesi, F., von Einem, J., Krapp, C., Jakobsen, M.R., Borgogna, C., Gariglio, M., De Andrea, M., Landolfo, S., 2018. Human cytomegalovirus tegument protein pp65 (pUL83) dampens type I interferon production by inactivating the DNA sensor cGAS without affecting STING. J. Virol. 92, e01774-17.

Brunette, R.L., Young, J.M., Whitley, D.G., Brodsky, I.E., Malik, H.S., Stetson, D.B., 2012. Extensive evolutionary and functional diversity among mammalian AIM2-like receptors. J. Exp. Med. 209, 1969–1983.

Burckstummer, T., Baumann, C., Bluml, S., Dixit, E., Durnberger, G., Jahn, H., Planyavsky, M., Bilban, M., Colinge, J., Bennett, K.L., Superti-Furga, G., 2009. An orthogonal proteomic-genomic screen identifies AIM2 as a cytoplasmic DNA sensor for the inflammasome. Nat. Immunol. 10, 266–272.

Caneparo, V., Pastorelli, L., Pisani, L.F., Bruni, B., Prodam, F., Boldorini, R., Roggenbuck, D., Vecchi, M., Landolfo, S., Gariglio, M., De Andrea, M., 2016. Distinct anti-IFI16 and anti-GP2 antibodies in inflammatory bowel disease and their variation with infliximab therapy. Inflamm. Bowel Dis. 22, 2977–2987.

Carlos, D., Costa, F.R., Pereira, C.A., Rocha, F.A., Yaochite, J.N., Oliveira, G.G., Carneiro, F.S., Tostes, R.C., Ramos, S.G., Zamboni, D.S., Camara, N.O., Ryffel, B., Silva, J.S., 2017. Mitochondrial DNA activates the NLRP3 inflammasome and predisposes to type 1 diabetes in murine model. Front. Immunol. 8, 164.

Chakrabarti, A., Banerjee, S., Franchi, L., Loo, Y.M., Gale Jr., M., Nunez, G., Silverman, R.H., 2015. RNase L activates the NLRP3 inflammasome during viral infections. Cell Host Microbe 17, 466–477.

Costa Franco, M.M., Marim, F., Guimaraes, E.S., Assis, N.R.G., Cerqueira, D.M., Alves-Silva, J., Harms, J., Splitter, G., Smith, J., Kanneganti, T.D., de Queiroz, N., Gutman, D., Barber, G.N., Oliveira, S.C., 2018. Brucella abortus triggers a cGAS-independent STING pathway to induce host protection that involves guanylate-binding proteins and inflammasome activation. J. Immunol. 200, 607–622.

Cridland, J.A., Curley, E.Z., Wykes, M.N., Schroder, K., Sweet, M.J., Roberts, T.L., Ragan, M.A., Kassahn, K.S., Stacey, K.J., 2012. The mammalian PYHIN gene family: phylogeny, evolution and expression. BMC Evol. Biol. 12, 140.

Cunha, L.D., Silva, A.L.N., Ribeiro, J.M., Mascarenhas, D.P.A., Quirino, G.F.S., Santos, L.L., Flavell, R.A., Zamboni, D.S., 2017. AIM2 engages active but unprocessed caspase-1 to induce noncanonical activation of the NLRP3 inflammasome. Cell Rep. 20, 794–805.

Dobbs, N., Burnaevskiy, N., Chen, D., Gonugunta, V.K., Alto, N.M., Yan, N., 2015. STING activation by translocation from the ER is associated with infection and autoinflammatory disease. Cell Host Microbe 18, 157–168.

Doitsh, G., Cavrois, M., Lassen, K.G., Zepeda, O., Yang, Z., Santiago, M.L., Hebbeler, A.M., Greene, W.C., 2010. Abortive HIV infection mediates CD4 T cell depletion and inflammation in human lymphoid tissue. Cell 143, 789–801.

Dorfleutner, A., Talbott, S.J., Bryan, N.B., Funya, K.N., Rellick, S.L., Reed, J.C., Shi, X., Rojanasakul, Y., Flynn, D.C., Stehlik, C., 2007. A Shope Fibroma virus PYRIN-only protein modulates the host immune response. Virus Genes 35, 685–694.

Eichholz, K., Bru, T., Tran, T.T., Fernandes, P., Welles, H., Mennechet, F.J., Manel, N., Alves, P., Perreau, M., Kremer, E.J., 2016. Immune-complexed adenovirus induce AIM2-mediated pyroptosis in human dendritic cells. PLoS Pathog. 12, e1005871.

Fang, R., Hara, H., Sakai, S., Hernandez-Cuellar, E., Mitsuyama, M., Kawamura, I., Tsuchiya, K., 2014. Type I interferon signaling regulates activation of the absent in melanoma 2 inflammasome during Streptococcus pneumoniae infection. Infect. Immun. 82, 2310–2317.

Fernandes-Alnemri, T., Yu, J.W., Datta, P., Wu, J., Alnemri, E.S., 2009. AIM2 activates the inflammasome and cell death in response to cytoplasmic DNA. Nature 458, 509–513.

Fernandes-Alnemri, T., Yu, J.W., Juliana, C., Solorzano, L., Kang, S., Wu, J., Datta, P., McCormick, M., Huang, L., McDermott, E., Eisenlohr, L., Landel, C.P., Alnemri, E.S., 2010. The AIM2 inflammasome is critical for innate immunity to Francisella tularensis. Nat. Immunol. 11, 385–393.

Finethy, R., Jorgensen, I., Haldar, A.K., de Zoete, M.R., Strowig, T., Flavell, R.A., Yamamoto, M., Nagarajan, U.M., Miao, E.A., Coers, J., 2015. Guanylate binding proteins enable rapid activation of canonical and noncanonical inflammasomes in Chlamydia-infected macrophages. Infect. Immun. 83, 4740–4749.

Gomes, M.T., Campos, P.C., Oliveira, F.S., Corsetti, P.P., Bortoluci, K.R., Cunha, L.D., Zamboni, D.S., Oliveira, S.C., 2013. Critical role of ASC inflammasomes and bacterial type IV secretion system in caspase-1 activation and host innate resistance to Brucella abortus infection. J. Immunol. 190, 3629–3638.

Gray, E.E., Winship, D., Snyder, J.M., Child, S.J., Geballe, A.P., Stetson, D.B., 2016. The AIM2-like receptors are dispensable for the interferon response to intracellular DNA. Immunity 45, 255–266.

Gugliesi, F., Bawadekar, M., De Andrea, M., Dell'Oste, V., Caneparo, V., Tincani, A., Gariglio, M., Landolfo, S., 2013. Nuclear DNA sensor IFI16 as circulating protein in autoimmune diseases is a signal of damage that impairs endothelial cells through high-affinity membrane binding. PLoS One 8, e63045.

Gupta, R., Ghosh, S., Monks, B., DeOliveira, R.B., Tzeng, T.C., Kalantari, P., Nandy, A., Bhattacharjee, B., Chan, J., Ferreira, F., Rathinam, V., Sharma, S., Lien, E., Silverman, N., Fitzgerald, K., Firon, A., Trieu-Cuot, P., Henneke, P., Golenbock, D.T., 2014. RNA and beta-hemolysin of group B Streptococcus induce interleukin-1beta (IL-1beta) by activating NLRP3 inflammasomes in mouse macrophages. J. Biol. Chem. 289, 13701–13705.

Han, Y., Chen, Z., Hou, R., Yan, D., Liu, C., Chen, S., Li, X., Du, W., 2015. Expression of AIM2 is correlated with increased inflammation in chronic hepatitis B patients. Virol. J. 12, 129.

Hanamsagar, R., Aldrich, A., Kielian, T., 2014. Critical role for the AIM2 inflammasome during acute CNS bacterial infection. J. Neurochem. 129, 704–711.

He, W.T., Wan, H., Hu, L., Chen, P., Wang, X., Huang, Z., Yang, Z.H., Zhong, C.Q., Han, J., 2015. Gasdermin D is an executor of pyroptosis and required for interleukin-1β secretion. Cell Res. 25, 1285–1298.

Hornung, V., Ablasser, A., Charrel-Dennis, M., Bauernfeind, F., Horvath, G., Caffrey, D.R., Latz, E., Fitzgerald, K.A., 2009. AIM2 recognizes cytosolic dsDNA and forms a caspase-1-activating inflammasome with ASC. Nature 458, 514–518.

Huang, Y., Liu, L., Ma, D., Liao, Y., Lu, Y., Huang, H., Qin, W., Liu, X., Fang, F., 2017a. Human cytomegalovirus triggers the assembly of AIM2 inflammasome in THP-1-derived macrophages. J. Med. Virol. 89, 2188–2195.

Huang, Y., Ma, D., Huang, H., Lu, Y., Liao, Y., Liu, L., Liu, X., Fang, F., 2017b. Interaction between HCMV pUL83 and human AIM2 disrupts the activation of the AIM2 inflammasome. Virol. J. 14, 34.

Jakobs, C., Perner, S., Hornung, V., 2015. AIM2 drives joint inflammation in a self-DNA triggered model of chronic polyarthritis. PLoS One 10, e0131702.

Jin, T., Perry, A., Jiang, J., Smith, P., Curry, J.A., Unterholzner, L., Jiang, Z., Horvath, G., Rathinam, V.A., Johnstone, R.W., Hornung, V., Latz, E., Bowie, A.G., Fitzgerald, K.A., Xiao, T.S., 2012. Structures of the HIN domain:DNA complexes reveal ligand binding and activation mechanisms of the AIM2 inflammasome and IFI16 receptor. Immunity 36, 561–571.

Johnson, K.E., Chikoti, L., Chandran, B., 2013. Herpes simplex virus 1 infection induces activation and subsequent inhibition of the IFI16 and NLRP3 inflammasomes. J. Virol. 87, 5005–5018.

Jones, J.W., Kayagaki, N., Broz, P., Henry, T., Newton, K., O'Rourke, K., Chan, S., Dong, J., Qu, Y., Roose-Girma, M., Dixit, V.M., Monack, D.M., 2010. Absent in melanoma 2 is required for innate immune recognition of Francisella tularensis. Proc. Natl. Acad. Sci. U. S. A. 107, 9771–9776.

Kailasan Vanaja, S., Rathinam, V.A., Atianand, M.K., Kalantari, P., Skehan, B., Fitzgerald, K.A., Leong, J.M., 2014. Bacterial RNA:DNA hybrids are activators of the NLRP3 inflammasome. Proc. Natl. Acad. Sci. U. S. A. 111, 7765–7770.

Kalantari, P., DeOliveira, R.B., Chan, J., Corbett, Y., Rathinam, V., Stutz, A., Latz, E., Gazzinelli, R.T., Golenbock, D.T., Fitzgerald, K.A., 2014. Dual engagement of the NLRP3 and AIM2 inflammasomes by plasmodium-derived hemozoin and DNA during malaria. Cell Rep. 6, 196–210.

Kang, R., Chen, R., Xie, M., Cao, L., Lotze, M.T., Tang, D., Zeh 3rd, H.J., 2016. The receptor for advanced glycation end products activates the AIM2 inflammasome in acute pancreatitis. J. Immunol. 196, 4331–4337.

Kanneganti, T.D., Body-Malapel, M., Amer, A., Park, J.H., Whitfield, J., Franchi, L., Taraporewala, Z.F., Miller, D., Patton, J.T., Inohara, N., Núñez, G., 2006a. Critical role for cryopyrin/Nalp3 in activation of caspase-1 in response to viral infection and double-stranded RNA. J. Biol. Chem. 281, 36560–36568.

Kanneganti, T.D., Ozoren, N., Body-Malapel, M., Amer, A., Park, J.H., Franchi, L., Whitfield, J., Barchet, W., Colonna, M., Vandenabeele, P., Bertin, J., Coyle, A., Grant, E.P., Akira, S., Nunez, G., 2006b. Bacterial RNA and small antiviral compounds activate caspase-1 through cryopyrin/Nalp3. Nature 440, 233–236.

Karki, R., Man, S.M., Malireddi, R.K.S., Gurung, P., Vogel, P., Lamkanfi, M., Kanneganti, T.D., 2015. Concerted activation of the AIM2 and NLRP3 inflammasomes orchestrates host protection against Aspergillus infection. Cell Host Microbe 17, 357–368.

Kawane, K., Ohtani, M., Miwa, K., Kizawa, T., Kanbara, Y., Yoshioka, Y., Yoshikawa, H., Nagata, S., 2006. Chronic polyarthritis caused by mammalian DNA that escapes from degradation in macrophages. Nature 443, 998–1002.

Kayagaki, N., Stowe, I.B., Lee, B.L., O'Rourke, K., Anderson, K., Warming, S., Cuellar, T., Haley, B., Roose-Girma, M., Phung, Q.T., Liu, P.S., Lill, J.R., Li, H., Wu, J., Kummerfeld, S., Zhang, J., Lee, W.P., Snipas, S.J., Salvesen, G.S., Morris, L.X., Fitzgerald, L., Zhang, Y., Bertram, E.M., Goodnow, C.C., Dixit, V.M., 2015. Caspase-11 cleaves gasdermin D for non-canonical inflammasome signalling. Nature 526, 666–671.

Kerur, N., Veettil, M.V., Sharma-Walia, N., Bottero, V., Sadagopan, S., Otageri, P., Chandran, B., 2011. IFI16 acts as a nuclear pathogen sensor to induce the inflammasome in response to Kaposi sarcoma-associated herpesvirus infection. Cell Host Microbe 9, 363–375.

Khare, S., Ratsimandresy, R.A., de Almeida, L., Cuda, C.M., Rellick, S.L., Misharin, A.V., Wallin, M.C., Gangopadhyay, A., Forte, E., Gottwein, E., Perlman, H., Reed, J.C., Greaves, D.R., Dorfleutner, A., Stehlik, C., 2014. The PYRIN domain-only protein POP3 inhibits ALR inflammasomes and regulates responses to infection with DNA viruses. Nat. Immunol. 15, 343–353.

Kim, S., Bauernfeind, F., Ablasser, A., Hartmann, G., Fitzgerald, K.A., Latz, E., Hornung, V., 2010. Listeria monocytogenes is sensed by the NLRP3 and AIM2 inflammasome. Eur. J. Immunol. 40, 1545–1551.

Kofoed, E.M., Vance, R.E., 2011. Innate immune recognition of bacterial ligands by NAIPs determines inflammasome specificity. Nature 477, 592–595.

Li, T., Diner, B.A., Chen, J., Cristea, I.M., 2012. Acetylation modulates cellular distribution and DNA sensing ability of interferon-inducible protein IFI16. Proc. Natl. Acad. Sci. U. S. A. 109, 10558–10563.

Li, T., Chen, J., Cristea, I.M., 2013. Human cytomegalovirus tegument protein pUL83 inhibits IFI16-mediated DNA sensing for immune evasion. Cell Host Microbe 14, 591–599.

Li, J., Hu, L., Liu, Y., Huang, L., Mu, Y., Cai, X., Weng, C., 2015. DDX19A senses viral RNA and mediates NLRP3-dependent inflammasome activation. J. Immunol. 195, 5732–5749.

Liu, S., Cai, X., Wu, J., Cong, Q., Chen, X., Li, T., Du, F., Ren, J., Wu, Y.T., Grishin, N.V., Chen, Z.J., 2015. Phosphorylation of innate immune adaptor proteins MAVS, STING, and TRIF induces IRF3 activation. Science 347, aaa2630.

Liu, T., Tang, Q., Liu, K., Xie, W., Liu, X., Wang, H., Wang, R.F., Cui, J., 2016a. TRIM11 suppresses AIM2 inflammasome by degrading AIM2 via p62-dependent selective autophagy. Cell Rep. 16, 1988–2002.

Liu, C., Yue, R., Yang, Y., Cui, Y., Yang, L., Zhao, D., Zhou, X., 2016b. AIM2 inhibits autophagy and IFN-beta production during M. bovis infection. Oncotarget 7, 46972–46987.

Liu, F., Niu, Q., Fan, X., Liu, C., Zhang, J., Wei, Z., Hou, W., Kanneganti, T.D., Robb, M.L., Kim, J.H., Michael, N.L., Sun, J., Soong, L., Hu, H., 2017. Priming and activation of inflammasome by canarypox virus vector ALVAC via the cGAS/IFI16-STING-type I IFN pathway and AIM2 sensor. J. Immunol. 199, 3293–3305.

Liu, B.C., Sarhan, J., Panda, A., Muendlein, H.I., Ilyukha, V., Coers, J., Yamamoto, M., Isberg, R.R., Poltorak, A., 2018. Constitutive interferon maintains GBP expression required for release of bacterial components upstream of pyroptosis and anti-DNA responses. Cell Rep. 24, 155.e5–168.e5.

Luan, Y., Yu, X.P., Yang, N., Frenkel, S., Chen, L., Liu, C.J., 2008. p204 protein overcomes the inhibition of core binding factor alpha-1-mediated osteogenic differentiation by Id helix-loop-helix proteins. Mol. Biol. Cell 19, 2113–2126.

Ludlow, L.E., Johnstone, R.W., Clarke, C.J., 2005. The HIN-200 family: more than interferon-inducible genes? Exp. Cell Res. 308, 1–17.

Lupfer, C., Anand, P.K., 2016. Integrating inflammasome signaling in sexually transmitted infections. Trends Immunol. 37, 703–714.

Man, S.M., Karki, R., Malireddi, R.K., Neale, G., Vogel, P., Yamamoto, M., Lamkanfi, M., Kanneganti, T.D., 2015. The transcription factor IRF1 and guanylate-binding proteins target activation of the AIM2 inflammasome by Francisella infection. Nat. Immunol. 16, 467–475.

Man, S.M., Karki, R., Sasai, M., Place, D.E., Kesavardhana, S., Temirov, J., Frase, S., Zhu, Q., Malireddi, R.K.S., Kuriakose, T., Peters, J.L., Neale, G., Brown, S.A., Yamamoto, M., Kanneganti, T.D., 2016. IRGB10 liberates bacterial ligands for sensing by the AIM2 and caspase-11-NLRP3 inflammasomes. Cell 167, 382–396.e317.

Maruzuru, Y., Ichinohe, T., Sato, R., Miyake, K., Okano, T., Suzuki, T., Koshiba, T., Koyanagi, N., Tsuda, S., Watanabe, M., Arii, J., Kato, A., Kawaguchi, Y., 2018. Herpes simplex virus 1 VP22 inhibits AIM2-dependent inflammasome activation to enable efficient viral replication. Cell Host Microbe 23, 254.e7–265.e7.

Meunier, E., Wallet, P., Dreier, R.F., Costanzo, S., Anton, L., Ruhl, S., Dussurgey, S., Dick, M.S., Kistner, A., Rigard, M., Degrandi, D., Pfeffer, K., Yamamoto, M., Henry, T., Broz, P., 2015. Guanylate-binding proteins promote activation of the AIM2 inflammasome during infection with Francisella novicida. Nat. Immunol. 16, 476–484.

Milora, K.A., Miller, S.L., Sanmiguel, J.C., Jensen, L.E., 2014. Interleukin-1alpha released from HSV-1-infected keratinocytes acts as a functional alarmin in the skin. Nat. Commun. 5, 5230.

Miraglia, M.C., Costa Franco, M.M., Rodriguez, A.M., Bellozi, P.M., Ferrari, C.C., Farias, M.I., Dennis, V.A., Barrionuevo, P., de Oliveira, A.C., Pitossi, F., Kim, K.S., Delpino, M.V., Oliveira, S.C., Giambartolomei, G.H., 2016. Glial cell-elicited activation of brain microvasculature in response to Brucella abortus infection requires ASC inflammasome-dependent IL-1beta production. J. Immunol. 196, 3794–3805.

Mitoma, H., Hanabuchi, S., Kim, T., Bao, M., Zhang, Z., Sugimoto, N., Liu, Y.J., 2013. The DHX33 RNA helicase senses cytosolic RNA and activates the NLRP3 inflammasome. Immunity 39, 123–135.

Monroe, K.M., Yang, Z., Johnson, J.R., Geng, X., Doitsh, G., Krogan, N.J., Greene, W.C., 2014. IFI16 DNA sensor is required for death of lymphoid CD4 T cells abortively infected with HIV. Science 343, 428–432.

Monteith, A.J., Kang, S., Scott, E., Hillman, K., Rajfur, Z., Jacobson, K., Costello, M.J., Vilen, B.J., 2016. Defects in lysosomal maturation facilitate the activation of innate sensors in systemic lupus erythematosus. Proc. Natl. Acad. Sci. U. S. A. 113, E2142–E2151.

Morrone, S.R., Wang, T., Constantoulakis, L.M., Hooy, R.M., Delannoy, M.J., Sohn, J., 2014. Cooperative assembly of IFI16 filaments on dsDNA provides insights into host defense strategy. Proc. Natl. Acad. Sci. U. S. A. 111, E62–E71.

Muruve, D.A., Petrilli, V., Zaiss, A.K., White, L.R., Clark, S.A., Ross, P.J., Parks, R.J., Tschopp, J., 2008. The inflammasome recognizes cytosolic microbial and host DNA and triggers an innate immune response. Nature 452, 103–107.

Nakahira, K., Haspel, J.A., Rathinam, V.A., Lee, S.J., Dolinay, T., Lam, H.C., Englert, J.A., Rabinovitch, M., Cernadas, M., Kim, H.P., Fitzgerald, K.A., Ryter, S.W., Choi, A.M., 2011. Autophagy proteins regulate innate immune responses by inhibiting the release of mitochondrial DNA mediated by the NALP3 inflammasome. Nat. Immunol. 12, 222–230.

Niemiec, M.J., Grumaz, C., Ermert, D., Desel, C., Shankar, M., Lopes, J.P., Mills, I.G., Stevens, P., Sohn, K., Urban, C.F., 2017. Dual transcriptome of the immediate neutrophil and Candida albicans interplay. BMC Genomics 18, 696.

Ogembo, J.G., Kannan, L., Ghiran, I., Nicholson-Weller, A., Finberg, R.W., Tsokos, G.C., Fingeroth, J.D., 2013. Human complement receptor type 1/CD35 is an Epstein-Barr virus receptor. Cell Rep. 3, 371–385.

Orzalli, M.H., DeLuca, N.A., Knipe, D.M., 2012. Nuclear IFI16 induction of IRF-3 signaling during herpesviral infection and degradation of IFI16 by the viral ICP0 protein. Proc. Natl. Acad. Sci. U. S. A. 109, E3008–E3017.

Paludan, S.R., Bowie, A.G., Horan, K.A., Fitzgerald, K.A., 2011. Recognition of herpesviruses by the innate immune system. Nat. Rev. Immunol. 11, 143–154.

Pan, J., Ou, Z., Cai, C., Li, P., Gong, J., Ruan, X.Z., He, K., 2018. Fatty acid activates NLRP3 inflammasomes in mouse Kupffer cells through mitochondrial DNA release. Cell. Immunol. 332, 111–120.

Park, E., Na, H.S., Song, Y.R., Shin, S.Y., Kim, Y.M., Chung, J., 2014. Activation of NLRP3 and AIM2 inflammasomes by Porphyromonas gingivalis infection. Infect. Immun. 82, 112–123.

Peng, K., Broz, P., Jones, J., Joubert, L.M., Monack, D., 2011. Elevated AIM2-mediated pyroptosis triggered by hypercytotoxic Francisella mutant strains is attributed to increased intracellular bacteriolysis. Cell. Microbiol. 13, 1586–1600.

Pierini, R., Juruj, C., Perret, M., Jones, C.L., Mangeot, P., Weiss, D.S., Henry, T., 2012. AIM2/ASC triggers caspase-8-dependent apoptosis in Francisella-infected caspase-1-deficient macrophages. Cell Death Differ. 19, 1709–1721.

Place, D.E., Kanneganti, T.D., 2018. Recent advances in inflammasome biology. Curr. Opin. Immunol. 50, 32–38.

Rathinam, V.A., Jiang, Z., Waggoner, S.N., Sharma, S., Cole, L.E., Waggoner, L., Vanaja, S.K., Monks, B.G., Ganesan, S., Latz, E., Hornung, V., Vogel, S.N., Szomolanyi-Tsuda, E., Fitzgerald, K.A., 2010. The AIM2 inflammasome is essential for host defense against cytosolic bacteria and DNA viruses. Nat. Immunol. 11, 395–402.

Rayamajhi, M., Zak, D.E., Chavarria-Smith, J., Vance, R.E., Miao, E.A., 2013. Cutting edge: mouse NAIP1 detects the type III secretion system needle protein. J. Immunol. 191, 3986–3989.

Saiga, H., Kitada, S., Shimada, Y., Kamiyama, N., Okuyama, M., Makino, M., Yamamoto, M., Takeda, K., 2012. Critical role of AIM2 in Mycobacterium tuberculosis infection. Int. Immunol. 24, 637–644.

Sauer, J.D., Witte, C.E., Zemansky, J., Hanson, B., Lauer, P., Portnoy, D.A., 2010. Listeria monocytogenes triggers AIM2-mediated pyroptosis upon infrequent bacteriolysis in the macrophage cytosol. Cell Host Microbe 7, 412–419.

Schattgen, S.A., Gao, G., Kurt-Jones, E.A., Fitzgerald, K.A., 2016. Cutting edge: DNA in the lung microenvironment during influenza virus infection tempers inflammation by engaging the DNA sensor AIM2. J. Immunol. 196, 29–33.

Schuster, S., Johnson, C.D., Hennebelle, M., Holtmann, T., Taha, A.Y., Kirpich, I.A., Eguchi, A., Ramsden, C.E., Papouchado, B.G., McClain, C.J., Feldstein, A.E., 2018. Oxidized linoleic acid metabolites induce liver mitochondrial dysfunction, apoptosis, and NLRP3 activation in mice. J. Lipid Res. 59, 1597–1609.

Sha, W., Mitoma, H., Hanabuchi, S., Bao, M., Weng, L., Sugimoto, N., Liu, Y., Zhang, Z., Zhong, J., Sun, B., Liu, Y.J., 2014. Human NLRP3 inflammasome senses multiple types of bacterial RNAs. Proc. Natl. Acad. Sci. U. S. A. 111, 16059–16064.

Shah, A.U., Sarwar, A., Orabi, A.I., Gautam, S., Grant, W.M., Park, A.J., Liu, J., Mistry, P.K., Jain, D., Husain, S.Z., 2009. Protease activation during in vivo pancreatitis is dependent on calcineurin activation. Am. J. Physiol. Gastrointest. Liver Physiol. 297, G967–G973.

Shah, S., Bohsali, A., Ahlbrand, S.E., Srinivasan, L., Rathinam, V.A., Vogel, S.N., Fitzgerald, K.A., Sutterwala, F.S., Briken, V., 2013. Cutting edge: Mycobacterium tuberculosis but not nonvirulent mycobacteria inhibits IFN-beta and AIM2 inflammasome-dependent IL-1beta production via its ESX-1 secretion system. J. Immunol. 191, 3514–3518.

Shi, J., Zhao, Y., Wang, K., Shi, X., Wang, Y., Huang, H., Zhuang, Y., Cai, T., Wang, F., Shao, F., 2015. Cleavage of GSDMD by inflammatory caspases determines pyroptotic cell death. Nature 526, 660–665.

Shimada, K., Crother, T.R., Karlin, J., Dagvadorj, J., Chiba, N., Chen, S., Ramanujan, V.K., Wolf, A.J., Vergnes, L., Ojcius, D.M., Rentsendorj, A., Vargas, M., Guerrero, C., Wang, Y., Fitzgerald, K.A., Underhill, D.M., Town, T., Arditi, M., 2012. Oxidized mitochondrial DNA activates the NLRP3 inflammasome during apoptosis. Immunity 36, 401–414.

Stratmann, S.A., Morrone, S.R., van Oijen, A.M., Sohn, J., 2015. The innate immune sensor IFI16 recognizes foreign DNA in the nucleus by scanning along the duplex. eLife 4, e11721.

Strittmatter, G.E., Sand, J., Sauter, M., Seyffert, M., Steigerwald, R., Fraefel, C., Smola, S., French, L.E., Beer, H.D., 2016. IFN-gamma primes keratinocytes for HSV-1-induced inflammasome activation. J. Invest. Dermatol. 136, 610–620.

Swanson, K.V., Junkins, R.D., Kurkjian, C.J., Holley-Guthrie, E., Pendse, A.A., El Morabiti, R., Petrucelli, A., Barber, G.N., Benedict, C.A., Ting, J.P., 2017. A noncanonical function of cGAMP in inflammasome priming and activation. J. Exp. Med. 214, 3611–3626.

Theil, D., Derfuss, T., Paripovic, I., Herberger, S., Meinl, E., Schueler, O., Strupp, M., Arbusow, V., Brandt, T., 2003. Latent herpesvirus infection in human trigeminal ganglia causes chronic immune response. Am. J. Pathol. 163, 2179–2184.

Torii, Y., Kawada, J.I., Murata, T., Yoshiyama, H., Kimura, H., Ito, Y., 2017. Epstein-Barr virus infection-induced inflammasome activation in human monocytes. PLoS One 12, e0175053.

Tsuchiya, K., Hara, H., Kawamura, I., Nomura, T., Yamamoto, T., Daim, S., Dewamitta, S.R., Shen, Y., Fang, R., Mitsuyama, M., 2010. Involvement of absent in melanoma 2 in inflammasome activation in macrophages infected with Listeria monocytogenes. J. Immunol. 185, 1186–1195.

Unterholzner, L., Keating, S.E., Baran, M., Horan, K.A., Jensen, S.B., Sharma, S., Sirois, C.M., Jin, T., Latz, E., Xiao, T.S., Fitzgerald, K.A., Paludan, S.R.,

Bowie, A.G., 2010. IFI16 is an innate immune sensor for intracellular DNA. Nat. Immunol. 11, 997–1004.

Vanhove, W., Peeters, P.M., Staelens, D., Schraenen, A., Van der Goten, J., Cleynen, I., De Schepper, S., Van Lommel, L., Reynaert, N.L., Schuit, F., Van Assche, G., Ferrante, M., De Hertogh, G., Wouters, E.F., Rutgeerts, P., Vermeire, S., Nys, K., Arijs, I., 2015. Strong upregulation of AIM2 and IFI16 inflammasomes in the mucosa of patients with active inflammatory bowel disease. Inflamm. Bowel Dis. 21, 2673–2682.

Wallet, P., Benaoudia, S., Mosnier, A., Lagrange, B., Martin, A., Lindgren, H., Golovliov, I., Michal, F., Basso, P., Djebali, S., Provost, A., Allatif, O., Meunier, E., Broz, P., Yamamoto, M., Py, B.F., Faudry, E., Sjostedt, A., Henry, T., 2017. IFN-gamma extends the immune functions of guanylate binding proteins to inflammasome-independent antibacterial activities during Francisella novicida infection. PLoS Pathog. 13, e1006630.

Warren, S.E., Armstrong, A., Hamilton, M.K., Mao, D.P., Leaf, I.A., Miao, E.A., Aderem, A., 2010. Cutting edge: cytosolic bacterial DNA activates the inflammasome via Aim2. J. Immunol. 185, 818–821.

Wellington, M., Koselny, K., Sutterwala, F.S., Krysan, D.J., 2014. Candida albicans triggers NLRP3-mediated pyroptosis in macrophages. Eukaryot. Cell 13, 329–340.

Weng, L., Mitoma, H., Trichot, C., Tricot, C., Bao, M., Liu, Y., Zhang, Z., Liu, Y.J., 2014. The E3 ubiquitin ligase tripartite motif 33 is essential for cytosolic RNA-induced NLRP3 inflammasome activation. J. Immunol. 193, 3676–3682.

Wu, J., Fernandes-Alnemri, T., Alnemri, E.S., 2010. Involvement of the AIM2, NLRC4, and NLRP3 inflammasomes in caspase-1 activation by Listeria monocytogenes. J. Clin. Immunol. 30, 693–702.

Xiao, T.S., 2015. The nucleic acid-sensing inflammasomes. Immunol. Rev. 265, 103–111.

Xin, H., Curry, J., Johnstone, R.W., Nickoloff, B.J., Choubey, D., 2003. Role of IFI 16, a member of the interferon-inducible p200-protein family, in prostate epithelial cellular senescence. Oncogene 22, 4831–4840.

Yang, J., Zhao, Y., Shi, J., Shao, F., 2013a. Human NAIP and mouse NAIP1 recognize bacterial type III secretion needle protein for inflammasome activation. Proc. Natl. Acad. Sci. U. S. A. 110, 14408–14413.

Yang, Y., Zhou, X., Kouadir, M., Shi, F., Ding, T., Liu, C., Liu, J., Wang, M., Yang, L., Yin, X., Zhao, D., 2013b. The AIM2 inflammasome is involved in macrophage activation during infection with virulent Mycobacterium bovis strain. J. Infect. Dis. 208, 1849–1858.

Yogarajah, T., Ong, K.C., Perera, D., Wong, K.T., 2017. AIM2 inflammasome-mediated pyroptosis in enterovirus A71-infected neuronal cells restricts viral replication. Sci. Rep. 7, 5845.

Yogarajah, T., Ong, K.C., Perera, D., Wong, K.T., 2018. RSAD2 and AIM2 modulate Coxsackievirus A16 and Enterovirus A71 replication in neuronal cells in different ways that may be associated with their 5' nontranslated regions. J. Virol. 92, e01914-17.

Zhang, Y., Kong, L., Carlson, C.S., Liu, C.J., 2008. Cbfa1-dependent expression of an interferon-inducible p204 protein is required for chondrocyte differentiation. Cell Death Differ. 15, 1760–1771.

Zhang, W., Cai, Y., Xu, W., Yin, Z., Gao, X., Xiong, S., 2013. AIM2 facilitates the apoptotic DNA-induced systemic lupus erythematosus via arbitrating macrophage functional maturation. J. Clin. Immunol. 33, 925–937.

Zhang, H., Luo, J., Alcorn, J.F., Chen, K., Fan, S., Pilewski, J., Liu, A., Chen, W., Kolls, J.K., Wang, J., 2017. AIM2 inflammasome is critical for influenza-induced lung injury and mortality. J. Immunol. 198, 4383–4393.

Zhao, Y., Yang, J., Shi, J., Gong, Y.N., Lu, Q., Xu, H., Liu, L., Shao, F., 2011. The NLRC4 inflammasome receptors for bacterial flagellin and type III secretion apparatus. Nature 477, 596–600.

Zhao, H., Gonzalezgugel, E., Cheng, L., Richbourgh, B., Nie, L., Liu, C., 2015. The roles of interferon-inducible p200 family members IFI16 and p204 in innate immune responses, cell differentiation and proliferation. Genes Dis. 2, 46–56.

Zhen, J., Zhang, L., Pan, J., Ma, S., Yu, X., Li, X., Chen, S., Du, W., 2014. AIM2 mediates inflammation-associated renal damage in hepatitis B virus-associated glomerulonephritis by regulating caspase-1, IL-1beta, and IL-18. Mediators Inflamm. 2014, 190860.

Zhong, Z., Liang, S., Sanchez-Lopez, E., He, F., Shalapour, S., Lin, X.J., Wong, J., Ding, S., Seki, E., Schnabl, B., Hevener, A.L., Greenberg, H.B., Kisseleva, T., Karin, M., 2018. New mitochondrial DNA synthesis enables NLRP3 inflammasome activation. Nature 560, 198–203.

Zhu, S., Ding, S., Wang, P., Wei, Z., Pan, W., Palm, N.W., Yang, Y., Yu, H., Li, H.B., Wang, G., Lei, X., de Zoete, M.R., Zhao, J., Zheng, Y., Chen, H., Zhao, Y., Jurado, K.A., Feng, N., Shan, L., Kluger, Y., Lu, J., Abraham, C., Fikrig, E., Greenberg, H.B., Flavell, R.A., 2017. Nlrp9b inflammasome restricts rotavirus infection in intestinal epithelial cells. Nature 546, 667–670.

CHAPTER SIX

Activating the Nucleic Acid-Sensing Machinery for Anticancer Immunity

Terry Medler[2], Jaina M. Patel[2], Alejandro Alice[2], Jason R. Baird[2], Hong-Ming Hu, Michael J. Gough[1]
Earle A. Chiles Research Institute, Robert W. Franz Cancer Center, Providence Portland Medical Center, Portland, OR, United States
[1]Corresponding author: e-mail address: michael.gough@providence.org

Contents

Abstract

Nucleic acid sensing pathways have likely evolved as part of a broad pathogen sensing strategy intended to discriminate infectious agents and initiate appropriate innate and adaptive controls. However, in the absence of infectious agents, nucleic acid sensing pathways have been shown to play positive and negative roles in regulating tumorigenesis, tumor progression and metastatic spread. Understanding the normal biology behind these pathways and how they are regulated in malignant cells and in the tumor immune environment can help us devise strategies to exploit nucleic acid sensing to manipulate anti-cancer immunity.

[2] Authors contributed equally.

International Review of Cell and Molecular Biology, Volume 344
ISSN 1937-6448
https://doi.org/10.1016/bs.ircmb.2018.08.006

1. INTRODUCTION

Cancer arises from dysregulated expansion of our own cells due to an accumulation of gene mutations that divert cells from their normal function. The progressive nature of tumorigenesis means that at various points, cancer cells have an array of mutations but are not yet malignant, and the final malignant clone may arise out of a premalignant state that may have been present for a prolonged period of time (Fearon and Vogelstein, 1990). Recently, with the rapid acceptance of immunotherapy, we have seen clinical validation of the power of the immune system to recognize and control tumors via targeting of mutated genes as antigenic targets (Tran et al., 2017). As an example, these antigenic targets can include mutated K-ras (Tran et al., 2016), which is considered an early common mutation in tumorigenesis (Fearon and Vogelstein, 1990). In this case, these data imply that there is an extended interaction between mutated cancer cells and a host immune system that has the capacity to recognize these mutations as antigens, yet cancer continues to progress and acquire further mutations until it develops into symptomatic disease.

This disconnect between the capacity of the immune system to recognize antigens and whether they actually do so is well known in immunology. Antigens are not sufficient to initiate immune responses. Immunology is all about the involvement of the right cells, in the right places, with the right signals to support immunity. While we continue to discover novel immune regulatory molecules, pathways and cells, certain features remain true. One truism is the core principle of antigen plus adjuvant being critical to initiate immune responses. This principal explains why systemic application of antigenic peptides alone results in antigen specific tolerance and eventual deletion of antigen specific T cells (Liblau et al., 1996). However, systemic application of a virus expressing this antigenic peptide or the peptide pulsed on an activated dendritic cell results in powerful antigen-specific responses. The difference lies in the context in which the antigen-specific T cells see their antigens, which is dominated by the influences of the cells that present the antigen, the adjuvant signals that influence the context of antigen presentation and the specific molecular interactions between antigen presenting cells and T cells. The role of nucleic acids in activating the immune system has been of interest to scientists ever since Isaacs et al. discovered that mouse cells infected with chicken nucleic acids produced interferons (Isaacs et al., 1963). As we will discuss, it is in this context that nucleic acid sensing mechanisms play a critical role in vaccination-based strategies to engender anti-cancer immunity.

In view of the capacity of immune cells to recognize and destroy cancer cells, it is now recognized that *negative* regulation of immune responses is a critical feature that allows mutated cells to develop into advanced malignancies (Hanahan and Weinberg, 2011). This requirement to suppress or control anti-cancer immunity is in addition to the positive role that inflammation can provide in driving tumorigenesis through constant remodeling of the epithelium (DeNardo et al., 2010). As we will discuss, nucleic acid sensors have been shown to influence tumorigenicity in animal models in part by regulating tumorigenic inflammation. These mechanisms remain influential throughout malignant progression, and as we will discuss, there are a range of autophagic and cell-death related processes that result in activation of nucleic acid sensors that activate inflammatory mechanisms to continuously remodel the tumor immune environment. However, the majority of the studies on nucleic acid sensing and cancer have focused on *treatment* of advanced malignancies. Here there are two dominant approaches. In the first, exogenous ligands for nucleic acid sensors are applied to tumors to activate inflammatory pathways in the tumor environment. In the second, treatments such as chemotherapy and radiation therapy influence endogenous nucleic acid processing mechanisms to activate nucleic acid sensors in the tumor environment. We will discuss the current status of each of these interventions and how they likely impact control of tumors by immune mechanisms. A list of abbreviations used in this review is provided in Table 1.

Table 1 Abbreviations

Abbreviation	Name
AIM2	Absent in melanoma 2
ALR	AIM2-like receptors
ATP	Adenosine triphosphate
BER	Base excision repair
CDA	Cyclic di-AMP
CDG	Cyclic di-GMP
CDN	Cyclic dinucleotide
cGAMP	Cyclic GMP-AMP
cGAS	Cyclic GMP-AMP synthase

Continued

Table 1 Abbreviations—cont'd

Abbreviation	Name
CLR	C-type lectin receptors
CpG	Cytosine triphosphate deoxynucleotide followed by a guanine triphosphate deoxynucleotide with a phosphodiester link
CTLA4	Cytotoxic T-lymphocyte-associated protein 4
DMBA	Dimethylbenz[*a*]anthracene
DRibble	DriPs- and SLiPs-filled autophagosomes
DRiPs	Defective ribosomal products
dsRNA	Double stranded RNA
ERAdP	Endoplasmic reticulum adaptor protein
GTP	Guanosine-5′-triphosphate
HNSCC	Head and neck squamous cell carcinoma
HPV	Human papilloma virus
HR	Homologous recombination
IFN	Interferon
IRF3	Interferon regulatory factor 3
M1 macrophage	Classically differentiated macrophage
M2 macrophage	Alternatively differentiated macrophage
MAVS	Mitochondrial antiviral-signaling protein
MDA5	Melanoma Differentiation-Associated protein 5
MMR	Mismatch repair
MyD88	Myeloid differentiation primary response 88
NAP	NF-κB-activating kinase associated protein
NER	Nucleotide excision repair
NFkB	Nuclear factor kappa-light-chain-enhancer of activated B cells
NHEJ	Non-homologous end joining
NLR	NOD-like receptor
NOD	Nucleotide-binding oligomerization domain-containing protein
ODN	Oligodeoxynucleotides
PD1	Programmed cell death protein 1

Table 1 Abbreviations—cont'd

Abbreviation	Name
RECON	Reductase controlling NF-κB
RIG-I	Retinoic acid-inducible gene I
RLR	RIG-I-like receptors
ROS	Reactive Oxygen Species
SHP	Src homology region 2 domain-containing phosphatase
SLiPs	Short-lived proteins
SSBR	Single-strand break repair
ssRNA	Single stranded RNA
STAT	Signal transducer and activator of transcription
STING	Stimulator of Interferon Genes
TBK1	TANK-binding kinase 1
Th1	Type 1 CD4 T helper cell
Th2	Type 2 CD4 T helper cell
TLR	Toll-like receptor
TNF	Tumor necrosis factor
TRAF	TNF receptor associated factor
TRAM	TRIF-related adaptor molecule
Treg	Regulatory CD4 T cell
TREX1	Three prime repair exonuclease 1
TRIF	TIR-domain-containing adapter-inducing interferon-β
VEGF	Vascular endothelial growth factor

2. OVERVIEW OF NUCLEIC ACID SENSORS

Nucleic acid sensing can be viewed as a subset of innate mechanisms to detect viral and bacterial infection. The prototypic innate sensors are the Toll-like receptors (TLRs) (Medzhitov et al., 1997), a hugely influential family of molecules that activate inflammatory mechanisms in response to a wide variety of targets, from lipopolysaccharide (TLR4) to unmethylated

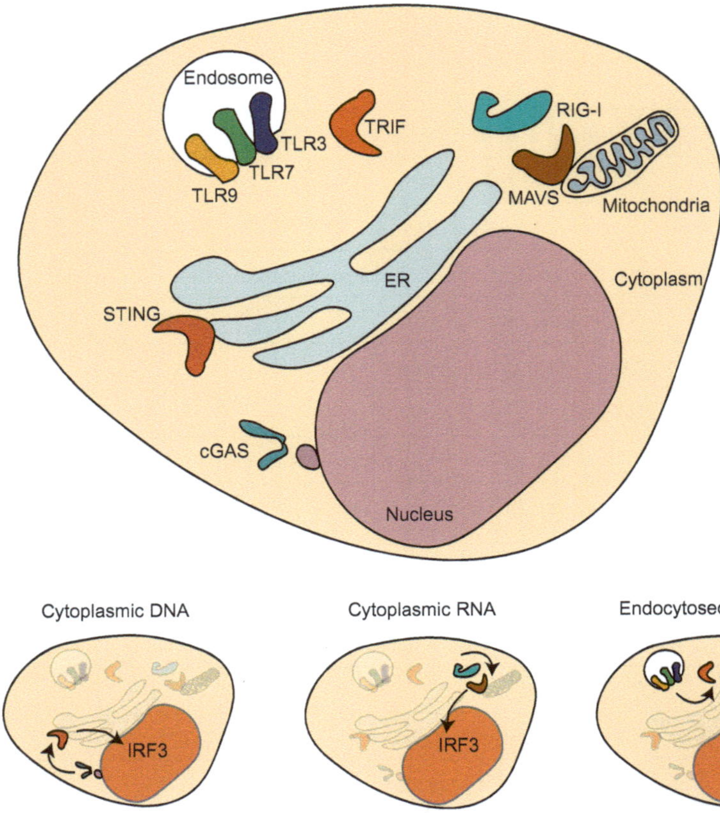

Fig. 1 Nucleic acid sensing mechanisms are dominantly located inside cells and consistently signal the presence of nucleic acids through activation and nuclear translocation of IRF3.

CpG dinucleotides (TLR9) (reviewed in Iwasaki and Medzhitov (2004); Yin et al. (2015)). In general, TLRs share a common cytoplasmic signaling domain, thought to oligomerize and recruit adaptor proteins, which include critical signaling molecules such as MyD88, TRIF, and TRAM (O'Neill and Bowie, 2007). It is notable that TLRs that recognize nucleic acids—TLR3, TLR7, and TLR9—have an intracellular localization, predominantly within endosomal compartments (reviewed in Blasius and Beutler (2010)) (Fig. 1). Broadly, TLR3 recognizes dsRNA, TLR7 recognizes ssRNA, and TLR9 recognizes unmethylated CpG links; however, each can recognize synthetic ligands that are important experimental cancer therapeutics. Not only do these receptors have an endosomal localization, but they also require endosomal acidification for activation (Häcker et al., 1998), which would correlate with destruction of internalized material and permit the host to distinguish an internalized infectious agent from inert material. While the signal transduction elicited by the TLRs that sense nucleic acids is not dissimilar from those sensing bacterial wall components—in that they utilize TIR domain-containing adaptors such as MyD88, TRIF, or TRAM—TLR3 is entirely Myd88-independent and only signals via TRIF (Hoebe et al., 2003). TRIF signaling is not unique to TLR3, but interestingly the other TLR members that use TRIF signaling may engage this pathway following receptor-mediated uptake and internalization of the TLR (Kagan et al., 2008; Tanimura et al., 2008). TRIF signal transduction differs somewhat from MyD88 signaling in that it recruits TRAF3 and TRAF6 to the signaling complex. While TRAF3 recruits TBK1 to the signaling complex and activates IRF3, TRAF6 recruits RIP-1 to the signaling complex, which activates TAK1 and subsequent NFκB and MAPK signaling (reviewed in Kawasaki and Kawai (2014)). TRAF6 also plays a role in Myd88 signaling, but Myd88-dependent signaling does not activate signal transduction pathways that lead to induction of IRF3 despite in some cases activating TBK1. The distinguishing mechanism is that IRF3 activation is dependent on its binding to the phosphorylated adaptor protein, in this case TRIF, allowing it to become phosphorylated by TBK1 (Liu et al., 2015). This mechanism is shared by the nucleic acid sensors MAVS and STING (Liu et al., 2015), such that one distinguishing feature of nucleic acid sensing pathways is IRF3 activation (Fig. 1).

MAVS and STING represent relatively recent additions to the signal transduction pathways activated following nucleic acid sensing in cells. MAVS (also known as IPS-1, Cardif, and VISA) is an adaptor protein that links the cytoplasmic sensors of viral RNA RIG-I, MDA5 and LGP2 to signaling pathways (Kawai et al., 2005; Meylan et al., 2005; Xu et al., 2005). STING functions similarly as an adaptor protein linked to signals from the DNA sensor cGAS (Liu et al., 2015), though as discussed later this can be bypassed via STING innate sensing of exogenously derived cyclic dinucleotides. In each case, IRF3 is activated, forms a homodimer and translocates to the nucleus to activate gene transcription. IRF3 is one of a family of transcription factors that participate in IFN gene transcription (reviewed in Honda et al. (2006)). IRF3 is most homologous to IRF7, and these proteins can form heterodimers. IRF3 is expressed at higher levels at baseline, is more stable than IRF7, and IRF3 homodimers are strong inducers of IFNβ gene transcription. By contrast IRF7 is induced by initial IRF3-induced gene expression and permits induction of a broader range of IFNα genes (Marié et al., 1998; Sato et al., 2000). In this way, continued detection of nucleic acids can generate a distinct early and late response in the vicinity of an infected cell.

While STING is commonly discussed as a nucleic acid sensor, it is not STING, but cGAS that directly binds DNA. Once cGAS binds to cytosolic double-stranded DNA, it catalyzes the formation of the second messenger molecule cyclic GMP-AMP (cGAMP) from ATP and GTP (Gao et al., 2013; Sun et al., 2013; Wu et al., 2013). cGAMP is an endogenous high-affinity ligand for the adaptor protein STING (also known as MITA, MPYS, ERIS, and TMEM173) which allows innate activation to endogenous nucleic acids as well as following viral or bacterial infection (Ishikawa and Barber, 2011; Ishikawa et al., 2009). cGAMP binding to STING homodimers embedded within the endoplasmic reticulum triggers a conformational change in STING, activating the receptor. Activated STING translocates to punctate structures in the cytoplasm via interaction with autophagy-related proteins (Saitoh et al., 2009), where it interacts with TBK1 and forms a complex for the activation of NFκB and IRF3 (Abe and Barber, 2014; Tanaka and Chen, 2012). This phosphorylation allows NFκB and IRF3 to translocate to the nucleus and activate target genes associated with cytokine production. In this way, endogenous DNA can be sensed in the cytoplasm and cross-activate the STING-based pathway of bacterial and viral sensing. Following this cascade, STING is rapidly degraded, which prevents sustained cytokine production (Prabakaran et al., 2018). In addition, alternate STING isoforms have been described that can sequester functional STING molecules and decrease responsiveness to cGAMP (Wang et al., 2018a). In addition to STING, alternate sensors for cyclic dinucleotides have been described. ERAdP is activated in macrophages infected by cyclic dinucleotides released by *Listeria monocytogenes* and independently results in NFκB activation and production of inflammatory cytokines (Xia et al., 2018). ERAdP has specificity for CDA (cyclic di-AMP) rather than cGAMP (cyclic GMP-AMP) (Xia et al., 2018), which means it is not able to respond to endogenous DNA sensed by cGAS. Nevertheless, this remains a potent cellular sensor of infection, which plays a significant response following bacterial infection (Xia et al., 2018). RECON is another recently described sensor for cyclic dinucleotides, and at steady state negatively regulates NFκB activation (McFarland et al., 2017). Following binding of cyclic dinucleotides, its activity is suppressed, resulting in increased NFκB activity in cells (McFarland et al., 2017). RECON has a higher affinity for CDA than

STING, but does not bind cGAMP (McFarland et al., 2017), meaning that like ERAdP, it will not be influenced by cGAS-mediated endogenous DNA sensing.

Cells of the immune system, as well as epithelial cells that come in contact with invading pathogens, have evolved these nucleic acid sensing mechanisms to combat foreign material present or mis-localized within cells. Pattern recognition receptors and damage-associated molecular pattern receptors expressed by these cell types include STING, TLRs, RIG-I-like receptors, DDX41, IFI16, IFI204, DAI, and components of the inflammasome (e.g., NLRs and AIM2) that regulate secretion of proinflammatory cytokines, type I and type II interferon responses, and pyroptosis. As we have discussed in a prior review, TLR and nucleic acid sensors are not ubiquitously expressed, but immune cells, and in particular myeloid cells, are extremely well provisioned with pattern recognition receptors and damage-associated molecular pattern sensors (Baird et al., 2017b). This would suggest that direct infection of myeloid cells would be required to detect infection, which does not support the concept of non-immune cells as early direct sensors of infection. However, myeloid sensors can be sufficient in many circumstances, since for example, TLR3 need not be expressed in infected cells to initiate an immune response following viral infection (Schulz et al., 2005). Cross-presenting dendritic cells expressing TLR3 can recognize viral nucleic acids following phagocytosis of infected cell material, resulting in improved dendritic cell activation, cross presentation and expansion of antiviral T cells (Schulz et al., 2005). By contrast, STING is much more widely expressed, but still not ubiquitous. For example, we have found that STING is expressed in basal cells of the epithelium, but is lost on differentiation into squamous cells (Baird et al., 2017a). This pattern is conserved during tumorigenesis, such that HPV^+ head and neck squamous cell carcinoma, which is thought to originate from infected basal cells, has a high level of STING expression, while HPV^- head and neck squamous cell carcinoma does not express STING (Baird et al., 2017a). In the pancreas, in a small set of samples we found that pancreatic acinar cells poorly express STING, but normal ductal cells and pancreatic ductal adenocarcinoma cells both express STING (Baird et al., 2016). In colorectal carcinoma, STING is expressed in both normal colorectal cells and in most colorectal carcinoma cell lines tested, though these cells respond poorly to endogenous DNA due to lack of additional components of the sensory and signaling cascades including lack of cGAS (Xia et al., 2016a). Epigenetic modulation was able to restore functional responses to DNA in some, but not all cell types (Xia et al., 2016a). These data would suggest that when considering the capacity of cancer cells to directly respond to nucleic acids, we can expect a very divergent response according to the cell type of origin and any epigenetic regulation of critical sensing and signaling components.

In view of the critical role of nucleic acid sensing in the response to infection, it seems surprising that some of our cells appear poorly protected. In the skin, it perhaps makes sense that differentiated squamous cells lack STING, since these are generally post-replicative and undergoing increased keratinization to form the skin barrier layer. Viral infection of these cells may occur, but would be of little consequence. However, viral infection of the basal cells is required for initiation of productive infection, which depends on access to the basal layer via a wound (reviewed in Stanley (2012)). In this setting, it is logical to focus antiviral protection in the basal cell. Interestingly, Human papilloma virus (HPV) E7 has been shown to antagonize STING activation, as does Adenovirus E1A (Lau et al., 2015). These data seem to confirm a functional role for STING in critical cell types to control viral infection, but also demonstrate a mechanism by which viruses can evade direct detection.

RIG-I, MDA5 and LGP2 family members are thought to be expressed by most cells and act as cytoplasmic sensors of viral RNA (reviewed in Loo and Gale (2011)). Under normal conditions, RIG-I has a conformation that results in auto-inhibition; however, following ligand binding interaction with ubiquitin, the formation of a tetramer structure results, which functions to recruit and activate MAVS (Peisley et al., 2014). The ligand preference for RIG-I is dsRNA containing 5′ triphosphates and biphosphates (Goubau et al., 2014). These dsRNA forms are present in the nucleus of cells early following transcription, but are removed before entry into the cytoplasm. Following viral infection, these unmodified RNA structures can be present in the cytoplasm, and thereby trigger the RIG-I sensor and induce type I IFN production (Goubau et al., 2013).

An alternative sensors for cytoplasmic RNA share the helicase activities of RIG-I-like receptors and can initiate anti-viral immunity (Zhang et al., 2011). Study of this pathway demonstrated that there is overlap in the function of the nucleic acid sensors. While STING itself functions as a direct sensor for cyclic dinucleotides generated endogenously or from infectious agents (Zhang et al., 2013), it also functions as an adapter protein downstream of DDX41 (Zhang et al., 2011). Similarly STING has been shown to be critical for signaling downstream of DAI (Takaoka et al., 2007), IFI16 (Unterholzner et al., 2010), and also RIG-I (Ishikawa and Barber, 2008).

In addition, while RIG-I has a specificity for 5′ triphosphorylated (50ppp) ends, these ends are not sufficient to activate RIG-I (Schlee et al., 2009). Additional motifs in the RNA backbone are required for functional activation of RIG-I, suggesting that RIG-I can serve to integrate recognition of foreign RNA structures (Loo and Gale, 2011). These data suggest that while we have the opportunity to target pathways individually through synthetic ligands, the signaling cascade in cells may be designed to integrate complex inputs through common pathways, or distinguish patterns of infection to generate appropriate responses. Information on how these regulatory pathways work with one another may be critical to effectively initiate and control immune responses to cancer.

This information is especially relevant since to generate anti-cancer immunity we are generally seeking cytotoxic CD8 T cells differentiated in the presence of Th1-type CD4 T cell responses to recognize endogenous tumor antigens presented on MHC class I. Stating a goal such as this is critical, since our immune system integrates information about the nature of its target and generates appropriate responses. For example, current anti-viral vaccines work by generating neutralizing antibodies that function to remove free virions, not to kill infected cells. This has proved a limitation in the usefulness of vaccination once virus is resident in our cells. In the case of HPV, the HPV vaccine is an effective prophylactic, but does not influence HPV integrated into infected cells. Once inside cells, CD8 T cells are required to recognize viral antigens presented on MHC class I and kill infected targets. So the optimum vaccine to generate protective antibody responses may not be the optimum vaccine to generate cytotoxic T cell mediated immunity, just as the appropriate immune response to a parasite is not the same as that to a virus. While effective antibody responses and T cell responses can be linked (Hulett et al., 2018), the nucleic acid sensors with their cytoplasmic localization specifically tell the immune system that the infectious agents are inside cells. These signals can form part of specific response algorithms that generate immune responses that can kill infected cells. So, if our goal is to generate immune responses that can recognize cancer cells with mutated antigens, these are much more likely to be presented as peptides on MHC class I rather than form extracellular structures that can be recognized by antibodies. To generate immune responses like those that kill virally-infected cells to instead target cancer, we likely need to recapitulate in tumors the kinds of signals associated with viral infection.

3. DNA SENSING IN TUMORIGENESIS VERSUS TREATMENT

Cells of the immune system, as well as epithelial cells that come in contact with invading pathogens have evolved these nucleic acid sensing mechanisms to combat foreign material present or mislocalized within cells. As discussed, pathogen-associated molecular pattern receptors and damage-associated molecular pattern receptors expressed by these cell types include STING, TLRs, RIG-I-like receptors, HMGB1, DDX41, IFI16, IFI204, DAI, and components of the inflammasome (e.g., NLRs and AIM2) that regulate secretion of proinflammatory cytokines, type I and type II interferon responses, and pyroptosis. While these nucleic acid sensors are critical for control and clearance of infections, we now appreciate that some of these processes also regulate tumorigenesis, as well as responses to cytotoxic therapies. As we will discuss, it is possible that cancer cells have a propensity to activate nucleic acid sensing mechanisms due to aberrant regulation of DNA repair and chromosome segregation, as well as necrotic release of nucleic acids into the tumor microenvironment, thereby regulating inflammation. This inflammation is generally discussed in the context of cancer treatment, but inflammation is also an extremely important feature of tumorigenesis.

As discussed above, the STING pathway is one of the primary mechanisms by which cytoplasmic DNA is sensed, either via direct activation by cGAS-generated cyclic dinucleotides (Zhang et al., 2013), or as an adapter protein downstream of DAI (Takaoka et al., 2007), RIG-I (Ishikawa and Barber, 2008), IFI16 (Unterholzner et al., 2010), and DDX41 (Zhang et al., 2011). Activation of STING results in activation of IRF3 and production of type I interferons, as well as activation of NF-κB and production of inflammatory mediators, such as TNFα (Ishikawa and Barber, 2008). Together, these downstream mediators function in maturation of APCs, as well as priming of antitumor $CD8^+$ T cell responses (Woo et al., 2014) that underlie the efficacy of chemotherapy (Sistigu et al., 2014), radiation therapy (Burnette et al., 2011; Deng et al., 2014; Lim et al., 2014), and anti-PD-L1 monoclonal antibody immunotherapy (Wang et al., 2017). Because activation of the STING pathway is important for type I interferon production and adaptive T cell responses, STING expression has been shown to correlate to clinical outcomes for several tumor types (reviewed in He et al. (2017)). STING has been shown to be decreased or defective by various mechanisms in breast cancer (Bhatelia et al., 2014), colorectal cancer (Xia et al., 2016a), melanoma (Xia et al., 2016b), gastric cancer (Song et al., 2017), and hepatocellular carcinoma (Bu et al., 2016) and decrease in STING is associated with poor clinical outcomes. Similarly, several other nucleic acid sensors have also been shown to limit oncogenesis. Decreased RIG-I expression was observed in hepatocellular carcinoma and led to poor outcomes, including shorter survival (Hou et al., 2014). While these studies implicate loss of nucleic acid sensors as a mechanism of immune escape, there are reports that STING can also contribute to cancer-associated inflammation in the chronic setting. In the DMBA mouse model of skin carcinogenesis, DMBA treatment leads to DNA adduct formation and nucleosome leakage of DNA into the cytoplasm where it activates the STING pathway and its downstream inflammatory cytokines, including TNFα (Ahn et al., 2014). In the context of chronic inflammation in skin carcinogenesis, TNFα has been shown to be a promoting force in squamous cell carcinoma development (Schioppa et al., 2011), and loss of STING rendered mice resistant to carcinogenesis (Ahn et al., 2014). As we will discuss, chromosomal instability is a common feature of tumors that results in production of micronuclei that easily rupture, and as we will discuss, this can release DNA into the cytosol (Hatch et al., 2013), thereby activating the cGAS-STING pathway (Bakhoum et al., 2018). However, in patients with tumors that have high chromosomal instability, instead of activating type I interferon and NF-κB signaling, a proinflammatory, non-canonical NF-κB signaling pathway can be activated and result in expression of genes associated with a mesenchymal phenotype and generate increased numbers of metastases (Bakhoum et al., 2018). This contrasts with another study, where it was found that decreased STING expression correlated with gastric cancer progression and more aggressive disease (Song et al., 2017). Knockdown of STING in gastric cancer cells resulted in increased migration, invasion, and colony formation, indicating its role in suppressing metastasis (Song et al., 2017). While the results of these studies on the metastatic properties of cancer cells are contradictory, it is evident that depending on the phenotype of the impacted cell (e.g., high chromosomal instability), there can be activation of either canonical or non-canonical NF-κB signaling, resulting in differential outcomes; however, the precise mechanism by which this occurs remains to be determined. In addition, prolonged exposure to IFN can engender an intrinsic resistance of cancer cells to immune control (Benci et al., 2016); thus a chronically inflamed site may permit outgrowth of cancer through a range of mechanisms.

4. EXOGENOUS ACTIVATION BY ADMINISTRATION OF LIGANDS

Given that the aforementioned studies demonstrate that nucleic acid sensing within tumors generally acts as a tumor suppressor, emerging studies are exploiting these pathways for their therapeutic potential. Synthetic cyclic dinucleotides have recently been developed for preclinical and clinical use as potent modulators of STING for anticancer therapy. Use of these agonists has been used successfully to induce tumor regression in preclinical models of pancreatic cancer (Baird et al., 2016), HNSCC (Baird et al., 2018; Moore et al., 2016), melanoma (Corrales et al., 2015) as well as in benign papilloma (Baird et al., 2018). Each of these models required intratumoral injection of cyclic dinucleotides as systemic administration fails to induce tumor regression, which could be problematic for widely metastatic tumors or tumors that are otherwise inaccessible. Given that intratumoral injection of cyclic dinucleotides induces rapid regression of tumors, several groups are studying the effects of cyclic dinucleotide administration on minimal residual disease at time of tumor resection (Park et al., 2018). This technique could be particularly useful in instances where the tumor is near critical structures and cannot be fully removed, or instead as a neoadjuvant therapy to induce tumor regression that can render the patient eligible for potentially curative resection. STING ligands have entered the clinical arena in a Phase Ib study whereby a synthetic cyclic dinucleotide is administered with anti-CTLA4 (NCT02675439) or with anti-PD-1 (NCT03172936). Additional preclinical studies are ongoing to identify clinical compounds that synergize with STING ligands in an effort to optimize the antitumor immune response while minimizing off-target immune effects.

In considering the role of nucleic acid sensors in cancer therapy, it is important to also discuss their role as vaccine adjuvants. Adjuvants help vaccines improve the antigen-specific immune response by providing signals that encourage presentation of antigens on activated professional antigen presenting cells. These adjuvants can include damage-associated molecular patterns and/or pathogen-associated molecular patterns that activate various pattern recognition receptors of innate immune cells including TLRs, STING—either directly or via cGAS, nucleotide-binding oligomerization domain (NOD)-like receptors (NLRs), absent in melanoma 2 (AIM2)-like receptors (ALRs), retinoic acid-inducible gene I (RIG-I)-like receptors (RLRs) or C-type lectin receptors (CLRs) (Kawai and Akira, 2010). Sensing of the damage-associated molecular patterns or pathogen-associated molecular patterns, such as microbial components (e.g., microbial DNA or LPS), by innate immune cells in the context of antigenic foreign proteins or peptides initiates a cascade of immune responses resulting in the elicitation of potent innate and adaptive immune responses against the invading pathogens. Exploitation of this process to target immune responses to tumor antigens is a well-established approach to direct the immune response to clear tumors (Mocellin et al., 2004). This comes with certain caveats, as thus far cancer vaccines have not fulfilled their promise in the clinic (Rosenberg et al., 2004).

In this context, nucleic acid-sensing is an essential strategy employed by the innate immune system to detect both pathogen-derived nucleic acids and self-DNA released by host apoptotic or necrotic cells. This exploits the fact that the presence of nucleic acids that gain access to the cytoplasm is perceived by mammalian cells as "stranger" or "danger" signals that trigger immunological responses. These can be exploited to generate non-specific antiviral states in cells treated with agents that stimulate STING (Ishikawa and Barber, 2011; Ishikawa et al., 2009; Lau et al., 2015; Sali et al., 2015), RIG-I (Goubau et al., 2014; Kato et al., 2005) or TRIF (Sali et al., 2015), but these can also be applied to improve responses to co-administered cancer-associated antigens. In this way, oligonucleotide or small-molecule agonists of nucleic acid sensors are being used in clinical trials to boost the immune response against poorly immunogenic cancers, as adjuvants in therapeutic immunizations against cancer, or in prophylactic vaccines against infections (reviewed in Coffman et al. (2010); Temizoz et al. (2016)). As we will discuss, one current limitation is that despite extensive study of vaccine adjuvants, there are few clinically approved options that are designed to direct Th1-type T cell responses. The majority of clinically-used vaccine adjuvants, such as alum, are designed to promote antibody-mediated neutralizing immunity to infectious diseases (Oleszycka and Lavelle, 2014). More efficacious cancer vaccine adjuvants can be developed both by developing totally new compounds with adjuvant activities that are optimum for cytotoxic T cell immunity, or by optimizing the formulation using various combinations of existing adjuvants. There are ongoing clinical studies aiming to characterize the types of immune responses elicited by these novel adjuvants to ensure they induce the optimal kind of anti-tumor immune responses in patients.

Poly I:C is a synthetic TLR3 ligand that has been reported to function as a potent type 1 adjuvant capable of activating antigen-specific antibody, CTL and Th1 type immune responses. TLR3 is an endosomal TLR that senses RNA virus infection in the cells by recognizing intracellular dsRNA. In addition to activating the TLR3-mediated signaling pathway, Poly I:C also was found to activate intracellular RNA sensors, RIG-I and MDA5, which is associated with its toxic effects, such as induction of systemic cytokine storms (Kato et al., 2006). Recently, ARNAX, a synthetic DNA–dsRNA hybrid molecule consisting of 140bp of measles virus vaccine strain-derived dsRNA with a 5′ GpC-type phosphorothioated oligodeoxynucleotides (ODNs) that binds to TLR3 but not to RIG-I or MDA5 in mouse, has been developed to avoid the toxic effects of Poly I:C (Matsumoto et al., 2015). This specific TLR3 ligand can induce potent NK cell- and CTL-mediated anti-tumor immune responses in mouse tumor models when provided along with a model tumor antigen (Matsumoto et al., 2015). This TLR3-aduvant-tumor-associated antigen vaccine approach synergizes well with anti-PD1 monoclonal antibody therapy to treat tumors that are resistant to vaccine therapy alone (Takeda et al., 2017). Polyinosinic–polycytidylic acid stabilized with poly-L-lysine and carboxymethylcellulose (Poly ICLC, Hiltonol) is another modified Poly I:C that is stabilized for protecting it from degradation by nucleases. In mice it showed that when used as an adjuvant, Poly I:C and Poly ICLC can induce potent tumor-specific CTL, NK and NKT cell responses providing significant tumor regression and prolonged survival of tumor-bearing mice (Damo et al., 2015). Clinical trials involving patients with different types of tumors showed that repetitive intramuscular injections of Poly ICLC without any antigen have low toxicity and promote anti-tumor immune responses (Salazar et al., 2014). An alternative modified nontoxic Poly I:C analogue is Poly (I:C_{12}C) (Ampligen) developed by introducing regular mismatching bases (G and U) into Poly I:C. These modifications make it more susceptible to hydrolysis and concomitantly decreasing its toxicity (reviewed in Jasani et al. (2009)). It was shown that systemic administration (intravenous) of Poly (I:C_{12}C) was nontoxic and well tolerated in HIV^+ patients. Furthermore, clinical trials in patients with metastatic malignancies, such as renal cancer, revealed that Poly (I:C_{12}C) can boost anti-tumor immune responses to provide clinical benefit and prolonged survival to patients via mechanisms activating potent NK and T-cell responses (Jasani et al., 2009).

As with TLR3, synthetic TLR9 ligands have been used as cancer vaccine adjuvants. CpG ODN is a synthetic TLR9 ligand capable of activating the TLR9–MyD88–IRF7 signaling pathway to induce type I interferons in addition to activating the TLR9–MyD88–NF-κB signaling pathway to induce pro-inflammatory cytokine production from immune cells (reviewed in Krieg (2006)). Clinical trials evaluating the adjuvant activities of CpG ODNs demonstrated that CpG ODNs can induce Th1-type immune responses, thereby becoming potential cancer vaccine adjuvants (reviewed in Temizoz et al. (2016)). Among the different types of CpG ODNs, D type ODN (class A ODN) can potently induce type I interferon production from pDCs but fails to activate B cells for antibody production. However, because of the presence of poly-G tails, D type ODN can form aggregates, limiting its applications for clinical use. On the other hand, K type CpG ODN (class B ODN), such as K3 CpG, does not form aggregates in solution, and it is capable of potently activating B cells for antibody and IL-6 production while only weakly inducing type I interferon production from pDCs (reviewed in Krieg (2006)). Therefore, the clinically available CpG ODN is a K type ODN. Clinical trials using CpG ODN as immunotherapeutic agents in cancer patients, such as melanoma and NSCLC, suggested that combination with chemotherapy or CpG ODN monotherapy can induce potent anti-tumor immune responses that correlate with clinical benefit (reviewed in Temizoz et al. (2016)).

Similarly, anti-cancer vaccines are being built around TLR7 ligands. TLR7 recognizes viral single-strand RNA and imidazoquinoline derivatives, including resiquimod (R848) and imiquimod. Although the mode of action of imiquimod is not completely understood, induction of type I interferons via the TLR7–MyD88–IRF7 pathway is known to play a major role for mediating its anti-viral and anti-tumor activities. An imiquimod based liquid formulation, TMX-101 (Vesimune), is currently being tested in a phase II clinical trial for noninvasive bladder cancer patients as an immunotherapeutic agent (reviewed in Temizoz et al. (2016); Vacchelli et al. (2012)).

One limitation on the application of adjuvants as cancer vaccines is the necessity to identify a suitable tumor associated antigen. While patients can exhibit diverse mutations that can be included in antigenic peptides, there are few mutated tumor antigens that are shared between different individuals, particularly in view of MHC diversity. Thus, more common approaches have tested vaccines against cancer testes antigens, such as MAGE antigens, those with differential expression, such as MUC1, or targets overexpressed in cancer, such as telomerase (reviewed in Vacchelli et al. (2012)). While these have generated antigen-specific responses, clinical success has been more elusive (Butts et al., 2014; Katakami et al., 2017). One issue is that they are not truly cancer specific, and we do not know whether responses to normal tissue blunt the anti-cancer immune response. When highly effective CAR T cells for melanoma-specific antigens were transferred to patients, they resulted in toxicity due to recognition of normal melanocytes (Johnson et al., 2009). These data suggest that when the response to antigens shared by normal cells is highly effective, toxicity becomes a limiting factor. Nevertheless, there are potential cancer specific shared antigens. EGFRvIII is a constitutively active and tumor-specific deletional mutant of EGFR found in a range of cancers including glioblastoma multiforme (GBM). The deletion of *EGFR* exons 2–7 results in a novel glycine at the junction and yields a tumor-specific antigen with demonstrated immunogenicity (Choi et al., 2009). Sampson et al. demonstrated that a peptide vaccine for the EGFRVIII antigenic peptide (CDX110) was safe and generated detectable peptide specific DTH responses in approximately half of the patients (Sampson et al., 2009). This approach was also effective in phase II clinical studies (Sampson et al., 2010), but did not show efficacy in a randomized phase III study (Weller et al., 2017). One caveat here is that the vaccine approach focused on humoral responses to EGFRvIII rather than T cell responses. By contrast, vaccine strategies based around engineered *L. monocytogenes* (Brockstedt et al., 2004, 2005), which generate strong CD8 T cell-mediated responses to expressed antigens (Pamer, 2004), can generate strong protective immunity to EGFRvIII-expressing tumors in preclinical models. Preliminary patient studies with CAR T cells targeting EGFRvIII have also shown promising results (O'Rourke et al., 2017). Therefore, common shared deletional mutants such as EGFRvIII may remain promising targets for cell-mediated approaches to control tumors. With the advent of large scale sequencing, it has become possible to identify patient-specific antigenic mutations using computational pipelines, and synthesis polyepitope vaccines for each patient (reviewed in Tran et al. (2017)). This is a potential solution to develop true tumor-specific cancer vaccines and will be dependent on an appropriate combination with immune adjuvants capable of driving Th1-type cytotoxic T cell responses, such as synthetic ligands of nucleic acid sensors.

5. ENDOGENOUS NUCLEIC ACID SENSOR ACTIVATION BY HOST LIGANDS

In view of our limited ability to develop cancer vaccines that apply to many different patients due to the great variation in antigenic mutations between people, an alternative approach is to use the patients own tumor as the nexus to initiate anti-tumor immunity (Crittenden et al., 2005). As discussed above, initial immune recognition of infection is often triggered through the recognition of conserved microbial patterns including those based on nucleic acids.

Of course, under normal conditions double-stranded DNA would not be anticipated to enter the cytosol. In healthy cells DNA is restricted to the nucleus and mitochondria, which avoids inappropriate activation of the immune system. The presence of both DNA damage responses and mechanisms to detect and destroy endogenous DNA plays an important role in normal cell behavior to protect the fidelity of an organism's genetic material. Thus, failure to correct damaged DNA can result in an array of diseases. Non-homologous end joining (NHEJ) and homologous recombination (HR) both remove double stranded breaks, most commonly caused by radiation damage, replication errors and certain mutagens (reviewed in West (2003)). Single-strand break repair (SSBR) ligates nicked DNA strands whereas mismatch repair (MMR) restores errors that occur during replication (reviewed in Caldecott (2008); Jiricny (2006)). Base excision repair (BER) and nucleotide excision repair (NER) reverse oxidative base modifications caused by free radicals and other oxidative stressors (reviewed in Bauer et al. (2015); Hoeijmakers (2009)). The interruption of these processes can result in abnormal accumulation and distribution of endogenous DNA fragments that have the potential to activate endogenous nucleic acid sensors. Thus, under stressful conditions modified endogenous DNA can trigger immune responses. In response to genomic stress yielding from a wide variety of sources: Reactive Oxygen Species (ROS) (Gehrke et al., 2013), redox-cycling events, ultraviolet light (Gao et al., 2013), ionizing radiation (Harding et al., 2017; Vanpouille-Box et al., 2017), and wide variety of chemical agents along with replication stress (Mackenzie et al., 2017), genomic DNA can accumulate in the cytoplasm.

The process by which genomic DNA gains access to the innate immune sensors residing in the cytoplasm has been recently addressed by a series of papers (Bartsch et al., 2017; Harding et al., 2017; Mackenzie et al., 2017). Chromosomal damage that occurs either by a normal cellular process such as mitosis or by exogenous physical or chemical stressors can result in fragmentation of chromosomes that are separated from the main chromatin mass. These can become enveloped by the nuclear membrane to form micronuclei that are separate from the primary nucleus. cGAS localizes to micronuclei and breakdown of the micronuclei membrane provides access of cGAS to DNA, resulting in activation of IFN pathways in these cells (Bartsch et al., 2017; Mackenzie et al., 2017). Therefore, this breakdown of the micronuclei membrane is a limiting factor in activation of cytoplasmic nucleic acid sensors. Closer examination of these micronuclei reveals that they contain an incomplete nuclear lamina, a fibrillar network composed of intermediate filaments and associated proteins that collectively function to give structure and maintain the structural integrity of the nucleus (Hatch et al., 2013). Fluorescence microscopy reveals sites of weakness in the micronuclear lamina, holes where the nuclear membrane can provide access to the chromatin within (Hatch et al., 2013). Thus, cGAS accumulates specifically in micronuclei with ruptured nuclear envelopes (Mackenzie et al., 2017), and does not necessarily require entry of DNA into the cytosol. However, ensuing micronuclear collapse can result in exposed chromatin in the cytoplasm (Hatch et al., 2013). These data indicate that while the primary nucleus remains intact, micronuclei formed following DNA damage can become disrupted, providing access to cGAS and endogenous synthesis of the STING ligand cGAMP. Similar mechanisms are necessary to explain how RIG-I, an endogenous sensor of viral RNA, can be activated in the absence of infection. It is plausible that micronuclei similarly carry triphosphorylated RNA ends that are present in the nucleus of cells early following transcription to the cytoplasm to activate RIG-I. RIG-I (Ranoa et al., 2016) and LGP2 (Widau et al., 2014), but not MDA5 (Sun et al., 2006), have been shown to detect endogenous nuclear material that has translocated to the cytoplasm following radiation therapy. Recent studies demonstrated that following radiation, cells produce type I IFN in a RIG-I dependent manner (Ranoa et al., 2016). LGP2 was shown to have an opposite effect and following radiation therapy LGP2 suppressed type I IFN induction (Widau et al., 2014). Interestingly, IFN induction was shown to contribute to the death of the irradiated cells, such that mice lacking RIG-I are protected against gastrointestinal epithelial cell death following total body radiation (Ranoa et al., 2016). However, it may not be necessary for RIG-I to directly recognize nucleic acids in order to contribute to local inflammation. RIG-I can physically associate with STAT1 and prevent SHP1-mediated inactivation of STAT1 (Hou et al., 2014). In this way, RIG-I can accentuate an initial IFN signal in cells, resulting in prolonged local inflammation without requiring interaction with viral RNA (Hou et al., 2014).

Despite the potential for DNA to enter the cytosol, cells have intrinsic mechanisms to degrade cytoplasmic DNA, which also exist in part as mechanisms to control infection. These include DNase I and DNase II, which degrade DNA in the cytoplasm. However, TREX1 (DNase III) appears as a critical molecule that regulates extranuclear DNA activation of immune pathways (Lehtinen et al., 2008; Stetson et al., 2008). Mice lacking TREX1 exhibit inflammatory diseases (Morita et al., 2004) and develop lupus-like autoimmunity (Grieves et al., 2015) via increased IFN production (Stetson et al., 2008). If not degraded by DNA exonucleases, cytosolic genomic DNA will be recognized by cGAS, leading to cyclic dinucleotide production, STING activation and ultimately the activation of the immune system via type I IFN (Hartlova et al., 2015). Thus, in cells lacking TREX1, cGAS is still required for IFN induction (Ablasser et al., 2014). Importantly, TREX-mediated destruction of cytoplasmic DNA has been shown to be a limiting factor in innate immune activation following irradiation of cancer cells, and while increasing doses of radiation result in increasing DNA in the cytosol, increased TREX expression counteracts any benefit of increased dose (Vanpouille-Box et al., 2017). The quality of cytoplasmic DNA also influences the consequence to the cells, since DNA that has undergone oxidative damage limits TREX1 degradation and potentiates sensing via cGAS and STING (Gehrke et al., 2013). These data indicate that despite careful regulatory pathways to control access of DNA to the cytoplasm, under unconventional conditions, during damage, or in transformed cells, the endogenous nucleic acid sensing can be activated, resulting in activation of immune mechanisms (Fig. 2). These can lead to pathologic autoimmunity, or be exploited to engender anti-cancer immunity.

6. DOES CANCER HAVE A PARTICULAR TENDENCY TO ACTIVATE NUCLEIC ACID SENSORS?

Since cancer cells fit into a number of the categories described above that can result in endogenous activation of nucleic acid sensors, and there is evidence that these sensors can regulate tumorigenesis and progression, it begs the question as to whether these sensors are constantly active in cancer cells. However, in addition to cell-intrinsic nuclear material, cancer can have an abundant supply of extrinsic nuclear material, since areas of necrosis are frequently detectable in growing tumors (Fig. 3). Necrotic material includes a range of endogenous adjuvants with varying ability to stimulate immune responses to associated proteins (Melcher et al., 1999). There is extensive literature that correlates prognosis of cancer patients with the presence of pathological necrosis in their tumor. While there can be conflicting results (reviewed in Gkogkou et al. (2014); Richards et al. (2011)), there is a propensity that the presence of necrosis is associated with poor outcomes across a range of malignancies

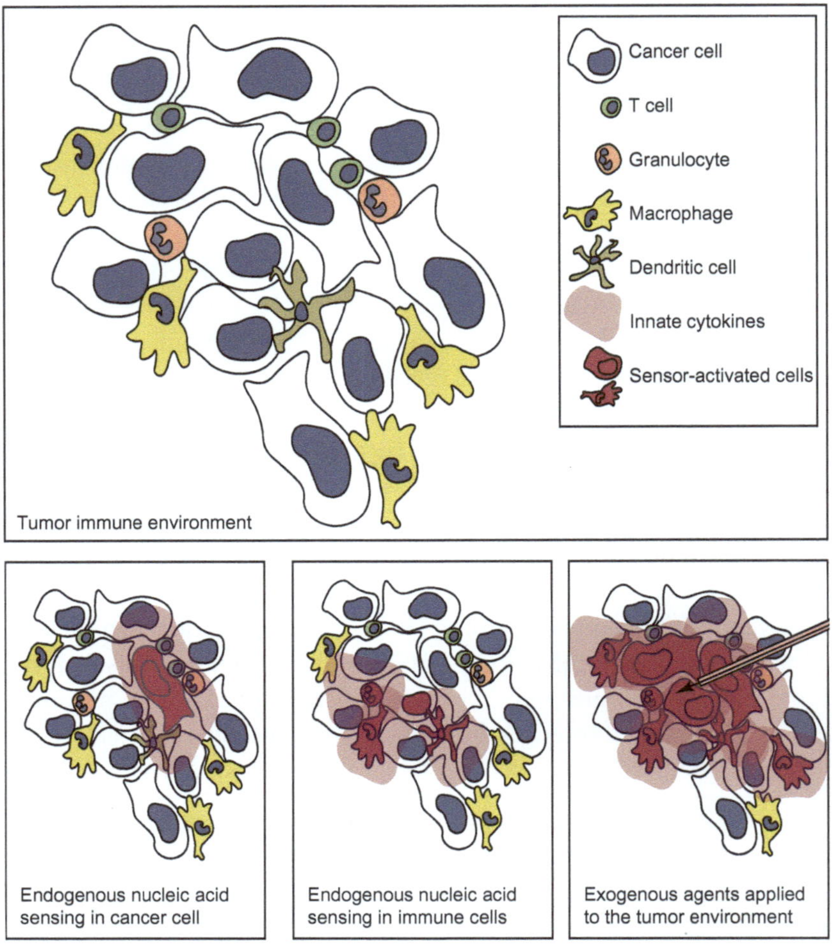

Fig. 2 The tumor can contain a heterogeneous cell population including cancer cells and immune cell types that can respond to nucleic acids. Nucleic acid sensing in cancer cells can result in inflammatory cytokine secretion to the microenvironment, or cancer cell material can be transferred to the neighboring immune cells, which can themselves sense nucleic acids and secrete inflammatory cytokines. Alternatively, exogenous administration of agents that activate nucleic acids can result in activation of multiple cell types in the tumor environment.

(Atanasov et al., 2017; Maiorano et al., 2010; Pichler et al., 2012; Pollheimer et al., 2010). If the release of cancer associated antigens and nucleic acids is a positive feature to generate anti-tumor immunity though associated activation of nucleic acid sensing pathways, these clinical data suggest the opposite. The presence of necrosis is often associated with larger

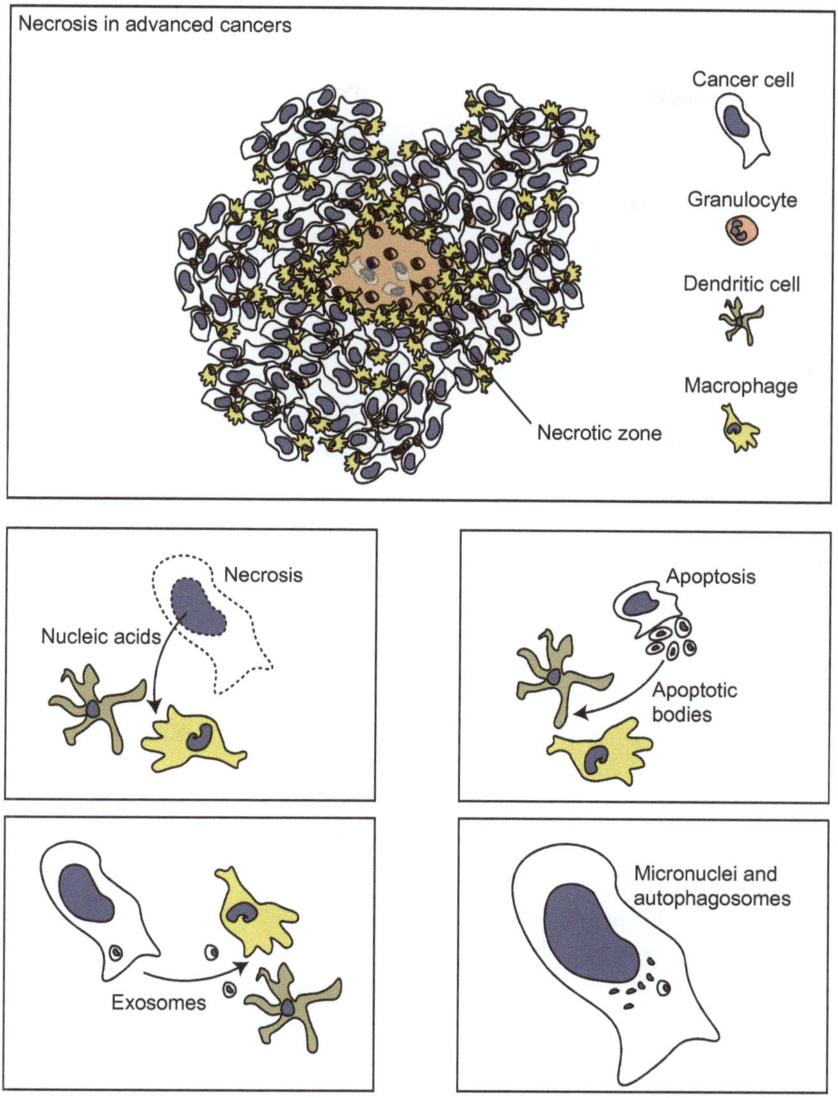

Fig. 3 Necrosis is a common feature in advanced cancers, due to aggressive growth of cancer cells outstripping the vascular supply. Necrosis can supply endogenous nucleic acids to surrounding immune cells. In addition, nucleic acids can be provided by apoptotic bodies, exosomes, or detected endogenously via micronuclei or autophagosomes.

tumors of higher grade, suggesting that these tumors have aggressively outstripped their supporting vasculature and stroma resulting in necrotic zones, but may also be sufficiently resistant to immune mechanisms so that innate immune activation either does not occur or does not impact tumor growth.

It is plausible that the response to necrotic material is limited by prior exposure of immune cells in the tumor environment to apoptotic cells. The presence of high levels of apoptotic cell death in the tumor has had mixed value as a prognostic tool to predict outcome across cancers (De Jong et al., 1999; Ghosh et al., 2001). The apoptotic index is a complicated gauge to interpret, since there are many ways to measure apoptosis. Broadly it would be assumed that more dying cancer cells are a benefit, and yet high levels of apoptosis often go along with high levels of cancer cell proliferation (Michael-Robinson et al., 2001). Therefore, as with measures of necrosis, patients with abundant dying cells are similarly not showing evidence of improved outcome, but instead associate with poor outcome. In view of these data, it is perhaps more critical to consider the cells that respond to endogenous adjuvants released from dying cancer cells, in addition to the presence or the quality of this antigen (Gough et al., 2001). While there are many phagocytes, only a subset of these can cross-present antigen to CD8 T cells. Similarly, while many cells can respond with pro-inflammatory signals to the presence of endogenous or exogenous adjuvants, phagocytes in tumors are commonly differentiated into phenotypes with a poor response to adjuvant signals. In mice, the progressive macrophage infiltration of tumors has been shown to be critical for neoangiogenesis, tumor progression and establishment of distant metastases (Lin et al., 2001, 2006) and larger macrophage infiltrates correlate with poor prognosis in patients (Leek et al., 1996). Macrophage infiltration is associated with necrotic areas in patient tumors (Leek et al., 1999) and the presence of these macrophages results in high levels of VEGF and increased microvessel density in patient tumors (Leek et al., 2000), likely as a biological attempt to overcome the hypoxia present in necrotic regions. Tumor macrophages are generally differentiated into phenotypes that can suppress T cell activation (Rodriguez et al., 2004), and treatment of these tumor macrophages with immunological adjuvants results in production of anti-inflammatory rather than pro-inflammatory cytokines (Crittenden et al., 2012). There is often a reciprocal relationship between T cell infiltration and macrophage infiltration in tumors, and where T cells are low and macrophages are high, the prognosis is especially poor (DeNardo et al., 2011; Mitchem et al., 2013). The mechanisms by which macrophages become suppressive to T cells may relate to the degree of cell death in the tumor. *In vitro*, apoptotic cells drive differentiation of macrophages into suppressive phenotypes that involve secretion of anti-inflammatory cytokines such as TGFβ and IL-10 and upregulation of suppressive molecules such as arginase I (Freire-de-Lima et al., 2006). *In vivo*, systemic administration of apoptotic cells is an efficient means to generate antigen-specific tolerance (A-Gonzalez et al., 2009) Thus, the data support a model where an aggressive tumor that outstrips its supportive structures can result in death of cancer cells, resulting in local necrosis. This, in turn, results in macrophage recruitment and following local interaction with dying cancer cells, these macrophages undergo M2 differentiation, secrete VEGF and precursors of collagen biosynthesis (Rodriguez et al., 2003, 2004), and direct revascularization and reorganization of the damaged tumor environment.

In this setting, the effect of tumor-derived material on nucleic acid sensors in the tumor environment may be unpredictable. However, nucleic acid sensors can also be activated by particles released by cells within the tumor microenvironment (Fig. 3). These particles include exosomes, microparticles/vesicles and autophagosomes that all vary in size. Exosomes are 60–100nm vesicles that originate from multivesicular bodies and fuse with the plasma membrane upon release into the extracellular space (Kurywchak et al., 2018; Zhang et al., 2015). Microparticles, on the other hand, are larger in size (100–1000nm) and are usually created in response to stress in which the parental cellular membrane encapsulates cytoplasm (Mause and Weber, 2010; Zhang et al., 2015). Autophagosomes vary in size (200–900nm) and are vesicles produced upon autophagy induction (Yi et al., 2012). Despite the differences in formation of these tumor-derived particles, they contain a variety of material from the parental cell from which they derive from, including many proteins, RNA and DNA (Kurywchak et al., 2018; Thakur et al., 2014; Yi et al., 2012). Exosomes and autophagosomes are also replete with mitochondrial nuclear acids (Medeiros et al., 2018; Thakur et al., 2014). Given the presence of RNA and DNA found within these vesicles, activation of nucleic acid sensors within recipient cells that phagocytose and engulf these particles is expected; however, the resulting outcome may lead contrastingly to tumor progression and metastasis, or antitumor immune responses that can control tumor progression and metastases.

Particles found within the tumor microenvironment have been associated with tumor progression and metastasis through nucleic acid sensing (Fig. 3). Exosomes derived from stromal cells were shown to contain RNA, which when transferred to breast cancer cells, activated RIG-I and antiviral signaling, leading to tumor progression that was resistant to therapy (Boelens et al., 2014). Nabet et al. further showed that the exosomes derived from stromal cells contained unshielded RNA, which did not have RNA-binding proteins attached as is found in the parental cells (Nabet et al., 2017). The unshielded RNA led to RIG-I activation in recipient breast cancer cells leading to inflammation that promoted metastases. In addition, if immune cells were the recipients of the exosomes with unshielded RNA, myeloid and dendritic cell subsets showed increased expression of activation markers and an inflammatory response was generated suggesting that the unshielded RNA found within the exosomes acted as a damage associated molecular pattern. Furthermore, Liu et al. showed that lung cancer-derived exosomes contain small nuclear RNAs which activate TLR3 in recipient alveolar type II cells (Liu et al., 2016). This TLR3 activation led to chemokine secretion that recruited neutrophils and promoted lung metastasis. Interestingly, only exosomal RNA, which was enriched with non-coding RNAs, and not tumor cell RNA, was responsible for the TLR3 activation (Antonopoulos et al., 2017; Liu et al., 2016). Moreover, microRNA found in tumor-derived exosomes can trigger murine TLR7 and human TLR8 in macrophages, also leading to tumor progression and metastases (Fabbri et al., 2012). Therefore, different forms of RNA found within exosomes have shown to promote tumor progression through activation of nucleic acid sensors in recipient cells, thus also promoting the use of exosomes as negative diagnostic biomarkers (Skog et al., 2008).

On the other hand, tumor-derived particles can also lead to antitumor immune responses. Kitai et al. used topotecan, a topoisomerase I inhibitor that triggers cell death, to treat breast cancer cells (Kitai et al., 2017). Topotecan treatment resulted in exosomes released from the breast cancer cells that contained DNA, which in turn activated the STING and IRF3 pathway when cultured with dendritic cells. Breast cancer-bearing mice treated with exosomes derived from topotecan-treated cells showed delayed tumor growth, which was abrogated in STING knockout mice showing a dependency on STING for antitumor immunity generated by the exosomes. In a similar manner, tumor cells exposed to ultraviolet radiation led to the release of tumor microparticles (Zhang et al., 2015). These tumor microparticles contained DNA, which when transferred to dendritic cells led to Type I IFN production through the cGAS-STING pathway. The DNA found within these tumor microparticles was of genomic and mitochondrial origin. The microparticles also contained RNA; however, Type I IFN production by dendritic cells was unaffected by RIG-I knockdown. Another pathway of IRF3 activation through exosomes was shown with exosomes derived from oral cancer cells (Wang et al., 2018b). The formation of these oral cancer-derived exosomes was not induced but purified from the supernatants of cultured oral cancer cells. These exosomes were found to be enriched with NAP1 (NF-κB-activating kinase associated protein 1), which acts upstream of IRF3. When NK cells internalized these exosomes, increased IRF3 phosphorylation and type I IFN production resulted, which further augmented NK cell cytotoxicity. In addition to exosomes and microparticles, tumor-derived autophagosomes or DRibbles (defective ribosomal products and short-lived protein-filled autophagosomes) have also shown to activate nucleic acid sensors, in particular TLR7 and TLR8, as well as TLR2, TLR4 and NOD2 (Xing et al., 2016). DRibbles are prepared by inducing autophagy and preventing lysosomal function in tumor cells resulting in the release of autophagosomes in the supernatant after mild sonication (Li et al., 2011). These resulting autophagosomes contain a wide variety of proteins and damage associated molecular patterns that induce pro-inflammatory cytokine production by immune cells (Xing et al., 2016), T cell activation (Li et al., 2008; Yi et al., 2012) and therapeutic efficacy in vaccinated tumor-bearing mice (Li et al., 2011; Su et al., 2015).

Therefore, exosomes or particles found within the tumor microenvironment, derived from either tumor cells or surrounding stromal cells, can also activate nucleic acid sensors. In this case, nucleic acid sensor activation can result in either a pro-tumor or anti-tumor response. The response generated depends on the type of cells to which the particles are transferred to as well as how the particle production is induced and released, by which immune stimulating anti-tumor properties usually result when particles are produced from cells exposed to stress (Chen et al., 2011; Thery et al., 2009).

7. ROLE OF DNA DAMAGING AGENTS IN CANCER TREATMENT

As discussed above, there is an abundance of data indicating that delivering agents that activate nucleic acid sensors in tumors can activate anticancer immunity in part by triggering nucleic acid sensing mechanisms within damaged cells. In view of these data, it would suggest that nucleic acid sensing mechanisms may contribute to immune activation following conventional cytotoxic therapies. Recent data suggest that this may be true (Harding et al., 2017; Vanpouille-Box et al., 2017). Radiation therapy was shown to activate STING in cancer cells via cGAS recognition of endogenous DNA in the cytoplasm (Harding et al., 2017; Vanpouille-Box et al., 2017). These effects were shown to be critical for immune-mediate control

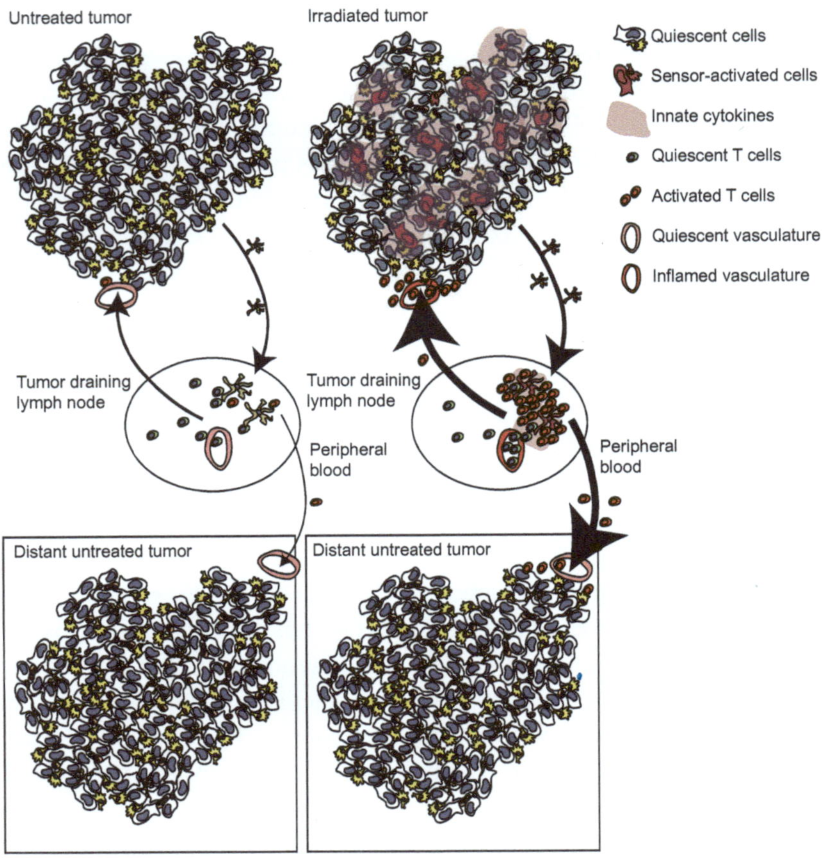

Fig. 4 Following radiation therapy, tumor associated antigens can be released to antigen-presenting cells that can travel to draining lymphatics and activate antigen-specific T cells. Activation of nucleic acid sensors can improve the function of antigen-presenting cells, permitting increased T cell expansion and improved recruitment of T cells to the draining lymphatics and back to the tumor via inflamed endothelia. Increased antigen specific T cells in the peripheral circulation can also travel to distant, untreated tumors, but their recruitment will be limited by poorly inflamed vasculature and ongoing negative regulation in the untreated site.

of residual cancer cells in the treated animals (Fig. 4). In view of these positive benefits of radiation therapy on anti-tumor immune responses it is puzzling that the overwhelming experience of treating millions of cancer patients over the course of decades is that radiation very very rarely results in immune control of untreated tumors (Abuodeh et al., 2015). The combination of radiation therapy with other immunotherapies has changed that situation, with multiple case studies (Postow et al., 2012; Stamell et al., 2012) or early phase clinical trials (Golden et al., 2015; Seung et al., 2012) demonstrating immune-mediated control of distant disease following radiation therapy of a limited subset of tumors. These data suggest that additional immunological interventions are needed to translate radiation-mediated activation of nucleic acid sensors in the tumor environment into distant responses. In addition, the contribution of nucleic acid sensors to tumor control varies according to the specific features of the tumor. In highly immunogenic tumors that are spontaneously rejected in mice with a competent immune system, the loss of STING in host dendritic cells allows these tumors to grow (Woo et al., 2014). In immunogenic tumors that grow in immune competent mice, but can be effectively cured by radiation therapy alone, endogenous activation of STING via the activity of cGAS in antigen-presenting cells is required for radiation-mediated local control (Deng et al., 2014). In our experience using poorly immunogenic tumors, the presence or absence of STING in the host does not affect tumor growth, but this could be substituted by local application of exogenous synthetic STING ligands in combination with radiation therapy (Baird et al., 2016). These data suggest that the immunogenicity of tumors—their ability to be rejected by immune mechanisms—is linked to endogenous nucleic acid sensing. However, with poorly immunogenic tumors the combination of exogenously applied STING ligands combined with radiation therapy had a limited ability to control distant tumors despite an increase in systemically circulating tumor antigen-specific T cells (Baird et al., 2016). This is consistent with the preclinical data showing that radiation therapy can activate cGAS-mediated activation of STING directly in irradiated cancer cells, since in this setting radiation therapy did not affect distant tumors without the addition of anti-CTLA4 monoclonal antibodies (Harding et al., 2017; Vanpouille-Box et al., 2017). Even without radiation therapy, the efficacy of checkpoint inhibitors has been shown to be dependent on nucleic acid sensing machinery for optimal tumor control (Wang et al., 2017). These data suggest that while nucleic acid sensing machinery can contribute to local control of tumors by radiation therapy, effective control of distant disease remains susceptible to the same checkpoint regulations that suppress adaptive immunity in tumors (Fig. 4). However, the critical event here is that local activation of nucleic acid sensing machinery by radiation therapy improves the generation of antigen specific T cells, and without these, checkpoint inhibitors may have no effectors to assist.

This dependence of checkpoint inhibitors on pre-existing immunity or immune responses generated by radiation therapy makes sense, since they act to block suppressive mechanisms rather than to generate *de novo* T cells. Consistent with this, we recently demonstrated that where pre-existing immunity is blocked, the combination of radiation therapy and checkpoint inhibitors is no longer able to control tumors (Crittenden et al., 2018). These data match prior studies using implantation of tumors as fragments rather than cell suspensions, which is an approach that can avoid implantation-mediated immunity to the tumor (Ochsenbein et al., 1999, 2001). Where tumors are implanted as fragments, radiation therapy plus checkpoint inhibition first required tumor antigen-specific vaccination before the combination was able to fully control of tumors (Zheng et al., 2016). In studies using tumors with lower levels of antigen that do not generate sufficient pre-existing immunity to allow tumor control, additional tumor antigen-specific vaccination was necessary for tumor control by radiation therapy (Hodge et al., 2012; Kudo-Saito et al., 2005). Therefore, nucleic acid sensing may be essential to initiate new immune responses as part of vaccines, as discussed above, or to propagate responses via IFN induction in irradiated cancer cells; however, it remains likely that additional interventions will be required to permit full immune-mediated control of residual cancer cells, and particularly where those residual cancer cells are outside the treatment field.

8. CONCLUSIONS

Nucleic acid sensors represent a particularly potent group of molecular pathways capable of initiating anti-bacterial and anti-viral immunity both via innate mechanisms as well as via induction of cell-mediated immunity. These pathways can also play a role in tumorigenesis, but triggering these pathways in advanced cancers can generate protective anti-tumor immune responses. There is a long history of combining innate adjuvants as part of anti-cancer vaccines (reviewed in Coffman et al. (2010); Temizoz et al. (2016)) or with cytotoxic therapies (reviewed in Baird et al. (2017b)); however, it has recently become clear that many of our conventional cancer therapies themselves trigger these innate immune mechanisms, and that this plays a role in their success. Understanding how best to manipulate these pathways to optimize local control for tumors, and amplify systemic control of metastatic disease, will likely require an improved understanding of the context of immune activation in the tumor environment, as well as the key cells that control positive and negative outcomes following combination therapies. In view of the key role of nucleic acid sensors in directing cell-mediated immunity, these likely remain potent targets for cancer therapy.

FUNDING

This work was funded by NCI R01CA182311 (MJG), and an American Cancer Society Postdoctoral Fellowship award (J.R.B.).

REFERENCES

Abe, T., Barber, G.N., 2014. Cytosolic-DNA-mediated, STING-dependent proinflammatory gene induction necessitates canonical NF-κB activation through TBK1. J. Virol. 88, 5328–5341.

Ablasser, A., Hemmerling, I., Schmid-Burgk, J.L., Behrendt, R., Roers, A., Hornung, V., 2014. TREX1 deficiency triggers cell-autonomous immunity in a cGAS-dependent manner. J. Immunol. 192, 5993–5997.

Abuodeh, Y., Venkat, P., Kim, S., 2015. Systematic review of case reports on the abscopal effect. Curr. Probl. Cancer 40, 25–37.

A-Gonzalez, N., Bensinger, S.J., Hong, C., Beceiro, S., Bradley, M.N., Zelcer, N., Deniz, J., Ramirez, C., Diaz, M., Gallardo, G., De Galarreta, C.R., Salazar, J., Lopez, F., Edwards, P., Parks, J., Andujar, M., Tontonoz, P., Castrillo, A., 2009. Apoptotic cells promote their own clearance and immune tolerance through activation of the nuclear receptor LXR. Immunity 31, 245–258.

Ahn, J., Xia, T., Konno, H., Konno, K., Ruiz, P., Barber, G.N., 2014. Inflammation-driven carcinogenesis is mediated through STING. Nat. Commun. 5, 5166.

Antonopoulos, D., Balatsos, N.A.A., Gourgoulianis, K.I., 2017. Cancer's smart bombs: tumor-derived exosomes target lung epithelial cells triggering pre-metastatic niche formation. J. Thorac. Dis. 9, 969–972.

Atanasov, G., Dietel, C., Feldbrugge, L., Benzing, C., Krenzien, F., Brandl, A., Mann, E., Englisch, J.P., Schierle, K., Robson, S.C., Splith, K., Morgul, M.H., Reutzel-Selke, A., Jonas, S., Pascher, A., Bahra, M., Pratschke, J., Schmelzle, M., 2017. Tumor necrosis and infiltrating macrophages predict survival after curative resection for cholangiocarcinoma. Oncoimmunology 6, e1331806.

Baird, J.R., Friedman, D., Cottam, B., Dubensky Jr., T.W., Kanne, D.B., Bambina, S., Bahjat, K., Crittenden, M.R., Gough, M.J., 2016. Radiotherapy combined with novel STING-targeting oligonucleotides results in regression of established tumors. Cancer Res. 76, 50–61.

Baird, J.R., Feng, Z., Xiao, H.D., Friedman, D., Cottam, B., Fox, B.A., Kramer, G., Leidner, R.S., Bell, R.B., Young, K.H., Crittenden, M.R., Gough, M.J., 2017a. STING expression and response to treatment with STING ligands in premalignant and malignant disease. PLoS One 12, e0187532.

Baird, J.R., Monjazeb, A.M., Shah, O., McGee, H., Murphy, W.J., Crittenden, M.R., Gough, M.J., 2017b. Stimulating innate immunity to enhance radiation therapy-induced tumor control. Int. J. Radiat. Oncol. Biol. Phys. 99, 362–373.

Baird, J.R., Feng, Z., Xiao, H.D., Friedman, D., Cottam, B., Fox, B.A., Kramer, G., Leidner, R.S., Bell, R.B., Young, K.H., Crittenden, M.R., Gough, M.J., 2018. Correction: STING expression and response to treatment with STING ligands in premalignant and malignant disease. PLoS One 13, e0192988.

Bakhoum, S.F., Ngo, B., Laughney, A.M., Cavallo, J.A., Murphy, C.J., Ly, P., Shah, P., Sriram, R.K., Watkins, T.B.K., Taunk, N.K., Duran, M., Pauli, C., Shaw, C., Chadalavada, K., Rajasekhar, V.K., Genovese, G., Venkatesan, S., Birkbak, N.J., McGranahan, N., Lundquist, M., Laplant, Q., Healey, J.H., Elemento, O., Chung, C.H., Lee, N.Y., Imielenski, M., Nanjangud, G., Pe'er, D., Cleveland, D.W., Powell, S.N., Lammerding, J., Swanton, C., Cantley, L.C., 2018. Chromosomal instability drives metastasis through a cytosolic DNA response. Nature 553, 467–472.

Bartsch, K., Knittler, K., Borowski, C., Rudnik, S., Damme, M., Aden, K., Spehlmann, M.E., Frey, N., Saftig, P., Chalaris, A., Rabe, B., 2017. Absence of RNase H2 triggers generation of immunogenic micronuclei removed by autophagy. Hum. Mol. Genet. 26, 3960–3972.

Bauer, N.C., Corbett, A.H., Doetsch, P.W., 2015. The current state of eukaryotic DNA base damage and repair. Nucleic Acids Res. 43, 10083–10101.

Benci, J.L., Xu, B., Qiu, Y., Wu, T.J., Dada, H., Twyman-Saint Victor, C., Cucolo, L., Lee, D.S.M., Pauken, K.E., Huang, A.C., Gangadhar, T.C., Amaravadi, R.K., Schuchter, L.M., Feldman, M.D., Ishwaran, H., Vonderheide, R.H., Maity, A., Wherry, E.J., Minn, A.J., 2016. Tumor interferon signaling regulates a multigenic resistance program to immune checkpoint blockade. Cell 167, 1540–1554.e12.

Bhatelia, K., Singh, A., Tomar, D., Singh, K., Sripada, L., Chagtoo, M., Prajapati, P., Singh, R., Godbole, M.M., Singh, R.;, 2014. Antiviral signaling protein MITA acts as a tumor suppressor in breast cancer by regulating NF-kappaB induced cell death. Biochim. Biophys. Acta 1842, 144–153.

Blasius, A.L., Beutler, B., 2010. Intracellular toll-like receptors. Immunity 32, 305–315.

Boelens, M.C., Wu, T.J., Nabet, B.Y., Xu, B., Qiu, Y., Yoon, T., Azzam, D.J., Twyman-Saint Victor, C., Wiemann, B.Z., Ishwaran, H., Ter Brugge, P.J., Jonkers, J., Slingerland, J., Minn, A.J., 2014. Exosome transfer from stromal to breast cancer cells regulates therapy resistance pathways. Cell 159, 499–513.

Brockstedt, D.G., Giedlin, M.A., Leong, M.L., Bahjat, K.S., Gao, Y., Luckett, W., Liu, W., Cook, D.N., Portnoy, D.A., Dubensky Jr., T.W., 2004. Listeria-based cancer vaccines that segregate immunogenicity from toxicity. Proc. Natl. Acad. Sci. U. S. A. 101, 13832–13837.

Brockstedt, D.G., Bahjat, K.S., Giedlin, M.A., Liu, W., Leong, M., Luckett, W., Gao, Y., Schnupf, P., Kapadia, D., Castro, G., Lim, J.Y., Sampson-Johannes, A., Herskovits, A.A., Stassinopoulos, A., Bouwer, H.G., Hearst, J.E., Portnoy, D.A., Cook, D.N., Dubensky Jr., T.W., 2005. Killed but metabolically active microbes: a new vaccine paradigm for eliciting effector T-cell responses and protective immunity. Nat. Med. 11, 853–860.

Bu, Y., Liu, F., Jia, Q.A., Yu, S.N., 2016. Decreased expression of TMEM173 predicts poor prognosis in patients with hepatocellular carcinoma. PLoS One 11, e0165681.

Burnette, B.C., Liang, H., Lee, Y., Chlewicki, L., Khodarev, N.N., Weichselbaum, R.R., Fu, Y.X., Auh, S.L., 2011. The efficacy of radiotherapy relies upon induction of type i interferon-dependent innate and adaptive immunity. Cancer Res. 71, 2488–2496.

Butts, C., Socinski, M.A., Mitchell, P.L., Thatcher, N., Havel, L., Krzakowski, M., Nawrocki, S., Ciuleanu, T.E., Bosquee, L., Trigo, J.M., Spira, A., Tremblay, L., Nyman, J., Ramlau, R., Wickart-Johansson, G., Ellis, P., Gladkov, O., Pereira, J.R., Eberhardt, W.E., Helwig, C., Schroder, A., Shepherd, F.A., 2014. Tecemotide (L-BLP25) versus placebo after chemoradiotherapy for stage III non-small-cell lung cancer (START): a randomised, double-blind, phase 3 trial. Lancet Oncol. 15, 59–68.

Caldecott, K.W., 2008. Single-strand break repair and genetic disease. Nat. Rev. Genet. 9, 619.

Chen, T., Guo, J., Yang, M., Zhu, X., Cao, X., 2011. Chemokine-containing exosomes are released from heat-stressed tumor cells via lipid raft-dependent pathway and act as efficient tumor vaccine. J. Immunol. 186, 2219–2228.

Choi, B.D., Archer, G.E., Mitchell, D.A., Heimberger, A.B., McLendon, R.E., Bigner, D.D., Sampson, J.H., 2009. EGFRvIII-targeted vaccination therapy of malignant glioma. Brain Pathol. 19, 713–723.

Coffman, R.L., Sher, A., Seder, R.A., 2010. Vaccine adjuvants: putting innate immunity to work. Immunity 33, 492–503.

Corrales, L., Glickman, L.H., McWhirter, S.M., Kanne, D.B., Sivick, K.E., Katibah, G.E., Woo, S.R., Lemmens, E., Banda, T., Leong, J.J., Metchette, K., Dubensky Jr., T.W., Gajewski, T.F., 2015. Direct activation of STING in the tumor microenvironment leads to potent and systemic tumor regression and immunity. Cell Rep. 11, 1018–1030.

Crittenden, M.R., Thanarajasingam, U., Vile, R.G., Gough, M.J., 2005. Intratumoral immunotherapy: using the tumour against itself. Immunology 114, 11–22.

Crittenden, M.R., Cottam, B., Savage, T., Nguyen, C., Newell, P., Gough, M.J., 2012. Expression of NF-kappaB p50 in tumor stroma limits the control of tumors by radiation therapy. PLoS One 7, e39295.

Crittenden, M., Zebertavage, L., Kramer, G., Bambina, S., Friedman, D., Troesch, V., Blair, T., Baird, J., Alice, A., Gough, M., 2018. Tumor cure by radiation therapy and checkpoint inhibitors depends on pre-existing immunity. Sci. Rep. 8, 7012.

Damo, M., Wilson, D.S., Simeoni, E., Hubbell, J.A., 2015. TLR-3 stimulation improves anti-tumor immunity elicited by dendritic cell exosome-based vaccines in a murine model of melanoma. Sci. Rep. 5, 17622.

De Jong, J.S., Diest, P.J.V., Baak, J.P.A., 1999. Number of apoptotic cells as a prognostic marker in invasive breast cancer. Br. J. Cancer 82, 368.

DeNardo, D.G., Andreu, P., Coussens, L.M., 2010. Interactions between lymphocytes and myeloid cells regulate pro- versus anti-tumor immunity. Cancer Metastasis Rev. 29, 309–316.

DeNardo, D.G., Brennan, D.J., Rexhepaj, E., Ruffell, B., Shiao, S.L., Madden, S.F., Gallagher, W.M., Wadhwani, N., Keil, S.D., Junaid, S.A., Rugo, H.S., Hwang, E.S., Jirstrom, K., West, B.L., Coussens, L.M., 2011. Leukocyte complexity predicts breast cancer survival and functionally regulates response to chemotherapy. Cancer Discov. 1, 54–67.

Deng, L., Liang, H., Xu, M., Yang, X., Burnette, B., Arina, A., Li, X.D., Mauceri, H., Beckett, M., Darga, T., Huang, X., Gajewski, T.F., Chen, Z.J., Fu, Y.X., Weichselbaum, R.R., 2014. STING-dependent cytosolic DNA sensing promotes radiation-induced Type I interferon-dependent antitumor immunity in immunogenic tumors. Immunity 41, 843–852.

Fabbri, M., Paone, A., Calore, F., Galli, R., Gaudio, E., Santhanam, R., Lovat, F., Fadda, P., Mao, C., Nuovo, G.J., Zanesi, N., Crawford, M., Ozer, G.H., Wernicke, D., Alder, H., Caligiuri, M.A., Nana-Sinkam, P., Perrotti, D., Croce, C.M., 2012. MicroRNAs bind to Toll-like receptors to induce prometastatic inflammatory response. Proc. Natl. Acad. Sci. U. S. A. 109, E2110–E2116.

Fearon, E.R., Vogelstein, B., 1990. A genetic model for colorectal tumorigenesis. Cell 61, 759–767.

Freire-de-Lima, C.G., Xiao, Y.Q., Gardai, S.J., Bratton, D.L., Schiemann, W.P., Henson, P.M., 2006. Apoptotic cells, through transforming growth factor-beta, coordinately induce anti-inflammatory and suppress pro-inflammatory eicosanoid and NO synthesis in murine macrophages. J. Biol. Chem. 281, 38376–38384.

Gao, P., Ascano, M., Wu, Y., Barchet, W., Gaffney, B.L., Zillinger, T., Serganov, A.A., Liu, Y., Jones, R.A., Hartmann, G., Tuschl, T., Patel, D.J., 2013. Cyclic [G(2',5')pA (3',5')p] is the metazoan second messenger produced by DNA-activated cyclic GMP-AMP synthase. Cell 153, 1094–1107.

Gehrke, N., Mertens, C., Zillinger, T., Wenzel, J., Bald, T., Zahn, S., Tuting, T., Hartmann, G., Barchet, W., 2013. Oxidative damage of DNA confers resistance to cytosolic nuclease TREX1 degradation and potentiates STING-dependent immune sensing. Immunity 39, 482–495.

Ghosh, M., Crocker, J., Morris, A., 2001. Apoptosis in squamous cell carcinoma of the lung: correlation with survival and clinicopathological features. J. Clin. Pathol. 54, 111–115.

Gkogkou, C., Frangia, K., Saif, M.W., Trigidou, R., Syrigos, K., 2014. Necrosis and apoptotic index as prognostic factors in non-small cell lung carcinoma: a review. Springerplus 3, 120.

Golden, E.B., Chhabra, A., Chachoua, A., Adams, S., Donach, M., Fenton-Kerimian, M., Friedman, K., Ponzo, F., Babb, J.S., Goldberg, J., Demaria, S., Formenti, S.C., 2015. Local radiotherapy and granulocyte-macrophage colony-stimulating factor to generate abscopal responses in patients with metastatic solid tumours: a proof-of-principle trial. Lancet Oncol. 16, 795–803.

Goubau, D., Deddouche, S., Reis e Sousa, C., 2013. Cytosolic sensing of viruses. Immunity 38, 855–869.

Goubau, D., Schlee, M., Deddouche, S., Pruijssers, A.J., Zillinger, T., Goldeck, M., Schuberth, C., Van der Veen, A.G., Fujimura, T., Rehwinkel, J., Iskarpatyoti, J.A., Barchet, W., Ludwig, J., Dermody, T.S., Hartmann, G., Reis e Sousa, C., 2014. Antiviral immunity via RIG-I-mediated recognition of RNA bearing 5'-diphosphates. Nature 514, 372–375.

Gough, M.J., Melcher, A.A., Ahmed, A., Crittenden, M.R., Riddle, D.S., Linardakis, E., Ruchatz, A.N., Emiliusen, L.M., Vile, R.G., 2001. Macrophages orchestrate the immune response to tumor cell death. Cancer Res. 61, 7240–7247.

Grieves, J.L., Fye, J.M., Harvey, S., Grayson, J.M., Hollis, T., Perrino, F.W., 2015. Exonuclease TREX1 degrades double-stranded DNA to prevent spontaneous lupus-like inflammatory disease. Proc. Natl. Acad. Sci. U. S. A. 112, 5117–5122.

Häcker, H., Mischak, H., Miethke, T., Liptay, S., Schmid, R., Sparwasser, T., Heeg, K., Lipford, G.B., Wagner, H., 1998. CpG-DNA-specific activation of antigen-presenting cells requires stress kinase activity and is preceded by non-specific endocytosis and endosomal maturation. EMBO J. 17, 6230–6240.

Hanahan, D., Weinberg, R.A., 2011. Hallmarks of cancer: the next generation. Cell 144, 646–674.

Harding, S.M., Benci, J.L., Irianto, J., Discher, D.E., Minn, A.J., Greenberg, R.A., 2017. Mitotic progression following DNA damage enables pattern recognition within micronuclei. Nature 548, 466–470.

Hartlova, A., Erttmann, S.F., Raffi, F.A., Schmalz, A.M., Resch, U., Anugula, S., Lienenklaus, S., Nilsson, L.M., Kroger, A., Nilsson, J.A., Ek, T., Weiss, S., Gekara, N.O., 2015. DNA damage primes the type I interferon system via the cytosolic DNA sensor STING to promote anti-microbial innate immunity. Immunity 42, 332–343.

Hatch, E.M., Fischer, A.H., Deerinck, T.J., Hetzer, M.W., 2013. Catastrophic nuclear envelope collapse in cancer cell micronuclei. Cell 154, 47–60.

He, L., Xiao, X., Yang, X., Zhang, Z., Wu, L., Liu, Z., 2017. STING signaling in tumorigenesis and cancer therapy: a friend or foe? Cancer Lett. 402, 203–212.

Hodge, J.W., Sharp, H.J., Gameiro, S.R., 2012. Abscopal regression of antigen disparate tumors by antigen cascade after systemic tumor vaccination in combination with local tumor radiation. Cancer Biother. Radiopharm. 27, 12–22.

Hoebe, K., Du, X., Georgel, P., Janssen, E., Tabeta, K., Kim, S.O., Goode, J., Lin, P., Mann, N., Mudd, S., Crozat, K., Sovath, S., Han, J., Beutler, B., 2003. Identification of Lps2 as a key transducer of MyD88-independent TIR signalling. Nature 424, 743.

Hoeijmakers, J.H., 2009. DNA damage, aging, and cancer. N. Engl. J. Med. 361, 1475–1485.

Honda, K., Takaoka, A., Taniguchi, T., 2006. Type I interferon [corrected] gene induction by the interferon regulatory factor family of transcription factors. Immunity 25, 349–360.

Hou, J., Zhou, Y., Zheng, Y., Fan, J., Zhou, W., Ng, I.O., Sun, H., Qin, L., Qiu, S., Lee, J.M., Lo, C.M., Man, K., Yang, Y., Yang, Y., Yang, Y., Zhang, Q., Zhu, X., Li, N., Wang, Z., Ding, G., Zhuang, S.M., Zheng, L., Luo, X., Xie, Y., Liang, A., Wang, Z., Zhang, M., Xia, Q., Liang, T., Yu, Y., Cao, X., 2014. Hepatic RIG-I

predicts survival and interferon-alpha therapeutic response in hepatocellular carcinoma. Cancer Cell 25, 49–63.
Hulett, T.W., Jensen, S.M., Wilmarth, P.A., Reddy, A.P., Ballesteros-Merino, C., Afentoulis, M.E., Dubay, C., David, L.L., Fox, B.A., 2018. Coordinated responses to individual tumor antigens by IgG antibody and CD8+ T cells following cancer vaccination. J. Immunother. Cancer 6, 27.
Isaacs, A., Cox, R.A., Rotem, Z., 1963. Foreign nucleic acids as the stimulus to make interferon. Lancet 2, 113–116.
Ishikawa, H., Barber, G.N., 2008. STING is an endoplasmic reticulum adaptor that facilitates innate immune signalling. Nature 455, 674–678.
Ishikawa, H., Barber, G.N., 2011. The STING pathway and regulation of innate immune signaling in response to DNA pathogens. Cell. Mol. Life Sci. 68, 1157–1165.
Ishikawa, H., Ma, Z., Barber, G.N., 2009. STING regulates intracellular DNA-mediated, type I interferon-dependent innate immunity. Nature 461, 788–792.
Iwasaki, A., Medzhitov, R., 2004. Toll-like receptor control of the adaptive immune responses. Nat. Immunol. 5, 987–995.
Jasani, B., Navabi, H., Adams, M., 2009. Ampligen: a potential toll-like 3 receptor adjuvant for immunotherapy of cancer. Vaccine 27, 3401–3404.
Jiricny, J., 2006. The multifaceted mismatch-repair system. Nat. Rev. Mol. Cell Biol. 7, 335–346.
Johnson, L.A., Morgan, R.A., Dudley, M.E., Cassard, L., Yang, J.C., Hughes, M.S., Kammula, U.S., Royal, R.E., Sherry, R.M., Wunderlich, J.R., Lee, C.C., Restifo, N.P., Schwarz, S.L., Cogdill, A.P., Bishop, R.J., Kim, H., Brewer, C.C., Rudy, S.F., Vanwaes, C., Davis, J.L., Mathur, A., Ripley, R.T., Nathan, D.A., Laurencot, C.M., Rosenberg, S.A., 2009. Gene therapy with human and mouse T-cell receptors mediates cancer regression and targets normal tissues expressing cognate antigen. Blood 114, 535–546.
Kagan, J.C., Su, T., Horng, T., Chow, A., Akira, S., Medzhitov, R., 2008. TRAM couples endocytosis of Toll-like receptor 4 to the induction of interferon-β. Nat. Immunol. 9, 361.
Katakami, N., Hida, T., Nokihara, H., Imamura, F., Sakai, H., Atagi, S., Nishio, M., Kashii, T., Satouchi, M., Helwig, C., Watanabe, M., Tamura, T., 2017. Phase I/II study of tecemotide as immunotherapy in Japanese patients with unresectable stage III non-small cell lung cancer. Lung Cancer 105, 23–30.
Kato, H., Sato, S., Yoneyama, M., Yamamoto, M., Uematsu, S., Matsui, K., Tsujimura, T., Takeda, K., Fujita, T., Takeuchi, O., Akira, S., 2005. Cell type-specific involvement of RIG-I in antiviral response. Immunity 23, 19–28.
Kato, H., Takeuchi, O., Sato, S., Yoneyama, M., Yamamoto, M., Matsui, K., Uematsu, S., Jung, A., Kawai, T., Ishii, K.J., Yamaguchi, O., Otsu, K., Tsujimura, T., Koh, C.-S., Reis e Sousa, C., Matsuura, Y., Fujita, T., Akira, S., 2006. Differential roles of MDA5 and RIG-I helicases in the recognition of RNA viruses. Nature 441, 101.
Kawai, T., Akira, S., 2010. The role of pattern-recognition receptors in innate immunity: update on Toll-like receptors. Nat. Immunol. 11, 373–384.
Kawai, T., Takahashi, K., Sato, S., Coban, C., Kumar, H., Kato, H., Ishii, K.J., Takeuchi, O., Akira, S., 2005. IPS-1, an adaptor triggering RIG-I- and Mda5-mediated type I interferon induction. Nat. Immunol. 6, 981–988.
Kawasaki, T., Kawai, T., 2014. Toll-like receptor signaling pathways. Front. Immunol. 5, 461.
Kitai, Y., Kawasaki, T., Sueyoshi, T., Kobiyama, K., Ishii, K.J., Zou, J., Akira, S., Matsuda, T., Kawai, T., 2017. DNA-containing exosomes derived from cancer cells treated with topotecan activate a STING-dependent pathway and reinforce antitumor immunity. J. Immunol. 198, 1649–1659.

Krieg, A.M., 2006. Therapeutic potential of Toll-like receptor 9 activation. Nat. Rev. Drug Discov. 5, 471–484.

Kudo-Saito, C., Schlom, J., Camphausen, K., Coleman, C.N., Hodge, J.W., 2005. The requirement of multimodal therapy (vaccine, local tumor radiation, and reduction of suppressor cells) to eliminate established tumors. Clin. Cancer Res. 11, 4533–4544.

Kurywchak, P., Tavormina, J., Kalluri, R., 2018. The emerging roles of exosomes in the modulation of immune responses in cancer. Genome Med. 10, 23.

Lau, L., Gray, E.E., Brunette, R.L., Stetson, D.B., 2015. DNA tumor virus oncogenes antagonize the cGAS-STING DNA-sensing pathway. Science 350, 568–571.

Leek, R.D., Lewis, C.E., Whitehouse, R., Greenall, M., Clarke, J., Harris, A.L., 1996. Association of macrophage infiltration with angiogenesis and prognosis in invasive breast carcinoma. Cancer Res. 56, 4625–4629.

Leek, R.D., Landers, R.J., Harris, A.L., Lewis, C.E., 1999. Necrosis correlates with high vascular density and focal macrophage infiltration in invasive carcinoma of the breast. Br. J. Cancer 79, 991–995.

Leek, R.D., Hunt, N.C., Landers, R.J., Lewis, C.E., Royds, J.A., Harris, A.L., 2000. Macrophage infiltration is associated with VEGF and EGFR expression in breast cancer. J. Pathol. 190, 430–436.

Lehtinen, D.A., Harvey, S., Mulcahy, M.J., Hollis, T., Perrino, F.W., 2008. The TREX1 double-stranded DNA degradation activity is defective in dominant mutations associated with autoimmune disease. J. Biol. Chem. 283, 31649–31656.

Li, Y., Wang, L.X., Yang, G., Hao, F., Urba, W.J., Hu, H.M., 2008. Efficient cross-presentation depends on autophagy in tumor cells. Cancer Res. 68, 6889–6895.

Li, Y., Wang, L.X., Pang, P., Cui, Z., Aung, S., Haley, D., Fox, B.A., Urba, W.J., Hu, H.M., 2011. Tumor-derived autophagosome vaccine: mechanism of cross-presentation and therapeutic efficacy. Clin. Cancer Res. 17, 7047–7057.

Liblau, R.S., Tisch, R., Shokat, K., Yang, X., Dumont, N., Goodnow, C.C., Mcdevitt, H.O., 1996. Intravenous injection of soluble antigen induces thymic and peripheral T-cells apoptosis. Proc. Natl. Acad. Sci. U. S. A. 93, 3031–3036.

Lim, J.Y., Gerber, S.A., Murphy, S.P., Lord, E.M., 2014. Type I interferons induced by radiation therapy mediate recruitment and effector function of CD8(+) T cells. Cancer Immunol. Immunother. 63, 259–271.

Lin, E.Y., Nguyen, A.V., Russell, R.G., Pollard, J.W., 2001. Colony-stimulating factor 1 promotes progression of mammary tumors to malignancy. J. Exp. Med. 193, 727–740.

Lin, E.Y., Li, J.F., Gnatovskiy, L., Deng, Y., Zhu, L., Grzesik, D.A., Qian, H., Xue, X.N., Pollard, J.W., 2006. Macrophages regulate the angiogenic switch in a mouse model of breast cancer. Cancer Res. 66, 11238–11246.

Liu, S., Cai, X., Wu, J., Cong, Q., Chen, X., Li, T., Du, F., Ren, J., Wu, Y.T., Grishin, N.V., Chen, Z.J., 2015. Phosphorylation of innate immune adaptor proteins MAVS, STING, and TRIF induces IRF3 activation. Science 347, aaa2630.

Liu, Y., Gu, Y., Han, Y., Zhang, Q., Jiang, Z., Zhang, X., Huang, B., Xu, X., Zheng, J., Cao, X., 2016. Tumor exosomal RNAs promote lung pre-metastatic niche formation by activating alveolar epithelial TLR3 to recruit neutrophils. Cancer Cell 30, 243–256.

Loo, Y.M., Gale Jr., M., 2011. Immune signaling by RIG-I-like receptors. Immunity 34, 680–692.

Mackenzie, K.J., Carroll, P., Martin, C.A., Murina, O., Fluteau, A., Simpson, D.J., Olova, N., Sutcliffe, H., Rainger, J.K., Leitch, A., Osborn, R.T., Wheeler, A.P., Nowotny, M., Gilbert, N., Chandra, T., Reijns, M.A.M., Jackson, A.P., 2017. cGAS surveillance of micronuclei links genome instability to innate immunity. Nature 548, 461–465.

Maiorano, E., Regan, M.M., Viale, G., Mastropasqua, M.G., Colleoni, M., Castiglione-Gertsch, M., Price, K.N., Gelber, R.D., Goldhirsch, A., Coates, A.S., 2010. Prognostic and predictive impact of central necrosis and fibrosis in early breast cancer: results from two International Breast Cancer Study Group randomized trials of chemoendocrine adjuvant therapy. Breast Cancer Res. Treat. 121, 211–218.

Marié, I., Durbin, J.E., Levy, D.E., 1998. Differential viral induction of distinct interferon-alpha genes by positive feedback through interferon regulatory factor-7. EMBO J. 17, 6660–6669.

Matsumoto, M., Tatematsu, M., Nishikawa, F., Azuma, M., Ishii, N., Morii-Sakai, A., Shime, H., Seya, T., 2015. Defined TLR3-specific adjuvant that induces NK and CTL activation without significant cytokine production in vivo. Nat. Commun. 6, 6280.

Mause, S.F., Weber, C., 2010. Microparticles: protagonists of a novel communication network for intercellular information exchange. Circ. Res. 107, 1047–1057.

McFarland, A.P., Luo, S., Ahmed-Qadri, F., Zuck, M., Thayer, E.F., Goo, Y.A., Hybiske, K., Tong, L., Woodward, J.J., 2017. Sensing of bacterial cyclic dinucleotides by the oxidoreductase RECON promotes NF-kappaB activation and shapes a proinflammatory antibacterial state. Immunity 46, 433–445.

Medeiros, T.C., Thomas, R.L., Ghillebert, R., Graef, M., 2018. Autophagy balances mtDNA synthesis and degradation by DNA polymerase POLG during starvation. J. Cell Biol. 217, 1601–1611.

Medzhitov, R., Preston-Hurlburt, P., Janeway Jr., C.A., 1997. A human homologue of the Drosophila Toll protein signals activation of adaptive immunity. Nature 388, 394–397.

Melcher, A., Gough, M., Todryk, S., Vile, R., 1999. Apoptosis or necrosis for tumor immunotherapy: what's in a name? J. Mol. Med. (Berl) 77, 824–833.

Meylan, E., Curran, J., Hofmann, K., Moradpour, D., Binder, M., Bartenschlager, R., Tschopp, J., 2005. Cardif is an adaptor protein in the RIG-I antiviral pathway and is targeted by hepatitis C virus. Nature 437, 1167–1172.

Michael-Robinson, J.M., Reid, L.E., Purdie, D.M., Biemer-Hüttmann, A.-E., Walsh, M.D., Pandeya, N., Simms, L.A., Young, J.P., Leggett, B.A., Jass, J.R., Radford-Smith, G.L., 2001. Proliferation, apoptosis, and survival in high-level microsatellite instability sporadic colorectal cancer. Clin. Cancer Res. 7, 2347–2356.

Mitchem, J.B., Brennan, D.J., Knolhoff, B.L., Belt, B.A., Zhu, Y., Sanford, D.E., Belaygorod, L., Carpenter, D., Collins, L., Piwnica-Worms, D., Hewitt, S., Udupi, G.M., Gallagher, W.M., Wegner, C., West, B.L., Wang-Gillam, A., Goedegebuure, P., Linehan, D.C., Denardo, D.G., 2013. Targeting tumor-infiltrating macrophages decreases tumor-initiating cells, relieves immunosuppression, and improves chemotherapeutic responses. Cancer Res. 73, 1128–1141.

Mocellin, S., Mandruzzato, S., Bronte, V., Lise, M., Nitti, D., 2004. Part I: vaccines for solid tumours. Lancet Oncol. 5, 681–689.

Moore, E., Clavijo, P.E., Davis, R., Cash, H., Van Waes, C., Kim, Y., Allen, C., 2016. Established T cell-inflamed tumors rejected after adaptive resistance was reversed by combination STING activation and PD-1 pathway blockade. Cancer Immunol. Res. 4, 1061–1071.

Morita, M., Stamp, G., Robins, P., Dulic, A., Rosewell, I., Hrivnak, G., Daly, G., Lindahl, T., Barnes, D.E., 2004. Gene-targeted mice lacking the Trex1 (DNase III) 3'->5' DNA exonuclease develop inflammatory myocarditis. Mol. Cell Biol. 24, 6719–6727.

Nabet, B.Y., Qiu, Y., Shabason, J.E., Wu, T.J., Yoon, T., Kim, B.C., Benci, J.L., Demichele, A.M., Tchou, J., Marcotrigiano, J., Minn, A.J., 2017. Exosome RNA unshielding couples stromal activation to pattern recognition receptor signaling in cancer. Cell 170, 352–366.e13.

O'Rourke, D.M., Nasrallah, M.P., Desai, A., Melenhorst, J.J., Mansfield, K., Morrissette, J.J.D., Martinez-Lage, M., Brem, S., Maloney, E., Shen, A., Isaacs, R.,

Mohan, S., Plesa, G., Lacey, S.F., Navenot, J.-M., Zheng, Z., Levine, B.L., Okada, H., June, C.H., Brogdon, J.L., Maus, M.V., 2017. A single dose of peripherally infused EGFRvIII-directed CAR T cells mediates antigen loss and induces adaptive resistance in patients with recurrent glioblastoma. Sci. Transl. Med. 9 (399). pii: eaaa0984. https://doi.org/10.1126/scitranslmed.aaa0984.

Ochsenbein, A.F., Klenerman, P., Karrer, U., Ludewig, B., Pericin, M., Hengartner, H., Zinkernagel, R.M., 1999. Immune surveillance against a solid tumor fails because of immunological ignorance. Proc. Natl. Acad. Sci. U. S. A. 96, 2233–2238.

Ochsenbein, A.F., Sierro, S., Odermatt, B., Pericin, M., Karrer, U., Hermans, J., Hemmi, S., Hengartner, H., Zinkernagel, R.M., 2001. Roles of tumour localization, second signals and cross priming in cytotoxic T-cell induction. Nature 411, 1058–1064.

Oleszycka, E., Lavelle, E.C., 2014. Immunomodulatory properties of the vaccine adjuvant alum. Curr. Opin. Immunol. 28, 1–5.

O'Neill, L.A., Bowie, A.G., 2007. The family of five: TIR-domain-containing adaptors in Toll-like receptor signalling. Nat. Rev. Immunol. 7, 353–364.

Pamer, E.G., 2004. Immune responses to Listeria monocytogenes. Nat. Rev. Immunol. 4, 812–823.

Park, C.G., Hartl, C.A., Schmid, D., Carmona, E.M., Kim, H.J., Goldberg, M.S., 2018. Extended release of perioperative immunotherapy prevents tumor recurrence and eliminates metastases. Sci. Transl. Med. 10 (433). pii: eaar1916. https://doi.org/10.1126/scitranslmed.aar1916.

Peisley, A., Wu, B., Xu, H., Chen, Z.J., Hur, S., 2014. Structural basis for ubiquitin-mediated antiviral signal activation by RIG-I. Nature 509, 110–114.

Pichler, M., Hutterer, G.C., Chromecki, T.F., Jesche, J., Kampel-Kettner, K., Rehak, P., Pummer, K., Zigeuner, R., 2012. Histologic tumor necrosis is an independent prognostic indicator for clear cell and papillary renal cell carcinoma. Am. J. Clin. Pathol. 137, 283–289.

Pollheimer, M.J., Kornprat, P., Lindtner, R.A., Harbaum, L., Schlemmer, A., Rehak, P., Langner, C., 2010. Tumor necrosis is a new promising prognostic factor in colorectal cancer. Hum. Pathol. 41, 1749–1757.

Postow, M.A., Callahan, M.K., Barker, C.A., Yamada, Y., Yuan, J., Kitano, S., Mu, Z., Rasalan, T., Adamow, M., Ritter, E., Sedrak, C., Jungbluth, A.A., Chua, R., Yang, A.S., Roman, R.A., Rosner, S., Benson, B., Allison, J.P., Lesokhin, A.M., Gnjatic, S., Wolchok, J.D., 2012. Immunologic correlates of the abscopal effect in a patient with melanoma. N. Engl. J. Med. 366, 925–931.

Prabakaran, T., Bodda, C., Krapp, C., Zhang, B.C., Christensen, M.H., Sun, C., Reinert, L., Cai, Y., Jensen, S.B., Skouboe, M.K., Nyengaard, J.R., Thompson, C.B., Lebbink, R.J., Sen, G.C., Van Loo, G., Nielsen, R., Komatsu, M., Nejsum, L.N., Jakobsen, M.R., Gyrd-Hansen, M., Paludan, S.R., 2018. Attenuation of cGAS-STING signaling is mediated by a p62/SQSTM1-dependent autophagy pathway activated by TBK1. EMBO J. 37, e97858.

Ranoa, D.R., Parekh, A.D., Pitroda, S.P., Huang, X., Darga, T., Wong, A.C., Huang, L., Andrade, J., Staley, J.P., Satoh, T., Akira, S., Weichselbaum, R.R., Khodarev, N.N., 2016. Cancer therapies activate RIG-I-like receptor pathway through endogenous non-coding RNAs. Oncotarget 7, 26496–26515.

Richards, C.H., Mohammed, Z., Qayyum, T., Horgan, P.G., McMillan, D.C., 2011. The prognostic value of histological tumor necrosis in solid organ malignant disease: a systematic review. Future Oncol. 7, 1223–1235.

Rodriguez, P.C., Zea, A.H., Desalvo, J., Culotta, K.S., Zabaleta, J., Quiceno, D.G., Ochoa, J.B., Ochoa, A.C., 2003. L-arginine consumption by macrophages modulates the expression of CD3 zeta chain in T lymphocytes. J. Immunol. 171, 1232–1239.

Rodriguez, P.C., Quiceno, D.G., Zabaleta, J., Ortiz, B., Zea, A.H., Piazuelo, M.B., Delgado, A., Correa, P., Brayer, J., Sotomayor, E.M., Antonia, S., Ochoa, J.B., Ochoa, A.C., 2004. Arginase I production in the tumor microenvironment by mature myeloid cells inhibits T-cell receptor expression and antigen-specific T-cell responses. Cancer Res. 64, 5839–5849.

Rosenberg, S.A., Yang, J.C., Restifo, N.P., 2004. Cancer immunotherapy: moving beyond current vaccines. Nat. Med. 10, 909–915.

Saitoh, T., Fujita, N., Hayashi, T., Takahara, K., Satoh, T., Lee, H., Matsunaga, K., Kageyama, S., Omori, H., Noda, T., Yamamoto, N., Kawai, T., Ishii, K., Takeuchi, O., Yoshimori, T., Akira, S., 2009. Atg9a controls dsDNA-driven dynamic translocation of STING and the innate immune response. Proc. Natl. Acad. Sci. U. S. A. 106, 20842–20846.

Salazar, A.M., Erlich, R.B., Mark, A., Bhardwaj, N., Herberman, R.B., 2014. Therapeutic in situ autovaccination against solid cancers with intratumoral poly-ICLC: case report, hypothesis, and clinical trial. Cancer Immunol. Res. 2, 720–724.

Sali, T.M., Pryke, K.M., Abraham, J., Liu, A., Archer, I., Broeckel, R., Staverosky, J.A., Smith, J.L., Al-Shammari, A., Amsler, L., Sheridan, K., Nilsen, A., Streblow, D.N., Defilippis, V.R., 2015. Characterization of a novel human-specific STING agonist that elicits antiviral activity against emerging alphaviruses. PLoS Pathog. 11, e1005324.

Sampson, J.H., Archer, G.E., Mitchell, D.A., Heimberger, A.B., Herndon 2nd, J.E., Lally-Goss, D., McGehee-Norman, S., Paolino, A., Reardon, D.A., Friedman, A.H., Friedman, H.S., Bigner, D.D., 2009. An epidermal growth factor receptor variant III-targeted vaccine is safe and immunogenic in patients with glioblastoma multiforme. Mol. Cancer Ther. 8, 2773–2779.

Sampson, J.H., Heimberger, A.B., Archer, G.E., Aldape, K.D., Friedman, A.H., Friedman, H.S., Gilbert, M.R., Herndon 2nd, J.E., Mclendon, R.E., Mitchell, D.A., Reardon, D.A., Sawaya, R., Schmittling, R.J., Shi, W., Vredenburgh, J.J., Bigner, D.D., 2010. Immunologic escape after prolonged progression-free survival with epidermal growth factor receptor variant III peptide vaccination in patients with newly diagnosed glioblastoma. J. Clin. Oncol. 28, 4722–4729.

Sato, M., Suemori, H., Hata, N., Asagiri, M., Ogasawara, K., Nakao, K., Nakaya, T., Katsuki, M., Noguchi, S., Tanaka, N., Taniguchi, T., 2000. Distinct and essential roles of transcription factors IRF-3 and IRF-7 in response to viruses for IFN-α/β gene induction. Immunity 13, 539–548.

Schioppa, T., Moore, R., Thompson, R.G., Rosser, E.C., Kulbe, H., Nedospasov, S., Mauri, C., Coussens, L.M., Balkwill, F.R., 2011. B regulatory cells and the tumor-promoting actions of TNF-alpha during squamous carcinogenesis. Proc. Natl. Acad. Sci. U. S. A. 108, 10662–10667.

Schlee, M., Roth, A., Hornung, V., Hagmann, C.A., Wimmenauer, V., Barchet, W., Coch, C., Janke, M., Mihailovic, A., Wardle, G., Juranek, S., Kato, H., Kawai, T., Poeck, H., Fitzgerald, K.A., Takeuchi, O., Akira, S., Tuschl, T., Latz, E., Ludwig, J., Hartmann, G., 2009. Recognition of 5' triphosphate by RIG-I helicase requires short blunt double-stranded RNA as contained in panhandle of negative-strand virus. Immunity 31, 25–34.

Schulz, O., Diebold, S.S., Chen, M., Naslund, T.I., Nolte, M.A., Alexopoulou, L., Azuma, Y.T., Flavell, R.A., Liljestrom, P., Reis e Sousa, C., 2005. Toll-like receptor 3 promotes cross-priming to virus-infected cells. Nature 433, 887–892.

Seung, S.K., Curti, B.D., Crittenden, M., Walker, E., Coffey, T., Siebert, J.C., Miller, W., Payne, R., Glenn, L., Bageac, A., Urba, W.J., 2012. Phase 1 study of stereotactic body radiotherapy and interleukin-2–tumor and immunological responses. Sci. Transl. Med. 4, 137ra74.

Sistigu, A., Yamazaki, T., Vacchelli, E., Chaba, K., Enot, D.P., Adam, J., Vitale, I., Goubar, A., Baracco, E.E., Remedios, C., Fend, L., Hannani, D., Aymeric, L., Ma, Y., Niso-Santano, M., Kepp, O., Schultze, J.L., Tuting, T., Belardelli, F., Bracci, L., La Sorsa, V., Ziccheddu, G., Sestili, P., Urbani, F., Delorenzi, M., Lacroix-Triki, M., Quidville, V., Conforti, R., Spano, J.P., Pusztai, L., Poirier-Colame, V., Delaloge, S., Penault-Llorca, F., Ladoire, S., Arnould, L., Cyrta, J., Dessoliers, M.C., Eggermont, A., Bianchi, M.E., Pittet, M., Engblom, C., Pfirschke, C., Preville, X., Uze, G., Schreiber, R.D., Chow, M.T., Smyth, M.J., Proietti, E., Andre, F., Kroemer, G., Zitvogel, L., 2014. Cancer cell-autonomous contribution of type I interferon signaling to the efficacy of chemotherapy. Nat. Med. 20, 1301–1309.

Skog, J., Wurdinger, T., Van Rijn, S., Meijer, D.H., Gainche, L., Sena-Esteves, M., Curry Jr., W.T., Carter, B.S., Krichevsky, A.M., Breakefield, X.O., 2008. Glioblastoma microvesicles transport RNA and proteins that promote tumour growth and provide diagnostic biomarkers. Nat. Cell Biol. 10, 1470–1476.

Song, S., Peng, P., Tang, Z., Zhao, J., Wu, W., Li, H., Shao, M., Li, L., Yang, C., Duan, F., Zhang, M., Zhang, J., Wu, H., Li, C., Wang, X., Wang, H., Ruan, Y., Gu, J., 2017. Decreased expression of STING predicts poor prognosis in patients with gastric cancer. Sci. Rep. 7, 39858.

Stamell, E.F., Wolchok, J.D., Gnjatic, S., Lee, N.Y., Brownell, I., 2012. The abscopal effect associated with a systemic anti-melanoma immune response. Int. J. Radiat. Oncol. Biol. Phys. 85, 293–295.

Stanley, M.A., 2012. Epithelial cell responses to infection with human papillomavirus. Clin. Microbiol. Rev. 25, 215–222.

Stetson, D.B., Ko, J.S., Heidmann, T., Medzhitov, R., 2008. Trex1 prevents cell-intrinsic initiation of autoimmunity. Cell 134, 587–598.

Su, H., Luo, Q., Xie, H., Huang, X., Ni, Y., Mou, Y., Hu, Q., 2015. Therapeutic antitumor efficacy of tumor-derived autophagosome (DRibble) vaccine on head and neck cancer. Int. J. Nanomedicine 10, 1921–1930.

Sun, Q., Sun, L., Liu, H.H., Chen, X., Seth, R.B., Forman, J., Chen, Z.J., 2006. The specific and essential role of MAVS in antiviral innate immune responses. Immunity 24, 633–642.

Sun, L., Wu, J., Du, F., Chen, X., Chen, Z.J., 2013. Cyclic GMP-AMP synthase is a cytosolic DNA sensor that activates the type I interferon pathway. Science 339, 786–791.

Takaoka, A., Wang, Z., Choi, M.K., Yanai, H., Negishi, H., Ban, T., Lu, Y., Miyagishi, M., Kodama, T., Honda, K., Ohba, Y., Taniguchi, T., 2007. DAI (DLM-1/ZBP1) is a cytosolic DNA sensor and an activator of innate immune response. Nature 448, 501–505.

Takeda, Y., Kataoka, K., Yamagishi, J., Ogawa, S., Seya, T., Matsumoto, M., 2017. A TLR3-specific adjuvant relieves innate resistance to PD-L1 blockade without cytokine toxicity in tumor vaccine immunotherapy. Cell Rep. 19, 1874–1887.

Tanaka, Y., Chen, Z.J., 2012. STING specifies IRF3 phosphorylation by TBK1 in the cytosolic DNA signaling pathway. Sci. Signal. 5, ra20.

Tanimura, N., Saitoh, S., Matsumoto, F., Akashi-Takamura, S., Miyake, K., 2008. Roles for LPS-dependent interaction and relocation of TLR4 and TRAM in TRIF-signaling. Biochem. Biophys. Res. Commun. 368, 94–99.

Temizoz, B., Kuroda, E., Ishii, K.J., 2016. Vaccine adjuvants as potential cancer immunotherapeutics. Int. Immunol. 28, 329–338.

Thakur, B.K., Zhang, H., Becker, A., Matei, I., Huang, Y., Costa-Silva, B., Zheng, Y., Hoshino, A., Brazier, H., Xiang, J., Williams, C., Rodriguez-Barrueco, R., Silva, J.M., Zhang, W., Hearn, S., Elemento, O., Paknejad, N., Manova-Todorova, K., Welte, K., Bromberg, J., Peinado, H., Lyden, D., 2014. Double-stranded DNA in exosomes: a novel biomarker in cancer detection. Cell Res. 24, 766–769.

Thery, C., Ostrowski, M., Segura, E., 2009. Membrane vesicles as conveyors of immune responses. Nat. Rev. Immunol. 9, 581–593.

Tran, E., Robbins, P.F., Lu, Y.C., Prickett, T.D., Gartner, J.J., Jia, L., Pasetto, A., Zheng, Z., Ray, S., Groh, E.M., Kriley, I.R., Rosenberg, S.A., 2016. T-cell transfer therapy targeting mutant KRAS in cancer. N. Engl. J. Med. 375, 2255–2262.

Tran, E., Robbins, P.F., Rosenberg, S.A., 2017. 'Final common pathway' of human cancer immunotherapy: targeting random somatic mutations. Nat. Immunol. 18, 255–262.

Unterholzner, L., Keating, S.E., Baran, M., Horan, K.A., Jensen, S.B., Sharma, S., Sirois, C.M., Jin, T., Latz, E., Xiao, T.S., Fitzgerald, K.A., Paludan, S.R., Bowie, A.G., 2010. IFI16 is an innate immune sensor for intracellular DNA. Nat. Immunol. 11, 997–1004.

Vacchelli, E., Martins, I., Eggermont, A., Fridman, W.H., Galon, J., Sautes-Fridman, C., Tartour, E., Zitvogel, L., Kroemer, G., Galluzzi, L., 2012. Trial watch: peptide vaccines in cancer therapy. Oncoimmunology 1, 1557–1576.

Vanpouille-Box, C., Alard, A., Aryankalayil, M.J., Sarfraz, Y., Diamond, J.M., Schneider, R.J., Inghirami, G., Coleman, C.N., Formenti, S.C., Demaria, S., 2017. DNA exonuclease Trex1 regulates radiotherapy-induced tumour immunogenicity. Nat. Commun. 8, 15618.

Wang, H., Hu, S., Chen, X., Shi, H., Chen, C., Sun, L., Chen, Z.J., 2017. cGAS is essential for the antitumor effect of immune checkpoint blockade. Proc. Natl. Acad. Sci. U. S. A. 114, 1637–1642.

Wang, P.-H., Fung, S.-Y., Gao, W.-W., Deng, J.-J., Cheng, Y., Chaudhary, V., Yuen, K.-S., Ho, T.-H., Chan, C.-P., Zhang, Y., Kok, K.-H., Yang, W., Chan, C.-P., Jin, D.-Y., 2018a. A novel transcript isoform of STING that sequesters cGAMP and dominantly inhibits innate nucleic acid sensing. Nucleic Acids Res. 46, 4054–4071.

Wang, Y., Qin, X., Zhu, X., Chen, W., Zhang, J., Chen, W., 2018b. Oral cancer-derived exosomal NAP1 enhances cytotoxicity of natural killer cells via the IRF-3 pathway. Oral Oncol. 76, 34–41.

Weller, M., Butowski, N., Tran, D.D., Recht, L.D., Lim, M., Hirte, H., Ashby, L., Mechtler, L., Goldlust, S.A., Iwamoto, F., Drappatz, J., O'rourke, D.M., Wong, M., Hamilton, M.G., Finocchiaro, G., Perry, J., Wick, W., Green, J., He, Y., Turner, C.D., Yellin, M.J., Keler, T., Davis, T.A., Stupp, R., Sampson, J.H., Investigators, A.I.T., 2017. Rindopepimut with temozolomide for patients with newly diagnosed, EGFRvIII-expressing glioblastoma (ACT IV): a randomised, double-blind, international phase 3 trial. Lancet Oncol. 18, 1373–1385.

West, S.C., 2003. Molecular views of recombination proteins and their control. Nat. Rev. Mol. Cell Biol. 4, 435–445.

Widau, R.C., Parekh, A.D., Ranck, M.C., Golden, D.W., Kumar, K.A., Sood, R.F., Pitroda, S.P., Liao, Z., Huang, X., Darga, T.E., Xu, D., Huang, L., Andrade, J., Roizman, B., Weichselbaum, R.R., Khodarev, N.N., 2014. RIG-I-like receptor LGP2 protects tumor cells from ionizing radiation. Proc. Natl. Acad. Sci. U. S. A. 111, E484–E491.

Woo, S.R., Fuertes, M.B., Corrales, L., Spranger, S., Furdyna, M.J., Leung, M.Y., Duggan, R., Wang, Y., Barber, G.N., Fitzgerald, K.A., Alegre, M.L., Gajewski, T.F., 2014. STING-dependent cytosolic DNA sensing mediates innate immune recognition of immunogenic tumors. Immunity 41, 830–842.

Wu, J., Sun, L., Chen, X., Du, F., Shi, H., Chen, C., Chen, Z.J., 2013. Cyclic GMP-AMP is an endogenous second messenger in innate immune signaling by cytosolic DNA. Science 339, 826–830.

Xia, T., Konno, H., Ahn, J., Barber, G.N., 2016a. Deregulation of STING signaling in colorectal carcinoma constrains DNA damage responses and correlates with tumorigenesis. Cell Rep. 14, 282–297.

Xia, T., Konno, H., Barber, G.N., 2016b. Recurrent loss of STING signaling in melanoma correlates with susceptibility to viral oncolysis. Cancer Res. 76, 6747–6759.

Xia, P., Wang, S., Xiong, Z., Zhu, X., Ye, B., Du, Y., Meng, S., Qu, Y., Liu, J., Gao, G., Tian, Y., Fan, Z., 2018. The ER membrane adaptor ERAdP senses the bacterial second messenger c-di-AMP and initiates anti-bacterial immunity. Nat. Immunol. 19, 141–150.

Xing, Y., Cao, R., Hu, H.M., 2016. TLR and NLRP3 inflammasome-dependent innate immune responses to tumor-derived autophagosomes (DRibbles). Cell Death Dis. 7, e2322.

Xu, L.G., Wang, Y.Y., Han, K.J., Li, L.Y., Zhai, Z., Shu, H.B., 2005. VISA is an adapter protein required for virus-triggered IFN-beta signaling. Mol. Cell 19, 727–740.

Yi, Y., Zhou, Z., Shu, S., Fang, Y., Twitty, C., Hilton, T.L., Aung, S., Urba, W.J., Fox, B.A., Hu, H.M., Li, Y., 2012. Autophagy-assisted antigen cross-presentation: autophagosome as the argo of shared tumor-specific antigens and DAMPs. Oncoimmunology 1, 976–978.

Yin, Q., Fu, T.M., Li, J., Wu, H., 2015. Structural biology of innate immunity. Annu. Rev. Immunol. 33, 393–416.

Zhang, Z., Yuan, B., Bao, M., Lu, N., Kim, T., Liu, Y.J., 2011. The helicase DDX41 senses intracellular DNA mediated by the adaptor STING in dendritic cells. Nat. Immunol. 12, 959–965.

Zhang, X., Shi, H., Wu, J., Zhang, X., Sun, L., Chen, C., Chen, Z.J., 2013. Cyclic GMP-AMP containing mixed phosphodiester linkages is an endogenous high-affinity ligand for STING. Mol. Cell 51, 226–235.

Zhang, H., Tang, K., Zhang, Y., Ma, R., Ma, J., Li, Y., Luo, S., Liang, X., Ji, T., Gu, Z., Lu, J., He, W., Cao, X., Wan, Y., Huang, B., 2015. Cell-free tumor microparticle vaccines stimulate dendritic cells via cGAS/STING signaling. Cancer Immunol. Res. 3, 196–205.

Zheng, W., Skowron, K.B., Namm, J.P., Burnette, B., Fernandez, C., Arina, A., Liang, H., Spiotto, M.T., Posner, M.C., Fu, Y.X., Weichselbaum, R.R., 2016. Combination of radiotherapy and vaccination overcomes checkpoint blockade resistance. Oncotarget 7, 43039–43051.

CHAPTER SEVEN

Cytosolic Nucleic Acid Sensors in Inflammatory and Autoimmune Disorders

Shahanshah Khan, Victoria Godfrey, Md. Hasan Zaki[1]
Department of Pathology, UT Southwestern Medical Center, Dallas, TX, United States
[1]Corresponding author: e-mail address: hasan.zaki@utsouthwestern.edu

Contents

Abstract

Innate immunity employs germline-encoded pattern recognition receptors (PRRs) to sense microbial pattern molecules. Recognition of pathogen-associated molecular patterns (PAMPs) by various PPRs located on the cell membrane or in the cytosol leads to the activation of cell signaling pathways and production of inflammatory mediators. Nucleic acids including DNA, RNA, and their derivatives are potent PAMPs which can

International Review of Cell and Molecular Biology, Volume 344
ISSN 1937-6448
https://doi.org/10.1016/bs.ircmb.2018.10.002

be recognized by multiple PRRs to induce inflammatory responses. While nucleic acid sensors can also sense endogenous nucleic acids, they are capable of discriminating self from non-self. However, defects in nucleic acid sensing PRRs or dysregulation of nucleic acid sensing signaling pathways may cause excessive activation of the immune system resulting in the development of inflammatory and autoimmune diseases. This review will discuss the major pathways for sensing intracellular nucleic acids and how defects in these nucleic acid sensing are associated with different kinds of autoimmune and inflammatory disorders.

1. INTRODUCTION

The innate immune response is a highly conserved host defense system which is equipped with germline-encoded pathogen sensors called pattern recognition receptors (PRRs) (Schlee and Hartmann, 2016). PPRs encounter pathogen or pathogen-associated molecular patterns (PAMPs) including DNA, RNA, proteins, glycans, lipids, and lipoproteins at the cell surface, cytosol, and endosome (Janeway, 1989; Medzhitov, 2009). PPRs also recognize danger/damage-associated molecular patterns (DAMPs), such as nucleic acids, intracellular proteins, and metabolites that are released by damaged cells or tissues (Venereau et al., 2015). The major classes of PRRs include Toll-like receptors (TLRs), retinoic acid-inducible gene (RIG)-I-like receptors (RLRs), nucleotide-binding oligomerization domain (NOD)-like receptors (NLRs), absent in melanoma 2 (AIM2)-like receptors (ALRs) and C-type lectin-like receptors (CLRs). Among all these receptors, the most extensively studied are the TLRs, which can detect a myriad of PAMPs from bacterial, viral, and parasitic infection on the cell surface or endosomal compartment. During a microbial infection, acidification and enzymatic digestion of the pathogen in the endosomal or lysosomal compartment lead to the release of the microbial cellular components into the cytosol where NLRs, ALRs, and other unclassified intracellular PRRs can recognize those PAMPs (Chow et al., 2015).

Sensing of PAMPs and DAMPs by diverse PRRs leads to the activation of cell signaling pathways involved in the production of inflammatory molecules. Inflammation is generally considered the body's protective response against infection and injury. However, excessive activation of inflammatory signaling pathways exerts pathological consequences causing inflammatory diseases. While infectious agents are the major triggers for inflammation, it can be initiated with the recognition of the body's own cellular materials or metabolites. Recognition of self-antigens by the immune system often causes the destruction of cells and tissues, a condition called autoimmune diseases. Nucleic acids such as RNA, DNA, and their derivatives are major PAMPs as well as DAMPs that are involved in multiple inflammatory and autoimmune diseases. Nucleic acids can be sensed by diverse PRRs including TLRs, RLR, NLRs, and ALRs. In addition, a major sensor for DNA is guanosine monophosphate-adenosine monophosphate synthase (cGAS). Activation of cGAS and other DNA sensing pathways leads to the production of cytokines, particularly interferons, which are associated with the pathogenesis of many inflammatory disorders. In this review, we will discuss our current understanding of major cytosolic nucleic acid sensors (Fig. 1) emphasizing their roles in inflammatory and autoimmune diseases.

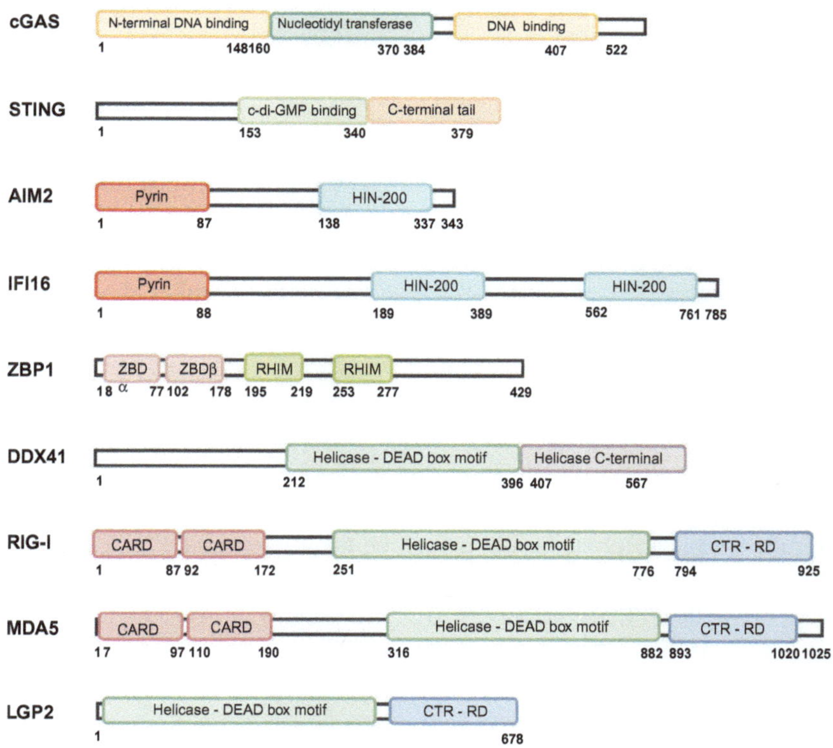

Fig. 1 Structures of nucleic acid sensors. Domain organization of major nucleic acid sensors.

2. THE cGAS-STING PATHWAY

Stimulator of interferon gene (STING) located in the endoplasmic reticulum (ER) is an adaptor protein involved in sensing nucleic acids and their derivatives. Since its discovery by Ishikawa and colleagues (2008 and 2009), STING is considered a major cytosolic DNA sensor involved in host defense, inflammation and various diseases (Barber, 2014; Burdette et al., 2011; Ishikawa and Barber, 2008; Ishikawa et al., 2009; Jin et al., 2013; Sauer et al., 2011; Sun et al., 2009; Zhong et al., 2008). In both human and mice, STING is a dimer consisting of 398 and 378 amino acids, respectively (Barber, 2014; Ishikawa and Barber, 2008; Ishikawa et al., 2009). STING was identified as a part of ER translocon system. Upon activation, STING migrates from the ER to the perinuclear membrane through interactions with the translocon associated protein (TRAP) component SSR2/TRAPβ, a key process involved in stimulating IFN-β expression (Barber, 2014; Hartmann et al., 1993; Ishikawa and Barber, 2008; Ishikawa et al., 2009; Menetret et al., 2005).

However, STING does not directly bind to microbial nucleic acids; instead it binds to cyclic dinucleotides (CDNs) or specifically cyclic guanosine monophosphate-adenosine monophosphate (cGAMP) (Sun et al., 2013; Wu et al., 2013). CDNs play critical roles in microbial pathogenesis including regulation of virulence genes, motility, and biofilm formation (Barker et al., 2013; Davies et al., 2012; Sun et al., 2013; Tamayo et al., 2007; Woodward et al., 2010). CDNs are produced either directly from intracellular microbes or through binding of exogenous or intracellular microbial DNA including self dsDNA leaked from the nucleus and mitochondria to the cyclic guanosine monophosphate-adenosine monophosphate synthase (cGAS), which acts as a second messenger (Ablasser et al., 2013; Burdette et al., 2011; Civril et al., 2013). Crystallographic studies revealed that cGAS does not bind to any bases of the DNA, but forms a complex with the sugar phosphate backbone independent of specific DNA sequences (Civril et al., 2013; Gao et al., 2013a; Li et al., 2013; Zhang et al., 2014).

Structurally, cGAS consists of a single domain of a nucleotidyl transferase and two dsDNA binding domains (Sun et al., 2013) (Fig. 1). Under normal physiological conditions, cGAS remains in an auto-inhibitory state (Civril et al., 2013; Gao et al., 2013a; Kranzusch et al., 2013; Sun et al., 2013; Zhang et al., 2014). However, in the presence of DNA, cGAS endogenously forms a 2:2 complex with DNA (Sun et al., 2013; Zhang et al., 2014). Once the microbial dsDNA binds to the active site of cGAS, a conformational change occurs in the C-terminal that contains the catalytic pocket, which catalyzes the formation of cGAMP from ATP and GTP (Sun et al., 2013; Wu et al., 2013; Zhang et al., 2014). The resulting cGAMP is consisted of two phosphodiester bonds—canonical 3′ OH-AMP plus 5′ phosphate-GMP and non-canonical 2′ OH-GMP plus 5′ phosphate-AMP, and therefore, termed as 2′-3′ cGAMP (Ablasser et al., 2013; Diner et al., 2013; Gao et al., 2013a; Zhang et al., 2013b). However, the secretory 3′-3′ cGAMP contains 3′-5′ phosphodiester bonds (Davies et al., 2012), which is known to be 10- to 1000-fold less potent than 2′-3′ cGAMP in STING activation (Diner et al., 2013; Gao et al., 2013b; Lolicato et al., 2014).

As a second messenger, cGAMP binds to the active site of STING and causes conformational changes in it that leads to the activation and transportation of STING from the ER to the Golgi apparatus through the ER-Golgi intermediate compartment (Dobbs et al., 2015; Ishikawa and Barber, 2008; Ishikawa et al., 2009; Saitoh et al., 2009; Wu et al., 2013; Zhang et al., 2013b; Zhong et al., 2008) (Fig. 2). During this trafficking process, STING is complexed with TANK-binding kinase 1 (TBK1) and then brought into the endolysosomal compartment (Ishikawa et al., 2009; Saitoh et al., 2009), where TBK1 phosphorylates interferon regulatory factor 3 (IRF3) and activates inhibitor of nuclear factor kappa-B kinase (IKK) (Fitzgerald et al., 2003; Sharma et al., 2003; Tanaka and Chen, 2012). Phosphorylated IRF3 undergoes dimerization and is then translocated into the nucleus (Fitzgerald et al., 2003; Sharma et al., 2003; Tanaka and Chen, 2012). On the other hand, activated IKK phosphorylates IκB which leads to the release and translocation of NF-κB into the nucleus, where NF-κB along with IRF3 binds to IFN-encoding genes and induces interferons and other inflammatory cytokines (Ishikawa and Barber, 2008) (Fig. 2).

2.1 The Role of cGAS-STING in Arthritis

Systemic lupus erythematous (SLE) and chronic polyarthritis are forms of autoimmune diseases, which are characterized by abnormal induction of inflammatory cytokine production that causes multiple organ injury (Crow, 2007; Helmick et al., 2008; Nagata, 2010). The link between mammalian DNA and arthritis was first reported with the observation that mice lacking DNase II developed polyarthritis, which resembles human rheumatoid arthritis (Kawane et al., 2006). DNase II, a DNA-degrading enzyme present in the lysosomal compartment of phagocytic cells, causes single-stranded DNA cleavage by hydrolyzing the phosphodiester backbone of

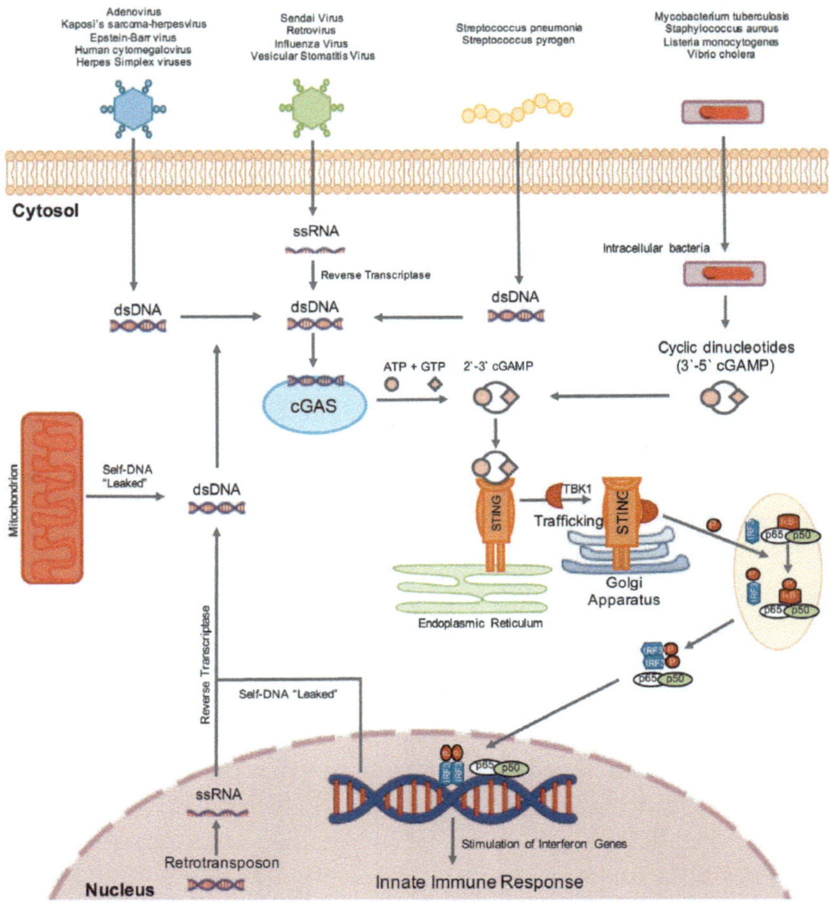

Fig. 2 The cGAS-STING signaling pathway. Cytosolic DNA derived from viruses, bacteria, and mitochondria is recognized by cGAS leading to the synthesis of cGAMP. Upon binding cGAMP, STING translocates from the ER to Golgi apparatus where it phosphorylates NF-κB and IRF3, resulting in the induction of type I interferons.

internalized DNA (Evans and Aguilera, 2003). Macrophages of DNase II-deficient mice are unable to degrade internalized DNA phagocytosed from red blood precursor cells (Fig. 3). The accumulation of internalized DNA in macrophages leads to an uncontrolled production of inflammatory cytokines such as TNFα and type I IFNs, which trigger chronic inflammation and development of arthritis (Henderson and Pettipher, 1989; Kawane et al., 2006; Magnusson et al., 2006; Yoshida et al., 2005). Although it was a milestone discovery in explaining the immunopathogenic mechanism of arthritis, the cellular sensor and molecular mechanism by which phagocytosed

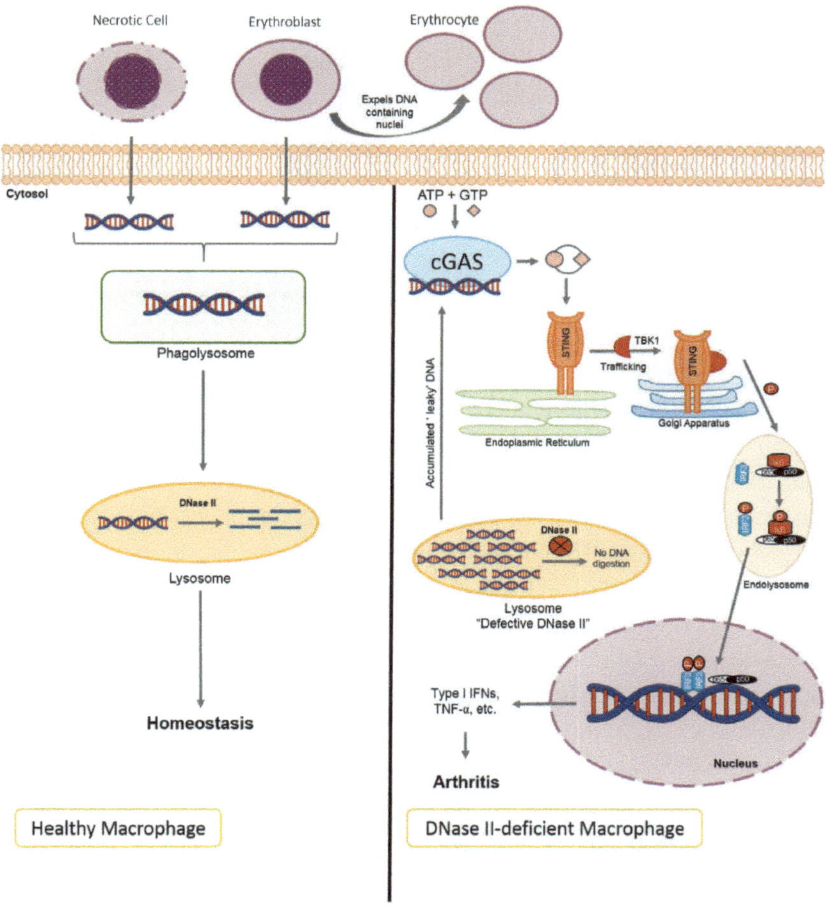

Fig. 3 cGAS and arthritis. Normally macrophages that take up DNA into the cell by the phagolysosome are degraded by DNase II. Whereas in DNase II-deficient macrophages, DNA accumulates in the lysosome and DNA that leaks out of the lysosome is subsequently detected by cGAS in the cytosol and activates the STING-dependent pathway resulting in the induction of cytokines, IFNs, and inflammation that promotes arthritis.

DNA induces inflammatory cytokine influx were unknown until recently. Ahn et al. identified STING as the cytoplasmic sensor for DNA released by red blood precursors, apoptotic and necrotic cells and demonstrated the importance of STING in the pathogenesis of polyarthritis (Ahn et al., 2012). Deletion of STING was shown to rescue mice from polyarthritis, suggesting that DNA sensing by STING is responsible for the induction of type I IFNs (Ahn et al., 2012). STING also plays a role in regulating cytoplasmic DNA internalization (Ahn et al., 2012; Barber, 2014; Ishikawa and Barber, 2008; Ishikawa et al., 2009). In DNase II-deficient mice, the DNA from necrotic and apoptotic cells is transported and accumulated into the lysosomal compartment of macrophages. Defective DNA degradation allows them to be released into the cytosol resulting in the activation of the STING pathway and upregulation of pro-inflammatory cytokines (Ahn et al., 2012; Kawane et al., 2001, 2006; Munoz et al., 2010; Yoshida et al., 2005).

2.2 Regulation of Intestinal Inflammatory Disorders by cGAS-STING

Ulcerative colitis (UC) and Crohn's disease (CD) are the two most common types of inflammatory bowel diseases (IBD), accounting 2.2 million and 1.5 million cases in Europe and the United States, respectively (Molodecky et al., 2012). It is commonly viewed that IBD is initiated with dysregulated immune response toward the gut microbiota (Kaser et al., 2010). Thus, one of the mechanisms of IBD induction involves dysregulation in the sensing of PAMPs by diverse PRRs in the gut. Indeed, several cytosolic sensors in the family of NLRs such as NOD2, NLRP3, NLRP6, and NLRP12 have been shown to play a central role in intestinal homeostasis and their defects are associated with colitis and colitis-associated colorectal cancer (Elinav et al., 2011; Hugot et al., 2001; Udden et al., 2017; Zaki et al., 2010, 2011). Since the cGAS-STING pathway senses microbial DNA and CDNs, it is considered that this pathway may play an important role in intestinal homeostasis. A study using $Il10^{-/-}$ mice, which spontaneously develop colitis, showed that defective STING ameliorates colitis with reduced inflammatory responses and pathological changes (Ahn et al., 2017). IL-10 is an immunosuppressive cytokine; therefore, loss of IL-10 leads to higher production of pro-inflammatory cytokines such as IL-1β, IL-18, and interferons (Ahn et al., 2017; Davidson et al., 1998). STING-mediated production of pro-inflammatory cytokines further enhances inflammatory responses via activation of myeloid differentiation primary response 88 (MyD88) dependent activation of NF-κB and mitogen-activated protein kinase (MAPK) pathways (Ahn et al., 2017; Hoshi et al., 2012; Rakoff-Nahoum et al., 2006; Salcedo et al., 2013). Therefore, defective STING pathways help to suppress inflammatory responses in the gut (Ahn et al., 2017). Notably, two other studies contradicted with this concept and underscored the beneficial role of the cGAS-STING pathway in intestinal homeostasis. Using an experimental injury model of colitis and colorectal tumorigenesis, it was demonstrated that sensing microbial DNA by cGAS-STING elicits protective immune responses in the gut

(Canesso et al., 2017; Zhu et al., 2014). These studies demonstrated that STING deficiency promotes colitis and colorectal tumorigenesis in mice (Canesso et al., 2017; Zhu et al., 2014). A possible reason of increased colitis susceptibility of STING-deficient mice is the alteration of gut microbiota, which reflects defective immune responses in the gut manifested by decreased intraepithelial lymphocytes and increased group 1 and 3 innate lymphoid cells (ILC1 and ILC3) (Canesso et al., 2017; Zhu et al., 2014). Activation of the cGAS-STING pathway by tumor-derived DNA also plays an essential role in anti-tumor immunity. STING-mediated sensing of DNA in dendritic cells leads to the production of type I IFNs which activate $CD8^+$T cells (Corrales et al., 2015; Deng et al., 2014; Woo et al., 2014). Colorectal cancer patients having low densities of $CD8^+$ T cells are more prone to tumor recurrence, suggesting a critical role for $CD8^+$ T cells in colorectal cancer immunity (Pages et al., 2009). Type I IFNs produced by RIG-I and STING pathways were shown to protect irradiation and immune-mediated intestinal tissue injury via promoting epithelial cell proliferation and production of antimicrobial peptide Reg3γ (Fischer et al., 2017). Thus, the cGAS-STING pathway plays both protective and detrimental roles in intestinal inflammatory disorders in a context-dependent manner. However, the relevance of type I IFNs and $CD8^+$ T cell-mediated anti-tumor immunity in increased colorectal cancer susceptibility of STING-deficient mice has not been examined. Future studies focusing on STING-dependent T cell priming in the context of colorectal cancer may provide a better understanding of the precise role of the cGAS-STING pathway in host defense against colorectal cancer.

2.3 cGAS-STING in Metabolic Disorders

Obesity and metabolic syndrome have emerged as a major global health concern. Despite the link between genetic defects and metabolic deficiencies, diet and lifestyle play a critical role in the pathogenesis of metabolic disorders via regulation of the body's inflammatory responses (Osborn and Olefsky, 2012; Roberto et al., 2015). Diet-derived free fatty acids induce endothelial inflammation, contributing to obesity (Li et al., 2009; Steinberg et al., 2000; Villaret et al., 2010; Wang et al., 2006). In addition, endothelial inflammation triggers cardiovascular and adipose tissue inflammation resulting in insulin resistance, a common condition observed in individuals with diabetes mellitus type 2 (T2D) and obesity (Gregor and Hotamisligil, 2011; Liao, 2013; Ouchi et al., 2011; Romeo et al., 2012). T2D, obesity, and insulin resistance are the main risk factors that increase the probability of nonalcoholic liver diseases and related metabolic disorders (Perry et al., 2014; Roberto et al., 2015; Yki-Jarvinen, 2014). Interestingly, STING was shown to be upregulated in high-fat diet (HFD)-fed mice which in turn activates interferon regulatory factor 3 (IRF3) by TBK-1 leading to hepatic inflammation and metabolic disorders through dysregulation of the NF-κB signaling pathway (Qiao et al., 2018). A separate study demonstrated that STING is involved in the early development of alcohol-induced acute and chronic liver disease in mice independent of type I IFNs (Petrasek et al., 2013). Biochemical analysis revealed that alcohol-induced acute liver disease pathogenesis is associated with ER stress with the activation of STING and IRF3. Activated IRF3 triggers hepatocyte apoptosis by interacting with B cell lymphoma 2-associated X protein (BAX) (Petrasek et al., 2013). All these studies suggest that endogenous DNA sensing via STING-cGAS play critical roles in multiple metabolic disorders and emphasize further investigation to dissect the precise mechanisms involved in these processes.

3. THE RNA SENSOR RIG-I-LIKE RECEPTORS (RLRs)

Cytosolic RNAs are recognized by a DExD/H-box RNA helicase family sensors including RIG-I, melanoma-differentiated gene 5 (MDA5), and laboratory of genetics and physiology 2 (LGP2) (Kang et al., 2002; Yoneyama et al., 2004). RIG-I and MDA5 share structurally similar domains, which include a N-terminal domain (NTD) consisting of two CARD domains, a RNA helicase DExD/H-box as its central domain (CD) and a C-terminal domain (CTD) lodged as a repressor domain (RD) (Yoneyama et al., 2004, 2005) (Fig. 1). However, LGP2 doesn't contain the N-terminal CARD domain (Fig. 1). The NTD of RIG-I serves as an adapter molecule for caspase activation, the central domain is involved in ATP hydrolysis and RNA unwinding, whereas, the C-terminal domain maintains RIG-I's autoinhibitory function (Hofmann et al., 1997; Johnson and Gale, 2006; Loo and Gale, 2011; Saito et al., 2007; Yoneyama et al., 2004, 2005).

RIG-I was originally identified as a viral sensor for cytoplasmic dsRNA and uncapped 5′-triphosphate-ssRNA (Habjan et al., 2008; Hornung et al., 2006; Kato et al., 2005, 2006; Loo et al., 2008; Luthra et al., 2011; McCartney et al., 2008; Plumet et al., 2007; Roth-Cross et al., 2008; Saito et al., 2007; Schlee, 2013; Sumpter et al., 2005; Yoneyama et al., 2004, 2005; Zhou et al., 2010). RIG-I recognizes a number of viruses in the family of Paramyxoviridae, Rhabdoviridae, Orthomyxoviridae, Filoviridae and Flaviviridae (Ori et al., 2017). RIG-I has also been shown to recognize

DNA viruses like Epstein–Barr virus, once their genomic DNA is converted into 5′triphosphate-ssRNA through RNA polymerase III (Samanta et al., 2006; Wang et al., 2008). Thus, 5′-triphosphate-ssRNA is considered to be RIG-I's main target to distinguish self RNA from non-self RNA (Schlee, 2013). Similar to RIG-I, MDA5 can recognize RNA from a number of viruses including Picornaviridae and Norovirus (Feng et al., 2012). Notably, several viruses such as Dengue virus, West Nile virus, Reovirus are recognized by both RIG-I and MDA5 (Errett et al., 2013; Fredericksen et al., 2008; Goubau et al., 2014; Kato et al., 2006; Loo et al., 2008). Although LGP2 can sense RNA, its involvement in detecting RNA during viral infection has yet to be identified.

Since its discovery as a sensor for dsRNA in 2004, RIG-I has been extensively studied to understand its role in antiviral immunity. RIG-I is primarily kept in an auto-inhibitory state through an indigenous phosphorylation of RD at residues Thr770 and Ser854/855 and CARDs at Ser8 and Thr170 by casein kinase II (CK II) and protein kinase-Cα/β (PKCα/β) (Maharaj et al., 2012; Sun et al., 2011). Upon binding and recognition of viral 5′-triphosphate-ssRNA within the helicase DExD/H-box, RIG-I undergoes conformational changes and simultaneously initiates antiviral responses through interaction with mitochondrial antiviral-signaling proteins (MAVS) (Bruns and Horvath, 2012; Gack, 2014; Loo and Gale, 2011). The conformational changes of activated RIG-I involve phosphatase PP1α and PP1γ, which are recruited to dephosphorylate RIG-I (Gack et al., 2007; Wies et al., 2013), leading to conformational changes within the CTD and Lys63-linked ubiquitination by ubiquitin ligase Riplet (Gack et al., 2007). Ubiquitination allows RIG-I to undergo oligomerization to expose its N-terminal CARDs, which are dephosphorylated by PP1α and PP1γ (Gack et al., 2007). Dephosphorylation of CARDs allows binding of RING-dependent ubiquitin E3 ligase (TRIM25) to CARD1 which further ubiquitinate CARD2 (Gack et al., 2007). However, these series of ubiquitination do not mark RIG-I to undergo degradation, instead, act as a scaffold for RIG-I to oligomerize and be released from RD autorepression allowing subsequent interaction with MAVS proteins to initiate antiviral responses through IFN production (Bruns and Horvath, 2012; Gack, 2014; Gack et al., 2007; Loo and Gale, 2011).

The binding of RIG-I to viral dsRNA or 5′-triphosphate containing RNA causes conformational changes in the helicase domains which allows recruitment of downstream MAVS through its N-terminal CARD domains. Thus, RIG-I in combination with MAVS prompts the activation of

TBK-1/IKKε and IKK kinases complexes that phosphorylate IRF3/7 and IκB, respectively. The phosphorylated IRF3 and IRF7 dimerize and translocate to the nucleus where it binds to the IFN-encoding gene and initiates the expression of IFNs. Phosphorylated IκB undergoes proteasomal degradation, allowing NF-κB to translocate into the nucleus and bind with κB element in the pro-inflammatory region that leads to the production of pro-inflammatory cytokines (Fig. 4). In line with this notion, other studies have revealed that RIG-I is a leading sensor for antiviral responses in the majority of cell populations except plasmacytoid dendritic cells (pDCs), which use TLR7 and TLR9 instead of RIG-I for detection of dsRNA (Kato et al., 2005; Wang et al., 2007).

3.1 RLRs in Autoimmune Diseases

Although induction of Type I IFNs is the body's protective response against viral infection, excessive production of these cytokines is associated with a number of autoimmune diseases (Gresser et al., 1975, 1980). Because of their involvement in inducing type I IFNs, RLRs are considered key immune molecules associated with autoimmune diseases. Genetic and clinical studies provide ample evidence suggesting that sensing self RNA or polymorphism of RLRs, particularly MDA5, is associated with several autoimmune diseases including type I diabetes, psoriasis, multiple sclerosis, and systemic lupus erythematosus (SLE) (Ferreira et al., 2010; Nejentsev et al., 2009; Zouk et al., 2014). Detection of anti-MDA5 antibodies along with elevated type I IFN responses has been implicated in dermatomyositis (Hattori et al., 2018; Sato et al., 2009; Zhang et al., 2018). Missense mutations have been detected in the MDA5 gene in patients with Singleton-Merten Syndrome (SMS), which is characterized by dental dysplasia, aortic calcification, skeletal abnormalities, glaucoma, psoriasis and other abnormalities (Rutsch et al., 2015). A recent study reported atypical SMS in 11 members of 2 families which is associated with mutations in RIG-I encoding gene DDX58 (Jang et al., 2015). A separate study showed that SMS-linked mutations in RIG-I are responsible for increased interaction of RIG-I with endogenous RNA (Lassig et al., 2015). RIG-I distinguishes viral dsRNA from self RNA by a mechanism that involves ATP hydrolysis by RIG-I's RNA translocase domain. Thus, mutations in RIG-I allow binding with endogenous RNA by preventing ATP hydrolysis (Lassig et al., 2015).

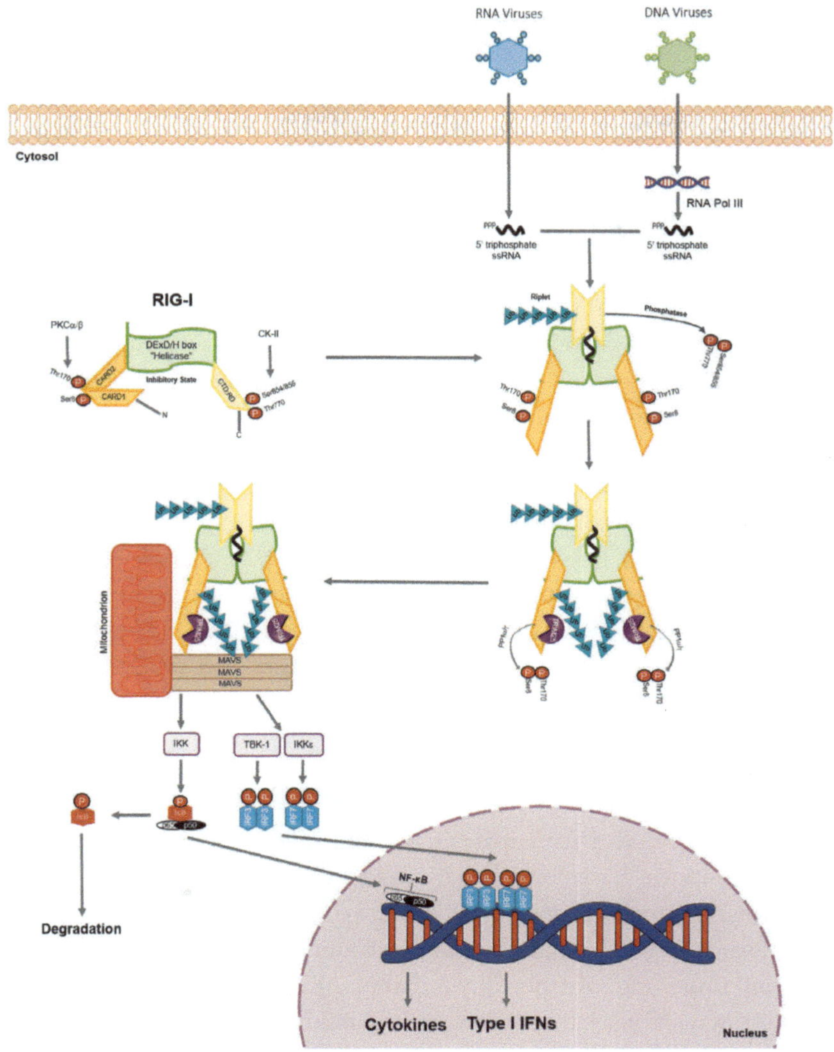

Fig. 4 Mechanism of RIG-I activation. RIG-I binds to 5′-triphosphate-ssRNA within its helicase DExD/H-box causing conformational changes due to being dephosphorylated by CK II and PKCα/β along with being ubiquinated by riplet. These structural changes lead to release of RIG-I from its auto-inhibitory state and simultaneously recruit MAVS which in combination with RIG-I, activate TBK-1/IKKe and IKK kinases complexes that phosphorylate IRF3/7 and IκB. Phosphorylated IRF3 and IRF7 dimerize and translocate to the nucleus where it binds to the IFN-encoding genes and initiates the expression of IFNs. Whereas, phosphorylated IκB undergoes degradation that allows NF-κB to translocate into the nucleus and bind with κB element in the pro-inflammatory region that promotes the induction of inflammatory cytokines.

3.2 RIG-I in Inflammatory Bowel Diseases and Colorectal Cancer

The role of RIG-I in inflammatory and autoimmune diseases is currently understudied. Wang et al. showed that Rig-I-deficient mice are susceptible to DSS-induced colitis (Wang et al., 2007). Supporting to this observation, the expression of RIG-I was seen downregulated in the intestinal epithelium of CD patients (Funke et al., 2011). Increased colitis pathogenesis in Rig-I-deficient mice was associated with reduced number and size of Peyer's patches and abnormal activation of peripheral T cells. Peyer's patches are one of the primary sites where initiation of the immune response against antigens occurs in the gut (Ohman et al., 2002). Lowered numbers of Peyer's patches contribute to the development of colitis (Spahn et al., 2002). Interestingly, Rig-I-deficient mice exhibited reduced expression of *Gαi2*, a heterotrimeric G protein subunit consisting of α, β, and γ subunits, which is considered a potential IBD susceptibility gene. Genetic studies showed that *Gαi2* is located within the IBD susceptible chromosome 3 at locus p21 (Hampe et al., 2001). In line with this notion, Gαi2 knockout mice were highly susceptible to IBD (Rudolph et al., 1995). Therefore, colitis susceptibility of Rig-I-deficient mice is linked with, at least in part, reduced expression of Gαi2 and dysregulated T responses (Solis et al., 2007; Wang et al., 2007).

3.3 Cross-Talk Between RIG-I and cGAS/STING

Since multiple PRRs participate in sensing nucleic acids, it is intriguing how the immune system copes with the activation of different PRRs and maintains immune homeostasis. It could be possible that different nucleic acid sensing PRRs and their signaling adapters cross-talk during microbial infection. Emerging evidence shows that cGAS/STING and RIG-I/MAVS pathways may show such a coordination in inducing inflammatory responses (Zevini et al., 2017). It was shown that STING serves as a co-adaptor for RIG-I to form a complex with MAVS (Castanier et al., 2010; Zhong et al., 2008). Consistently, STING and RIG-I double knockout macrophages and dendritic cells failed to produce type I IFNs in response to poly (dA:dT) (Brunette et al., 2012; Ishikawa and Barber, 2008; Zhong et al., 2008). It was also seen that STING is regulated by RIG-I at both the transcriptional and post-transcriptional levels (Liu et al., 2016b; Wu et al., 2017). In addition to direct cross-talk, these two pathways may participate in innate immunity in a synergistic fashion. A recent study showed that radiation-induced tissue injury leads to activation of both MAVS and

STING leading to the production of type I IFNs, which contributes to protection against inflammation and injury via promoting epithelial integrity and expression of antimicrobial peptides (Fischer et al., 2017). Thus, these two pathways can act simultaneously in diverse pathophysiological conditions for sensing nucleic acids to elicit type-I IFNs and T cell-mediated immunity.

4. THE DNA SENSOR AIM2

AIM2 is another major cytosolic DNA sensor belonging to the family interferon-inducible HIN-200 which can sense both endogenous and exogenous dsDNA (Burckstummer et al., 2009; Fernandes-Alnemri et al., 2009; Hornung et al., 2009). AIM2 was first discovered by a cDNA screening of malignant melanoma cell lines (Deyoung et al., 1997). AIM2 localizes in human chromosome 1q22 and chromosome 1q21 in mice (Deschamps et al., 2003; Ludlow et al., 2005). AIM2 was found to be expressed in the spleen, small intestine, and small peripheral blood leukocytes (Han et al., 2015). AIM2 was initially studied as a putative tumor suppressor due to its ability to inhibit cell proliferation as seen in human breast cancer cells and fibroblasts (Chen et al., 2006). But breakthrough studies in 2009 from several groups discovered AIM2's role as a cytosolic PRR for dsDNA and an essential initiator for inflammasome activation (Burckstummer et al., 2009; Fernandes-Alnemri et al., 2009; Hornung et al., 2009; Roberts et al., 2009). Since then AIM2 has been shown to play a key role in inflammasome activation during many bacterial and viral infections including *Listeria monocytogenes*, *Streptococcus pneumonia*, *Staphylococcus aureus*, *Brucella abortus*, *Porphyromons gingivalis*, *Chlamydia muridarum*, *Mycobacterium tuberculosis*, and *M. bovis Francisella novicida*, *F. tularensis*, mouse cytomegalovirus (MCMV), and vaccinia viruses (Fang et al., 2011, 2014; Fernandes-Alnemri et al., 2010; Gomes et al., 2013; Hanamsagar et al., 2014; Jones et al., 2010; Kim et al., 2010; Man et al., 2015a; Park et al., 2014; Reinholz et al., 2013; Saiga et al., 2012; Wassermann et al., 2015; Yang et al., 2013).

Structural analysis revealed AIM2 to have an N-terminal pyrin domain (PYD) and a C-terminal HIN-200 domain which contains two dsDNA binding domains. The PYD of AIM2 recruits adaptor protein apoptosis-associated speck-like protein (ASC), which interacts with pro-caspase-1 via its caspase recruitment domain (CARD) leading to the formation of an inflammasome complex (Jin et al., 2012). In addition to AIM2, several members of the NLR family including NLRP1, NLRP3, NLRC4, and

NLRP6 are known to form an inflammasome complex (Lamkanfi and Dixit, 2009; Latz et al., 2013). Activated caspase-1, processed by the inflammasome, participates in cleaving pro-IL-1β and pro-IL-18 along with gasdermin D to produce their mature forms (Fig. 5) (Latz et al., 2013). Cleavage of gasdermin D allows the release of its N-terminal fragment which initiates pyroptosis, inflammatory cell death, and the release of the

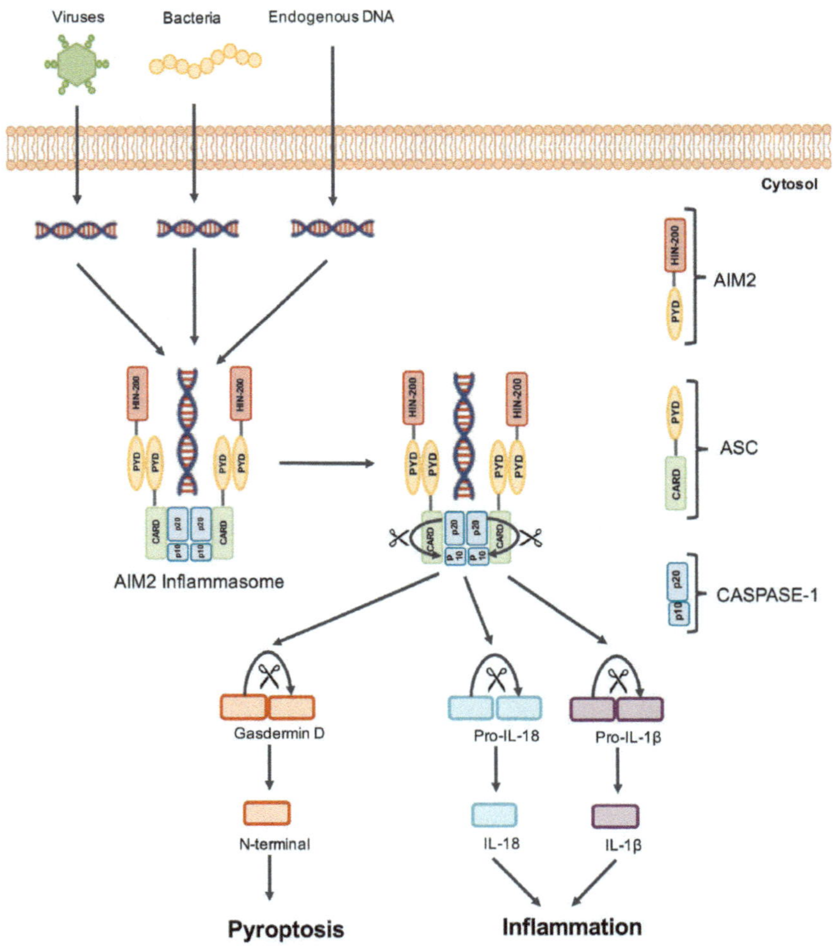

Fig. 5 The AIM2 signaling pathway. AIM2 can recognize a myriad of dsDNA via its HIN-200 domain. Once bound to the dsDNA, the PYD of AIM2 recruits ASC which further interacts with pro-caspase-1 to form the inflammasome complex. The inflammasome formation leads to the activation of caspase-1 which cleaves gasdermin D, pro-IL1β, and pro-IL-18 into their mature forms. The activation of these proinflammatory molecules triggers inflammation and cell death.

mature cytokines into the cell (Kayagaki et al., 2015; Liu et al., 2016a; Man et al., 2016a; Shi et al., 2015). Recent studies demonstrated that *F. novicida* activates AIM2 via activation of cGAS-STING pathway which induces interferon regulatory factor 1(IRF1) (Man et al., 2015a). Among the IRF1 inducible genes, guanylate binding proteins (GBPs) and IRGB10 were shown to be involved in *F. novicida*-induced AIM2 activation (Man et al., 2015a, 2016b; Meunier et al., 2015). These two proteins facilitate release of bacterial DNA into the cytosol. However, IRF1 and its downstream proteins are not required for AIM2 inflammasome by transfection of DNA (polydA:dT) and other bacterial and viral DNA (Fernandes-Alnemri et al., 2010; Man et al., 2015a; Meunier et al., 2015; Muruve et al., 2008; Rathinam et al., 2010). Thus, AIM2 may be activated by different bacterial DNA through distinct mechanisms and our understanding of the molecular mechanisms of AIM2 activation during different bacterial infections is still poor.

4.1 AIM2 and Intestinal Inflammatory Disorders

The inflammasome has recently emerged as a critical regulator of intestinal homeostasis. Previous studies depicted the beneficial role of the inflammasome in protection against colitis in mice (Allen et al., 2010; Elinav et al., 2011; Zaki et al., 2010). However, the role of AIM2 in colitis and colorectal tumorigenesis was unknown until recently. Consistent to other inflammasome activating sensors including NLRP3 and NLRP6, AIM2 was shown to protect mice from DSS-induced colitis (Hu et al., 2015). Colitis susceptibility of Aim2-deficient mice was shown to be associated with altered gut microbiota, particularly Enterobacteriaceae. Clinical studies documented ample evidence on the association of higher abundance of *Escherichia coli* with IBD pathogenesis (Martin et al., 2004; Martinez-Medina et al., 2014). Higher *E. coli* burden in Aim2-deficient mice was associated with reduced activation of caspase-1 and subsequent production of IL-18. The role of IL-18 in suppressing *E. coli* growth was evident by dramatic reduction of *E. coli* burden in $Aim2^{-/-}$ mice upon administration of recombinant IL-18 (Hu et al., 2015). Mechanistically, IL-18 was shown to trigger the production of metal chelators lipocaline 2 (Lcn2), S100A8, and S100A9 and antimicrobial peptides Reg3β and Reg3γ (Hu et al., 2015). A separate study supported this protective function of the Aim2 inflammasome against colitis and showed that AIM2-mediated activation of IL-18 induces IL-22 and signal transducer and activator of transcription 3 (STAT3) activation, which regulates Reg3β and Reg3γ (Ratsimandresy et al., 2017). These studies underscore a crucial role for AIM2 in regulating intestinal inflammation and preventing dysbiosis in the gut.

Interestingly, two separate studies reported that AIM2 protects mice against colorectal tumorigenesis induced by mutagen azoxymethane (AOM) with repeated exposures of DSS in an inflammasome-independent manner (Man et al., 2015b; Wilson et al., 2015). While increased tumorigenesis in *Aim2*$^{-/-}$ mice was associated with increased epithelial cell proliferation, there was no apparent defect in the levels of IL-1β and IL-18 in them (Man et al., 2015b; Wilson et al., 2015). A lack of defective inflammasome activation in *Aim2*$^{-/-}$ mice could be due to altered gut microbiota which may activate other inflammasome pathways. These studies suggested that higher proliferation of Aim2-deficient colon epithelial cells during tumorigenesis was associated with increased activation of alpha serine/threonine-protein kinase (AKT). Overexpression of AIM2 in mouse embryonic fibroblast cells or colonic epithelial cells was shown to inhibit AKT via interaction with DNA-dependent protein kinase (DNA-PK) (Wilson et al., 2015). Interestingly, such AIM2-mediated inhibition of AKT was not observed in myeloid cells (Wilson et al., 2015). Notably, the background of *Aim2*$^{-/-}$ mice used in these studies that failed to observe AIM2-dependent inflammasome activation during colorectal tumorigenesis (Man et al., 2015b, Wilson et al., 2015) is not same as of mice used in other studies showing defective inflammasome function (Hu et al., 2015; Ratsimandresy et al., 2017). Future studies should define the precise role of inflammasome-dependent and independent functions of AIM2 in the regulation of tumorigenesis.

4.2 Role of AIM2 in Inflammatory Skin Diseases

Psoriasis is a chronic skin disease that causes rapid accumulation of skin cells that form white-silver patches on the skin that are dry and irritating (Nestle et al., 2009). IL-1β is key for cutaneous inflammation in psoriasis progression and is produced mainly by keratinocytes in the epidermis (Renne et al., 2010; Shepherd et al., 2004). Due to psoriasis causing uncontrolled proliferation of the epidermis there is an overabundance of host DNA from neighboring dead cells and additional accumulation of DNA in the cytosol hyperactivating AIM2. Elevated levels of cytosolic DNA and AIM2 expression have been detected in psoriatic lesions compared to normal skin in humans (Dombrowski et al., 2011). Expression studies also identified AIM2 induction in keratinocytes through in vitro epidermis models as seen under inflammatory conditions whereas in normal conditions AIM2 expression is restricted to melanocytes and Langerhans cells (de Koning et al., 2012). AIM2 has a role as the first line of defense after disruption or damage to the skin barrier where initial AIM2 activation wards off invading pathogens via inflammasome activation and pyroptosis. However, prolonged skin damage and aberrant AIM2 activation can lead to the development of chronic inflammatory skin diseases such as psoriasis, rosacea, and atopic dermatitis (de Koning et al., 2012; Dombrowski et al., 2011).

4.3 Role of AIM2 in Liver Inflammation

Chronic liver inflammation and innate immune activation is key in the pathogenesis of various conditions such as hepatic fibrosis, cirrhosis, nonalcoholic fatty liver disease (NAFLD) and nonalcoholic steatohepatitis (NASH). Liver injury and hepatic cell death lead to the release of cellular components such as DNA and RNA which can activate the immune system. Clinically, AIM2 inflammasome activation is associated with liver cirrhosis (Lozano-Ruiz et al., 2015). AIM2 inflammasome activation was seen to be highly induced in macrophages isolated from ascitic fluid (AF), accumulated fluid in the abdominal cavity, of patients with liver cirrhosis compared to healthy patients. Isolated AF-derived macrophages did not need a priming signal for AIM2 activation indicating the pre-activated state of AIM2 in patients with extensive hepatic cirrhosis (Lozano-Ruiz et al., 2015). AIM2 also contributes to dietary steatohepatitis in a TLR9-MyD88 dependent manner through upregulated expression and activation by TLR9 ligands and damaged host DNA (Csak et al., 2014).

4.4 Role of AIM2 in Other Autoimmune and Autoinflammatory Diseases

AIM2 has been implicated in systemic lupus erythematosus (SLE) development. It was shown that AIM2 is essential for expression of *Fcgr2b*, a gene that encodes for the inhibitory FcγRIIB receptor which is a major lupus susceptibility gene (Panchanathan et al., 2011). Monozygotic twin studies revealed the reduction of DNA methylation in the gene encoding AIM2 in the twin with SLE versus their healthy sibling counterpart (Javierre et al., 2010). Interestingly, there is also a high sexual disparity in development of SLE with a female to male ratio of 8:1 and males that do develop SLE has significantly more aggressive disease development (Tedeschi et al., 2013).

The hallmarks of SLE pathogenesis are the production of antibodies against host DNA and atypical macrophage activation (Kimkong et al., 2009; Tedeschi et al., 2013). AIM2 inflammasome could contribute to this sexual disparity in SLE due to mRNA expression of unstimulated macrophages is decreased in females and increased in male SLE patients (Yang et al., 2015).

AIM2 has been implicated in various other autoinflammatory diseases such as acute pancreatitis (Algaba-Chueca et al., 2017; Vanhove et al., 2015). It is also seen that AIM2 activation can cause disease progression in a tissue-specific manner as seen in polyarthritis utilizing the DNase II mouse model, expression of pro-inflammatory cytokines and inflammation of diseased joints was AIM2 dependent, whereas systemic inflammation was AIM2 independent showing AIM2 as a key endogenous self-DNA receptor in autoinflammatory diseases (Jakobs et al., 2015). Similar to AIM2's role in psoriasis, the inability to eliminate self-DNA and deviant activation of AIM2 is crucial for the development of disease pathogenesis and inflammation. As an extensive dsDNA sensor, AIM2 is crucial for regulating the innate immune responses to pat'hogens and inflammation in order to prevent autoinflammatory diseases.

5. DNA SENSOR IFI16

IFNγ-inducible protein 16 (IFI16) or its mouse orthologue p204 is another important DNA binding protein in the HIN200 family. Structurally, IFI16 is similar to AIM2 in that it contains a N-terminal pyrin domain but contains two HIN domains whereas AIM2 only has one HIN domain (Dell'oste et al., 2015). Initial findings described that IFI16 is expressed in cells of myeloid origins and localized to the nucleus (Dawson and Trapani, 1995; Dawson et al., 1995). However, recent studies demonstrated that IFI16 is also present in the cytoplasm in different cell types and shuttles between nucleus and cytoplasm at different physiological contexts (Bawadekar et al., 2015). It has been shown that IFI16 interacts with human cytomegalovirus (HCMV) DNA within the nucleus but later translocates to the cytoplasm and localizes with virus assembly complexes as a part of viral virulence strategy (Dell'oste et al., 2014; Gariano et al., 2012). Similarly, during Kaposi's Sarcoma-associated Herpesvirus (KSHV) infection, IFI16 senses its DNA and forms an inflammasome complex in association with ASC and caspase-1, which is later translocated to the cytoplasm for activation of IL-1β and IL-18 (Singh et al., 2013). UV irradiation leads to relocalization of IFI16 from nucleus to cytoplasm in keratinocytes and this process may involve post-translational modification of IFI16 (Costa et al., 2011).

Sensing viral DNA by IFI16 elicits host defense responses by inducing interferons and activation of the inflammasome (Ansari et al., 2015; Dell'oste et al., 2015; Gariano et al., 2012; Johnson et al., 2013, 2014; Singh et al., 2013). It was observed that vaccinia virus induces IFN-β in a TLR-independent manner in monocytes. Proteins that were pulled down with the viral DNA interaction included IFI16 (Unterholzner et al., 2010). IFI16 has been implicated to have a role in p53-mediated tumor suppression, transcriptional regulation, and apoptosis (Deyoung et al., 1997; Duan et al., 2011; Johnstone et al., 1998, 2000; Liao et al., 2011). IFI16 has the ability to scan long stretches of exposed dsDNA in an one-dimensional manner that allows it to distinguish between self and foreign DNA. It has also been seen that chromatization and nucleosomes can also inhibit IFI16 scanning protecting host DNA (Stratmann et al., 2015).

5.1 IFI16 and Autoimmune Diseases

Prevalence of autoantibodies against IFI16 has been seen in patients with multiple autoimmune disorders. Similar to AIM2, IFI16 is also known to have a role in SLE; over 29% of SLE patients develop autoantibodies due to IFI16's high specificity for antinuclear antibodies (Seelig et al., 1994). As IFI16 is critical for processes like apoptosis and cell proliferation, dysfunction can heighten development for autoimmune disease and the same gene family *Ifi-200* (including AIM2) has been seen to contribute to SLE susceptibility (Choubey and Panchanathan, 2008). Multiple groups in 2015 discovered IFI16 to be associated with primary Sjogren's syndrome (pSS), which is a chronic autoimmune disease primarily affecting the exocrine glands and is also characterized by increased lymphocyte infiltration in the minor salivary glands (MSGs) (Alunno et al., 2015a; Baer et al., 2016). Patients with pSS had higher levels of IFI16 and anti-IFI16 in the serum compared to healthy individuals (Alunno et al., 2015b). MSGs do not constitutively express IFI16 whereas pSS-MSGs had marked expression, suggesting a role in glandular inflammation that occurs in pSS and functioning as markers of severe disease. IFI16 is believed to be associated with rheumatoid arthritis since patients exhibited increased expression of IFI16 in the serum. IFI16 has also been identified as a susceptibility gene for Behçet disease, a chronic inflammatory disease of blood vessels distinguished by oral ulcers (Alunno et al., 2016; Ortiz-Fernandez et al., 2015).

6. ADDITIONAL CYTOPLASMIC NUCLEIC ACID SENSORS

6.1 ZBP1

Z-DNA-binding protein 1 (ZBP1) was initially discovered as the novel gene, DLM-1, associated with anti-tumor host response due to upregulated expression in tumor and surrounding stromal tissue. It was also upregulated in macrophages during stimulation with LPS and IFN-γ, characterizing it as an interferon-inducible gene (Fu et al., 1999). Structurally, ZBP1 is consisted of two N-terminal Z-DNA binding domains (ZBDs), which has been reported to bind Z-DNA, B-DNA and RNA, followed by two receptor-interacting protein homotypic interaction motifs (RHIM) domains (Fig. 1) (Ha et al., 2006, 2008; Schwartz et al., 2001). When ZBP1 was found to bind Z-DNA, left-handed dsDNA conformation, through its ZBD it inspired its current name, ZBP1 (Herbert and Rich, 1996; Kuriakose and Kanneganti, 2018). Taniguchi and colleagues discovered ZBP1's ability to initiate DNA-mediated activation of innate immunity via induction of type I IFNs in a-ssociation with IRF3 and TBK1 that led them to propose ZBP1 as a cytosolic DNA sensor with a proposed name of DNA-dependent activator of IFN-regulatory factors (DAI) (Herbert and Rich, 1996; Takaoka et al., 2007). More recently, studies have focused on the two RHIM domains which allow interactions with TIR-domain-containing adapter-inducing interferon-β (TRIF), receptor-interacting serine/threonine-protein kinase 1 (RIPK1), RIPK3 and other RIP kinases (Kaiser et al., 2008; Sun et al., 2002). RHIM-dependent interactions of ZBP1 are essential for the regulation of necroptosis in response to viral infection and inflammation via induction of NF-κB (Rebsamen et al., 2009; Upton et al., 2012). ZBP1 has also been reported to sense both viral RNA, influenza A virus (IAV), and murine cytomegalovirus (Kuriakose et al., 2016; Sridharan et al., 2017; Thapa et al., 2016). Recognition of IAV by ZBP1 leads to NLRP3 inflammasome activation and induction of cell death (Kuriakose et al., 2016). ZBP-1 induced cell death during IAV infection involves the production of type I IFNs. Since, Type I IFNs are also induced by RIG-I and TLR signaling, Kesavardhana et al. recently investigated the contribution of these pathways in IAV-induced cell death and explored that sensing IAV by RIG-I triggers cell death via ZBP-1 (Kesavardhana et al., 2017). Further biochemical studies revealed that ZBP1 detects viral ribonucleoprotein (vRNP) and undergoes ubiquitination as a part of its involvement in innate immune response and apoptosis during IAV infection (Kesavardhana et al., 2017).

There have been several reports associating ZBP1 with autoinflammatory diseases, specifically to SLE like other cytosolic sensors due to correlation with aberrant DNA recognition and its corresponding autoimmune responses (Walport, 2000). ZBP1 expression was notably increased in SLE patients and lupus mice models promoting the development of SLE nephritis (Finke et al., 2007; Qiao et al., 2005). ZBP1 also has been identified to be involved with lymphocyte-derived self-apoptotic DNA (ALD-DNA)-induced macrophage activation and polarization via induction of the IRF3 and NF-κB pathways resulting in calcium signaling (Liu et al., 2008; Zhang et al., 2010). Knockdown of ZBP1 in ALD-DNA induced macrophages halted activation, whereas inhibition of ZBP1 was seen to alleviate SLE syndrome and decrease inflammatory responses and macrophage activation in lupus mice (Zhang et al., 2013a). At this moment, our knowledge on the physiological function of ZBP1 in different inflammatory diseases is limited. But, being a cytosolic DNA sensor involved in cell death and inflammatory responses, ZBP1 may potentially be involved in other autoimmune and inflammatory diseases.

6.2 DDX41

DExD/H-box protein 41 (DDX41) is a part of the RIG-I-like receptor (RLR) family, like RIG-I it carries a DExD/H-box domain containing RNA helicase that senses viral PAMPs (Onoguchi et al., 2011). RLRs are best known to induce type I IFNs and antiviral genes in response to viral infections. DDX41 was discovered by a siRNA screening of the DExD/H-box helicase family and has been reported to sense DNA in a STING-dependent manner in response to DNA viruses, HSV-1 and adenovirus (Zhang et al., 2011). Parvatiyar et al. showed that DDX41 directly interacts with CDNs. Deletion of DDX41 resulted in reduced activation of the STING-TBK1-IRF3 pathway (Parvatiyar et al., 2012). These observations lead to a proposed model where DDX41 synergistically enhances CDN-mediated STING activation and IFN responses via its interaction with CDNs and STING (Parvatiyar et al., 2012). Recent studies reported that inherited and acquired mutations in DDX41 are associated with acute myeloid leukemia (AML) (Cardoso et al., 2016; Ding et al., 2012). Mutations in DDX41 were associated with altered pre-mRNA splicing and RNA processing (Polprasert et al., 2015). Thus, DDX41 could be targeted for AML therapy, but the precise mechanism through which DDX41 suppresses tumorigenesis is yet to be discovered (Jiang et al., 2017).

7. CONCLUSION

There has been a significant progress in understanding the role of nucleic acids in innate immunity and diverse disease pathogenesis. In general, nucleic acid sensing by multiple PRRs can trigger activation of type I IFNs, inflammasome, and other inflammatory signaling pathways. As seen with many other receptors, dysregulation in the recognition of nucleic acids can promote the pathogenesis of autoimmune and inflammatory diseases such as arthritis, lupus, colitis, etc. In the last decade, many breakthrough studies have introduced the importance of nucleic acid sensing receptors for a healthy life showing that many inflammatory and autoimmune diseases are linked to the defective sensing of nucleic acids. There are still many unresolved issues. For example, how different PRRs coordinate and cross-talk when they recognize nucleic acids in the cytoplasm. It is also not clear, why several PRRs recognize nucleic acids from a particular group of pathogens but not others. Furthermore, the precise mechanisms through which PRRs distinguish between self and non-self nucleic acids are not very clear. Addressing these concerns may open means to developing therapeutics against autoimmune and inflammatory diseases.

ACKNOWLEDGEMENT

This work was supported by Cancer Prevention and Research Institute of Texas (CPRIT) Individual Investigator Award (RP160169), and UT Southwestern Start-up funding given to M.H.Z.

REFERENCES

Ablasser, A., Goldeck, M., Cavlar, T., Deimling, T., Witte, G., Rohl, I., Hopfner, K.P., Ludwig, J., Hornung, V., 2013. cGAS produces a 2′-5′-linked cyclic dinucleotide second messenger that activates STING. Nature 498, 380–384.

Ahn, J., Gutman, D., Saijo, S., Barber, G.N., 2012. STING manifests self DNA-dependent inflammatory disease. Proc. Natl. Acad. Sci. U. S. A. 109, 19386–19391.

Ahn, J., Son, S., Oliveira, S.C., Barber, G.N., 2017. STING-dependent signaling underlies IL-10 controlled inflammatory colitis. Cell Rep. 21, 3873–3884.

Algaba-Chueca, F., de -Madaria, E., Lozano-Ruiz, B., Martinez-Cardona, C., Quesada-Vazquez, N., Bachiller, V., Tarin, F., Such, J., Frances, R., Zapater, P., Gonzalez-Navajas, J.M., 2017. The expression and activation of the AIM2 inflammasome correlates with inflammation and disease severity in patients with acute pancreatitis. Pancreatology 17, 364–371.

Allen, I.C., Tekippe, E.M., Woodford, R.M., Uronis, J.M., Holl, E.K., Rogers, A.B., Herfarth, H.H., Jobin, C., TING, J.P., 2010. The NLRP3 inflammasome functions as a negative regulator of tumorigenesis during colitis-associated cancer. J. Exp. Med. 207, 1045–1056.

Alunno, A., Caneparo, V., Carubbi, F., Bistoni, O., Caterbi, S., Bartoloni, E., Giacomelli, R., Gariglio, M., Landolfo, S., Gerli, R., 2015a. Interferon gamma-inducible protein 16 in primary sjogren's syndrome: a novel player in disease pathogenesis? Arthritis Res. Ther. 17, 208.

Alunno, A., Caneparo, V., Carubbi, F., Bistoni, O., Caterbi, S., Gariglio, M., Bartoloni, E., Landolfo, S., Gerli, R., 2015b. Interferon gamma-inducible protein 16 (IFI16) and anti-IFI16 antibodies in primary sjogren's syndrome: findings in serum and minor salivary glands. Reumatismo 67, 85–90.

Alunno, A., Caneparo, V., Bistoni, O., Caterbi, S., Terenzi, R., Gariglio, M., Bartoloni, E., Manzo, A., Landolfo, S., Gerli, R., 2016. Circulating interferon-inducible protein IFI16 correlates with clinical and serological features in rheumatoid arthritis. Arthritis Care Res. (Hoboken) 68, 440–445.

Ansari, M.A., Dutta, S., Veettil, M.V., Dutta, D., Iqbal, J., Kumar, B., Roy, A., Chikoti, L., Singh, V.V., Chandran, B., 2015. Herpesvirus genome recognition induced acetylation of nuclear IFI16 is essential for its cytoplasmic translocation, inflammasome and IFN-beta responses. PLoS Pathog. 11, e1005019.

Baer, A.N., Petri, M., Sohn, J., Rosen, A., Casciola-Rosen, L., 2016. Association of antibodies to interferon-inducible protein-16 with markers of more severe disease in primary sjogren's syndrome. Arthritis Care Res. (Hoboken) 68, 254–260.

Barber, G.N., 2014. STING-dependent cytosolic DNA sensing pathways. Trends Immunol. 35, 88–93.

Barker, J.R., Koestler, B.J., Carpenter, V.K., Burdette, D.L., Waters, C.M., Vance, R.E., Valdivia, R.H., 2013. STING-dependent recognition of cyclic di-AMP mediates type I interferon responses during *Chlamydia trachomatis* infection. MBio 4, e00018-13.

Bawadekar, M., de Andrea, M., Gariglio, M., Landolfo, S., 2015. Mislocalization of the interferon inducible protein IFI16 by environmental insults: implications in autoimmunity. Cytokine Growth Factor Rev. 26, 213–219.

Brunette, R.L., Young, J.M., Whitley, D.G., Brodsky, I.E., Malik, H.S., Stetson, D.B., 2012. Extensive evolutionary and functional diversity among mammalian AIM2-like receptors. J. Exp. Med. 209, 1969–1983.

Bruns, A.M., Horvath, C.M., 2012. Activation of RIG-I-like receptor signal transduction. Crit. Rev. Biochem. Mol. Biol. 47, 194–206.

Burckstummer, T., Baumann, C., Bluml, S., Dixit, E., Durnberger, G., Jahn, H., Planyavsky, M., Bilban, M., Colinge, J., Bennett, K.L., Superti-Furga, G., 2009. An orthogonal proteomic-genomic screen identifies AIM2 as a cytoplasmic DNA sensor for the inflammasome. Nat. Immunol. 10, 266–272.

Burdette, D.L., Monroe, K.M., Sotelo-Troha, K., Iwig, J.S., Eckert, B., Hyodo, M., Hayakawa, Y., Vance, R.E., 2011. STING is a direct innate immune sensor of cyclic di-GMP. Nature 478, 515–518.

Canesso, M.C.C., Lemos, L., Neves, T.C., Marim, F.M., Castro, T.B.R., Veloso, E.S., Queiroz, C.P., Ahn, J., Santiago, H.C., Martins, F.S., Alves-Silva, J., Ferreira, E., Cara, D.C., Vieira, A.T., Barber, G.N., Oliveira, S.C., Faria, A.M.C., 2017. The cytosolic sensor STING is required for intestinal homeostasis and control of inflammation. Mucosal Immunol. 11, 820–834.

Cardoso, S.R., Ryan, G., Walne, A.J., Ellison, A., Lowe, R., Tummala, H., Rio-Machin, A., Collopy, L., Al Seraihi, A., Wallis, Y., Page, P., Akiki, S., Fitzgibbon, J., Vulliamy, T., Dokal, I., 2016. Germline heterozygous Ddx41 variants in a subset of familial myelodysplasia and acute myeloid leukemia. Leukemia 30, 2083–2086.

Castanier, C., Garcin, D., Vazquez, A., Arnoult, D., 2010. Mitochondrial dynamics regulate the RIG-I-like receptor antiviral pathway. EMBO Rep. 11, 133–138.

Chen, I.F., Ou-Yang, F., Hung, J.Y., Liu, J.C., Wang, H., Wang, S.C., Hou, M.F., Hortobagyi, G.N., Hung, M.C., 2006. AIM2 suppresses human breast cancer cell proliferation in vitro and mammary tumor growth in a mouse model. Mol. Cancer Ther. 5, 1–7.

Choubey, D., Panchanathan, R., 2008. Interferon-inducible Ifi200-family genes in systemic lupus erythematosus. Immunol. Lett. 119, 32–41.
Chow, J., Franz, K.M., Kagan, J.C., 2015. PRRs are watching you: localization of innate sensing and signaling regulators. Virology 479–480, 104–109.
Civril, F., Deimling, T., de Oliveira Mann, C.C., Ablasser, A., Moldt, M., Witte, G., Hornung, V., Hopfner, K.P., 2013. Structural mechanism of cytosolic DNA sensing by cGAS. Nature 498, 332–337.
Corrales, L., Glickman, L.H., Mcwhirter, S.M., Kanne, D.B., Sivick, K.E., Katibah, G.E., Woo, S.R., Lemmens, E., Banda, T., Leong, J.J., Metchette, K., Dubensky Jr., T.W., Gajewski, T.F., 2015. Direct activation of STING in the tumor microenvironment leads to potent and systemic tumor regression and immunity. Cell Rep. 11, 1018–1030.
Costa, S., Borgogna, C., Mondini, M., de Andrea, M., Meroni, P.L., Berti, E., Gariglio, M., Landolfo, S., 2011. Redistribution of the nuclear protein IFI16 into the cytoplasm of ultraviolet B-exposed keratinocytes as a mechanism of autoantigen processing. Br. J. Dermatol. 164, 282–290.
Crow, M.K., 2007. Type I interferon in systemic lupus erythematosus. Curr. Top. Microbiol. Immunol. 316, 359–386.
Csak, T., Pillai, A., Ganz, M., Lippai, D., Petrasek, J., Park, J.-K., Kodys, K., Dolganiuc, A., Kurt-Jones, E.A., Szabo, G., 2014. Both bone marrow-derived and non-bone marrow-derived cells contribute to AIM2 and NLRP3 inflammasome activation in a MyD88-dependent manner in dietary steatohepatitis. Liver Int. 34, 1402–1413.
Davidson, N.J., Hudak, S.A., Lesley, R.E., Menon, S., Leach, M.W., Rennick, D.M., 1998. IL-12, but not IFN-gamma, plays a major role in sustaining the chronic phase of colitis in IL-10-deficient mice. J. Immunol. 161, 3143–3149.
Davies, B.W., Bogard, R.W., Young, T.S., Mekalanos, J.J., 2012. Coordinated regulation of accessory genetic elements produces cyclic di-nucleotides for *V. cholerae* virulence. Cell 149, 358–370.
Dawson, M.J., Trapani, J.A., 1995. Ifi 16 gene encodes a nuclear protein whose expression is induced by interferons in human myeloid leukaemia cell lines. J. Cell. Biochem. 57, 39–51.
Dawson, M.J., Trapani, J.A., Briggs, R.C., Nicholl, J.K., Sutherland, G.R., Baker, E., 1995. The closely linked genes encoding the myeloid nuclear differentiation antigen (MNDA) and IFI16 exhibit contrasting haemopoietic expression. Immunogenetics 41, 40–43.
de Koning, H.D., Bergboer, J.G., van den Bogaard, E.H., van Vlijmen-Willems, I.M., Rodijk-Olthuis, D., Simon, A., Zeeuwen, P.L., Schalkwijk, J., 2012. Strong induction of AIM2 expression in human epidermis in acute and chronic inflammatory skin conditions. Exp. Dermatol. 21, 961–964.
Dell'oste, V., Gatti, D., Gugliesi, F., de Andrea, M., Bawadekar, M., Lo Cigno, I., Biolatti, M., Vallino, M., Marschall, M., Gariglio, M., Landolfo, S., 2014. Innate nuclear sensor IFI16 translocates into the cytoplasm during the early stage of in vitro human cytomegalovirus infection and is entrapped in the egressing virions during the late stage. J. Virol. 88, 6970–6982.
Dell'oste, V., Gatti, D., Giorgio, A.G., Gariglio, M., Landolfo, S., de Andrea, M., 2015. The interferon-inducible DNA-sensor protein IFI16: a key player in the antiviral response. New Microbiol. 38, 5–20.
Deng, L., Liang, H., Xu, M., Yang, X., Burnette, B., Arina, A., Li, X.D., Mauceri, H., Beckett, M., Darga, T., Huang, X., Gajewski, T.F., Chen, Z.J., Fu, Y.X., Weichselbaum, R.R., 2014. STING-dependent cytosolic DNA sensing promotes radiation-induced type I interferon-dependent antitumor immunity in immunogenic tumors. Immunity 41, 843–852.
Deschamps, S., Meyer, J., Chatterjee, G., Wang, H., Lengyel, P., Roe, B.A., 2003. The mouse Ifi200 gene cluster: genomic sequence, analysis, and comparison with the human HIN-200 gene cluster. Genomics 82, 34–46.

Deyoung, K.L., Ray, M.E., Su, Y.A., Anzick, S.L., Johnstone, R.W., Trapani, J.A., Meltzer, P.S., Trent, J.M., 1997. Cloning a novel member of the human interferon-inducible gene family associated with control of tumorigenicity in a model of human melanoma. Oncogene 15, 453–457.

Diner, E.J., Burdette, D.L., Wilson, S.C., Monroe, K.M., Kellenberger, C.A., Hyodo, M., Hayakawa, Y., Hammond, M.C., Vance, R.E., 2013. The innate immune DNA sensor cGAS produces a noncanonical cyclic dinucleotide that activates human STING. Cell Rep. 3, 1355–1361.

Ding, L., Ley, T.J., Larson, D.E., Miller, C.A., Koboldt, D.C., Welch, J.S., Ritchey, J.K., Young, M.A., Lamprecht, T., McLellan, M.D., McMichael, J.F., Wallis, J.W., Lu, C., Shen, D., Harris, C.C., Dooling, D.J., Fulton, R.S., Fulton, L.L., Chen, K., Schmidt, H., Kalicki-Veizer, J., Magrini, V.J., Cook, L., Mcgrath, S.D., Vickery, T.L., Wendl, M.C., Heath, S., Watson, M.A., Link, D.C., Tomasson, M.H., Shannon, W.D., Payton, J.E., Kulkarni, S., Westervelt, P., Walter, M.J., Graubert, T.A., Mardis, E.R., Wilson, R.K., Dipersio, J.F., 2012. Clonal evolution in relapsed acute myeloid leukaemia revealed by whole-genome sequencing. Nature 481, 506–510.

Dobbs, N., Burnaevskiy, N., Chen, D., Gonugunta, V.K., Alto, N.M., Yan, N., 2015. STING activation by translocation from the ER is associated with infection and autoinflammatory disease. Cell Host Microbe 18, 157–168.

Dombrowski, Y., Peric, M., Koglin, S., Kammerbauer, C., Goss, C., Anz, D., Simanski, M., Glaser, R., Harder, J., Hornung, V., Gallo, R.L., Ruzicka, T., Besch, R., Schauber, J., 2011. Cytosolic DNA triggers inflammasome activation in keratinocytes in psoriatic lesions. Sci. Transl. Med. 3, 82ra38.

Duan, X., Ponomareva, L., Veeranki, S., Choubey, D., 2011. IFI16 induction by glucose restriction in human fibroblasts contributes to autophagy through activation of the ATM/AMPK/p53 pathway. PLoS One 6, e19532.

Elinav, E., Strowig, T., Kau, A.L., Henao-Mejia, J., Thaiss, C.A., Booth, C.J., Peaper, D.R., Bertin, J., Eisenbarth, S.C., Gordon, J.I., Flavell, R.A., 2011. NLRP6 inflammasome regulates colonic microbial ecology and risk for colitis. Cell 145, 745–757.

Errett, J.S., Suthar, M.S., McMillan, A., Diamond, M.S., Gale Jr., M., 2013. The essential, nonredundant roles of RIG-I and MDA5 in detecting and controlling West Nile virus infection. J. Virol. 87, 11416–11425.

Evans, C.J., Aguilera, R.J., 2003. DNase II: genes, enzymes and function. Gene 322, 1–15.

Fang, R., Tsuchiya, K., Kawamura, I., Shen, Y., Hara, H., Sakai, S., Yamamoto, T., Fernandes-Alnemri, T., Yang, R., Hernandez-Cuellar, E., Dewamitta, S.R., Xu, Y., Qu, H., Alnemri, E.S., Mitsuyama, M., 2011. Critical roles of ASC inflammasomes in caspase-1 activation and host innate resistance to *Streptococcus pneumoniae* infection. J. Immunol. 187, 4890–4899.

Fang, R., Hara, H., Sakai, S., Hernandez-Cuellar, E., Mitsuyama, M., Kawamura, I., Tsuchiya, K., 2014. Type I interferon signaling regulates activation of the absent in melanoma 2 inflammasome during *Streptococcus pneumoniae* infection. Infect. Immun. 82, 2310–2317.

Feng, Q., Hato, S.V., Langereis, M.A., Zoll, J., Virgen-Slane, R., Peisley, A., Hur, S., Semler, B.L., van Rij, R.P., van Kuppeveld, F.J., 2012. MDA5 detects the double-stranded RNA replicative form in picornavirus-infected cells. Cell Rep. 2, 1187–1196.

Fernandes-Alnemri, T., Yu, J.W., Datta, P., Wu, J., Alnemri, E.S., 2009. Aim2 activates the inflammasome and cell death in response to cytoplasmic DNA. Nature 458, 509–513.

Fernandes-Alnemri, T., Yu, J.W., Juliana, C., Solorzano, L., Kang, S., Wu, J., Datta, P., McCormick, M., Huang, L., McDermott, E., Eisenlohr, L., Landel, C.P., Alnemri, E.S., 2010. The AIM2 inflammasome is critical for innate immunity to *Francisella tularensis*. Nat. Immunol. 11, 385–393.

Ferreira, R.C., Pan-Hammarstrom, Q., Graham, R.R., Gateva, V., Fontan, G., Lee, A.T., Ortmann, W., Urcelay, E., Fernandez-Arquero, M., Nunez, C., Jorgensen, G.,

Ludviksson, B.R., Koskinen, S., Haimila, K., Clark, H.F., Klareskog, L., Gregersen, P.K., Behrens, T.W., Hammarstrom, L., 2010. Association of IFIH1 and other autoimmunity risk alleles with selective IgA deficiency. Nat. Genet. 42, 777–780.

Finke, D., Randers, K., Hoerster, R., Hennig, H., Zawatzky, R., Marion, T., Brockmann, C., Klempt-Giessing, K., Jacobsen, K., Kirchner, H., Goerg, S., 2007. Elevated levels of endogenous apoptotic DNA and IFN-alpha in complement C4-deficient mice: implications for induction of systemic lupus erythematosus. Eur. J. Immunol. 37, 1702–1709.

Fischer, J.C., Bscheider, M., Eisenkolb, G., Lin, C.C., Wintges, A., Otten, V., Lindemans, C.A., Heidegger, S., Rudelius, M., Monette, S., Porosnicu Rodriguez, K.A., Calafiore, M., Liebermann, S., Liu, C., Lienenklaus, S., Weiss, S., Kalinke, U., Ruland, J., Peschel, C., Shono, Y., Docampo, M., Velardi, E., Jenq, R.R., Hanash, A.M., Dudakov, J.A., Haas, T., van den Brink, M.R.M., Poeck, H., 2017. RIG-I/MAVS and STING signaling promote gut integrity during irradiation- and immune-mediated tissue injury. Sci. Transl. Med. 9, eaag2513.

Fitzgerald, K.A., Mcwhirter, S.M., Faia, K.L., Rowe, D.C., Latz, E., Golenbock, D.T., Coyle, A.J., Liao, S.M., Maniatis, T., 2003. IKKepsilon and TBK1 are essential components of the IRF3 signaling pathway. Nat. Immunol. 4, 491–496.

Fredericksen, B.L., Keller, B.C., Fornek, J., Katze, M.G., Gale Jr., M., 2008. Establishment and maintenance of the innate antiviral response to West Nile virus involves both RIG-I and MDA5 signaling through IPS-1. J. Virol. 82, 609–616.

Fu, Y., Comella, N., Tognazzi, K., Brown, L.F., Dvorak, H.F., Kocher, O., 1999. Cloning of DLM-1, a novel gene that is up-regulated in activated macrophages, using RNA differential display. Gene 240, 157–163.

Funke, B., Lasitschka, F., Roth, W., Penzel, R., Meuer, S., Saile, M., Gretz, N., Sido, B., Schirmacher, P., Autschbach, F., 2011. Selective downregulation of retinoic acid-inducible gene I within the intestinal epithelial compartment in Crohn's disease. Inflamm. Bowel Dis. 17, 1943–1954.

Gack, M.U., 2014. Mechanisms of RIG-I-like receptor activation and manipulation by viral pathogens. J. Virol. 88, 5213–5216.

Gack, M.U., Shin, Y.C., Joo, C.H., Urano, T., Liang, C., Sun, L., Takeuchi, O., Akira, S., Chen, Z., Inoue, S., Jung, J.U., 2007. TRIM25 RING-finger E3 ubiquitin ligase is essential for RIG-I-mediated antiviral activity. Nature 446, 916–920.

Gao, P., Ascano, M., Wu, Y., Barchet, W., Gaffney, B.L., Zillinger, T., Serganov, A.A., Liu, Y., Jones, R.A., Hartmann, G., Tuschl, T., Patel, D.J., 2013a. Cyclic [G(2′,5′)pA (3′,5′)p] is the metazoan second messenger produced by DNA-activated cyclic GMP-AMP synthase. Cell 153, 1094–1107.

GAO, P., Ascano, M., Zillinger, T., Wang, W., Dai, P., Serganov, A.A., Gaffney, B.L., Shuman, S., Jones, R.A., Deng, L., Hartmann, G., Barchet, W., Tuschl, T., Patel, D.J., 2013b. Structure-function analysis of STING activation by c[G(2′,5′)pA (3′,5′)p] and targeting by antiviral DMXAA. Cell 154, 748–762.

Gariano, G.R., Dell'oste, V., Bronzini, M., Gatti, D., Luganini, A., de Andrea, M., Gribaudo, G., Gariglio, M., Landolfo, S., 2012. The intracellular DNA sensor IFI16 gene acts as restriction factor for human cytomegalovirus replication. PLoS Pathog. 8, e1002498.

Gomes, M.T., Campos, P.C., Oliveira, F.S., Corsetti, P.P., Bortoluci, K.R., Cunha, L.D., Zamboni, D.S., Oliveira, S.C., 2013. Critical role of ASC inflammasomes and bacterial type IV secretion system in caspase-1 activation and host innate resistance to *Brucella abortus* infection. J. Immunol. 190, 3629–3638.

Goubau, D., Schlee, M., Deddouche, S., Pruijssers, A.J., Zillinger, T., Goldeck, M., Schuberth, C., van der Veen, A.G., Fujimura, T., Rehwinkel, J., Iskarpatyoti, J.A., Barchet, W., Ludwig, J., Dermody, T.S., Hartmann, G., Reis, E.S.C., 2014. Antiviral immunity via RIG-I-mediated recognition of RNA bearing 5′-diphosphates. Nature 514, 372–375.

Gregor, M.F., Hotamisligil, G.S., 2011. Inflammatory mechanisms in obesity. Annu. Rev. Immunol. 29, 415–445.

Gresser, I., Tovey, M.G., Maury, C., Chouroulinkov, I., 1975. Lethality of interferon preparations for newborn mice. Nature 258, 76–78.

Gresser, I., Morel-Maroger, L., Riviere, Y., Guillon, J.C., Tovey, M.G., Woodrow, D., Sloper, J.C., Moss, J., 1980. Interferon-induced disease in mice and rats. Ann. N. Y. Acad. Sci. 350, 12–20.

Ha, S.C., van Quyen, D., Hwang, H.Y., Oh, D.B., Brown 2nd, B.A., Lee, S.M., Park, H.J., Ahn, J.H., Kim, K.K., Kim, Y.G., 2006. Biochemical characterization and preliminary X-ray crystallographic study of the domains of human ZBP1 bound to left-handed Z-DNA. Biochim. Biophys. Acta 1764, 320–323.

Ha, S.C., Kim, D., Hwang, H.Y., Rich, A., Kim, Y.G., Kim, K.K., 2008. The crystal structure of the second Z-DNA binding domain of human DAI (ZBP1) in complex with Z-DNA reveals an unusual binding mode to Z-DNA. Proc. Natl. Acad. Sci. U. S. A. 105, 20671–20676.

Habjan, M., Andersson, I., Klingstrom, J., Schumann, M., Martin, A., Zimmermann, P., Wagner, V., Pichlmair, A., Schneider, U., Muhlberger, E., Mirazimi, A., Weber, F., 2008. Processing of genome 5′ termini as a strategy of negative-strand RNA viruses to avoid RIG-I-dependent interferon induction. PLoS One 3, e2032.

Hampe, J., Lynch, N.J., Daniels, S., Bridger, S., Macpherson, A.J., Stokkers, P., Forbes, A., Lennard-Jones, J.E., Mathew, C.G., Curran, M.E., Schreiber, S., 2001. Fine mapping of the chromosome 3p susceptibility locus in inflammatory bowel disease. Gut 48, 191–197.

Han, Y., Chen, Z., Hou, R., Yan, D., Liu, C., Chen, S., Li, X., Du, W., 2015. Expression of AIM2 is correlated with increased inflammation in chronic hepatitis B patients. Virol. J. 12, 129.

Hanamsagar, R., Aldrich, A., Kielian, T., 2014. Critical role for the AIM2 inflammasome during acute CNS bacterial infection. J. Neurochem. 129, 704–711.

Hartmann, E., Gorlich, D., Kostka, S., Otto, A., Kraft, R., Knespel, S., Burger, E., Rapoport, T.A., Prehn, S., 1993. A tetrameric complex of membrane proteins in the endoplasmic reticulum. Eur. J. Biochem. 214, 375–381.

Hattori, Y., Matsuyama, K., Takahashi, T., Shu, E., Kanoh, H., Seishima, M., 2018. Anti-MDA5 antibody-positive dermatomyositis presenting with cellulitis-like erythema on the mandible as an initial symptom. Case Rep. Dermatol. 10, 110–114.

Helmick, C.G., Felson, D.T., Lawrence, R.C., Gabriel, S., Hirsch, R., Kwoh, C.K., Liang, M.H., Kremers, H.M., Mayes, M.D., Merkel, P.A., Pillemer, S.R., Reveille, J.D., Stone, J.H., National Arthritis Data Workgroup, 2008. Estimates of the prevalence of arthritis and other rheumatic conditions in the United States. Part I. Arthritis Rheum. 58, 15–25.

Henderson, B., Pettipher, E.R., 1989. Arthritogenic actions of recombinant IL-1 and tumour necrosis factor alpha in the rabbit: evidence for synergistic interactions between cytokines in vivo. Clin. Exp. Immunol. 75, 306–310.

Herbert, A., Rich, A., 1996. The biology of left-handed Z-DNA. J. Biol. Chem. 271, 11595–11598.

Hofmann, K., Bucher, P., Tschopp, J., 1997. The CARD domain: a new apoptotic signalling motif. Trends Biochem. Sci. 22, 155–156.

Hornung, V., Ellegast, J., Kim, S., Brzozka, K., Jung, A., Kato, H., Poeck, H., Akira, S., Conzelmann, K.K., Schlee, M., Endres, S., Hartmann, G., 2006. 5′-triphosphate RNA is the ligand for RIG-I. Science 314, 994–997.

Hornung, V., Ablasser, A., Charrel-Dennis, M., Bauernfeind, F., Horvath, G., Caffrey, D.R., Latz, E., Fitzgerald, K.A., 2009. AIM2 recognizes cytosolic dsDNA and forms a caspase-1-activating inflammasome with ASC. Nature 458, 514–518.

Hoshi, N., Schenten, D., Nish, S.A., Walther, Z., Gagliani, N., Flavell, R.A., Reizis, B., Shen, Z., Fox, J.G., Iwasaki, A., Medzhitov, R., 2012. MyD88 signalling in colonic mononuclear phagocytes drives colitis in IL-10-deficient mice. Nat. Commun. 3, 1120.

Hu, S., Peng, L., Kwak, Y.-T., Tekippe, E.M., Pasare, C., Malter, J.S., Hooper, L.V., Zaki, M.H., 2015. The DNA sensor AIM2 maintains intestinal homeostasis via regulation of epithelial antimicrobial host defense. Cell Rep. 13, 1922–1936.

Hugot, J.P., Chamaillard, M., Zouali, H., Lesage, S., Cezard, J.P., Belaiche, J., Almer, S., Tysk, C., O'morain, C.A., Gassull, M., Binder, V., Finkel, Y., Cortot, A., Modigliani, R., Laurent-Puig, P., Gower-Rousseau, C., Macry, J., Colombel, J.F., Sahbatou, M., Thomas, G., 2001. Association of NOD2 leucine-rich repeat variants with susceptibility to crohn's disease. Nature 411, 599–603.

Ishikawa, H., Barber, G.N., 2008. STING is an endoplasmic reticulum adaptor that facilitates innate immune signalling. Nature 455, 674–678.

Ishikawa, H., Ma, Z., Barber, G.N., 2009. STING regulates intracellular DNA-mediated, type I interferon-dependent innate immunity. Nature 461, 788–792.

Jakobs, C., Perner, S., Hornung, V., 2015. AIM2 drives joint inflammation in a self-DNA triggered model of chronic polyarthritis. PLoS One 10, e0131702.

Janeway Jr., C.A., 1989. Approaching the asymptote? Evolution and revolution in immunology. Cold Spring. Harb. Symp. Quant. Biol. 54 (Pt. 1), 1–13.

Jang, M.A., Kim, E.K., Now, H., Nguyen, N.T., Kim, W.J., Yoo, J.Y., Lee, J., Jeong, Y.M., Kim, C.H., Kim, O.H., Sohn, S., Nam, S.H., Hong, Y., Lee, Y.S., Chang, S.A., Jang, S.Y., Kim, J.W., Lee, M.S., Lim, S.Y., Sung, K.S., Park, K.T., Kim, B.J., Lee, J.H., Kim, D.K., Kee, C., Ki, C. S., 2015. Mutations in DDX58, which encodes RIG-I, cause atypical singleton-merten syndrome. Am. J. Hum. Genet. 96, 266–274.

Javierre, B.M., Fernandez, A.F., Richter, J., Al-Shahrour, F., Martin-Subero, J.I., Rodriguez-Ubreva, J., Berdasco, M., Fraga, M.F., O'Hanlon, T.P., Rider, L.G., Jacinto, F.V., Lopez-Longo, F.J., Dopazo, J., Forn, M., Peinado, M.A., Carreno, L., Sawalha, A.H., Harley, J.B., Siebert, R., Esteller, M., Miller, F.W., Ballestar, E., 2010. Changes in the pattern of DNA methylation associate with twin discordance in systemic lupus erythematosus. Genome Res. 20, 170–179.

Jiang, Y., Zhu, Y., Liu, Z.-J., Ouyang, S., 2017. The emerging roles of the DDX41 protein in immunity and diseases. Protein Cell 8, 83–89.

Jin, T., Perry, A., Jiang, J., Smith, P., Curry, J.A., Unterholzner, L., JIANG, Z., Horvath, G., Rathinam, V., Johnstone, R.W., Hornung, V., Latz, E., Bowie, A.G., Fitzgerald, K.A., Xiao, T.S., 2012. Structures of the HIN domain:DNA complexes reveal ligand binding and activation mechanisms of the AIM2 inflammasome and IFI16 receptor. Immunity 36, 561–571.

Jin, L., Getahun, A., Knowles, H.M., Mogan, J., Akerlund, L.J., Packard, T.A., Perraud, A.L., Cambier, J.C., 2013. STING/MPYS mediates host defense against *Listeria monocytogenes* infection by regulating Ly6C(hi) monocyte migration. J. Immunol. 190, 2835–2843.

Johnson, C.L., Gale Jr., M., 2006. CARD games between virus and host get a new player. Trends Immunol. 27, 1–4.

Johnson, K.E., Chikoti, L., Chandran, B., 2013. Herpes simplex virus 1 infection induces activation and subsequent inhibition of the IFI16 and NLRP3 inflammasomes. J. Virol. 87, 5005–5018.

Johnson, K.E., Bottero, V., Flaherty, S., Dutta, S., Singh, V.V., Chandran, B., 2014. IFI16 restricts HSV-1 replication by accumulating on the hsv-1 genome, repressing HSV-1 gene expression, and directly or indirectly modulating histone modifications. PLoS Pathog. 10, e1004503.

Johnstone, R.W., Kerry, J.A., Trapani, J.A., 1998. The human interferon-inducible protein, IFI 16, is a repressor of transcription. J. Biol. Chem. 273, 17172–17177.

Johnstone, R.W., Wei, W., Greenway, A., Trapani, J.A., 2000. Functional interaction between p53 and the interferon-inducible nucleoprotein IFI 16. Oncogene 19, 6033–6042.
Jones, J.W., Kayagaki, N., Broz, P., Henry, T., Newton, K., O'rourke, K., Chan, S., Dong, J., Qu, Y., Roose-Girma, M., Dixit, V.M., Monack, D.M., 2010. Absent in melanoma 2 is required for innate immune recognition of *Francisella tularensis*. Proc. Natl. Acad. Sci. U. S. A. 107, 9771–9776.
Kaiser, W.J., Upton, J.W., Mocarski, E.S., 2008. Receptor-interacting protein homotypic interaction motif-dependent control of NF-kappa B activation via the DNA-dependent activator of IFN regulatory factors. J. Immunol. 181, 6427–6434.
Kang, D.C., Gopalkrishnan, R.V., Wu, Q., Jankowsky, E., Pyle, A.M., Fisher, P.B., 2002. mda-5: an interferon-inducible putative RNA helicase with double-stranded RNA-dependent ATPase activity and melanoma growth-suppressive properties. Proc. Natl. Acad. Sci. U. S. A. 99, 637–642.
Kaser, A., Zeissig, S., Blumberg, R.S., 2010. Inflammatory bowel disease. Annu. Rev. Immunol. 28, 573–621.
Kato, H., Sato, S., Yoneyama, M., Yamamoto, M., Uematsu, S., Matsui, K., Tsujimura, T., Takeda, K., Fujita, T., Takeuchi, O., Akira, S., 2005. Cell type-specific involvement of RIG-I in antiviral response. Immunity 23, 19–28.
Kato, H., Takeuchi, O., Sato, S., Yoneyama, M., Yamamoto, M., Matsui, K., Uematsu, S., Jung, A., Kawai, T., Ishii, K.J., Yamaguchi, O., Otsu, K., Tsujimura, T., Koh, C.S., Reis e Sousa, C., Matsuura, Y., Fujita, T., Akira, S., 2006. Differential roles of MDA5 and RIG-I helicases in the recognition of RNA viruses. Nature 441, 101–105.
Kawane, K., Fukuyama, H., Kondoh, G., Takeda, J., Ohsawa, Y., Uchiyama, Y., Nagata, S., 2001. Requirement of DNase II for definitive erythropoiesis in the mouse fetal liver. Science 292, 1546–1549.
Kawane, K., Ohtani, M., Miwa, K., Kizawa, T., Kanbara, Y., Yoshioka, Y., Yoshikawa, H., Nagata, S., 2006. Chronic polyarthritis caused by mammalian DNA that escapes from degradation in macrophages. Nature 443, 998–1002.
Kayagaki, N., Stowe, I.B., Lee, B.L., O'rourke, K., Anderson, K., Warming, S., Cuellar, T., Haley, B., Roose-Girma, M., Phung, Q.T., Liu, P.S., Lill, J.R., Li, H., Wu, J., Kummerfeld, S., Zhang, J., Lee, W.P., Snipas, S.J., Salvesen, G.S., Morris, L.X., Fitzgerald, L., Zhang, Y., Bertram, E.M., Goodnow, C.C., Dixit, V.M., 2015. Caspase-11 cleaves gasdermin D for non-canonical inflammasome signalling. Nature 526, 666–671.
Kesavardhana, S., Kuriakose, T., Guy, C.S., Samir, P., Malireddi, R.K.S., Mishra, A., Kanneganti, T.D., 2017. ZBP1/DAI ubiquitination and sensing of influenza vRNPs activate programmed cell death. J. Exp. Med. 214, 2217–2229.
Kim, S., Bauernfeind, F., Ablasser, A., Hartmann, G., Fitzgerald, K.A., Latz, E., Hornung, V., 2010. *Listeria monocytogenes* is sensed by the NLRP3 and AIM2 inflammasome. Eur. J. Immunol. 40, 1545–1551.
Kimkong, I., Avihingsanon, Y., Hirankarn, N., 2009. Expression profile of HIN200 in leukocytes and renal biopsy of SLE patients by real-time RT-PCR. Lupus 18, 1066–1072.
Kranzusch, P.J., Lee, A.S., Berger, J.M., Doudna, J.A., 2013. Structure of human cGAS reveals a conserved family of second-messenger enzymes in innate immunity. Cell Rep. 3, 1362–1368.
Kuriakose, T., Kanneganti, T.-D., 2018. ZBP1: innate sensor regulating cell death and inflammation. Trends Immunol. 39, 123–134.
Kuriakose, T., Man, S.M., Malireddi, R.K.S., Karki, R., Kesavardhana, S., Place, D.E., Neale, G., Vogel, P., Kanneganti, T.-D., 2016. ZBP1/DAI is an innate sensor of influenza virus triggering the NLRP3 inflammasome and programmed cell death pathways. Sci. Immunol. 1, aag2045.

Lamkanfi, M., Dixit, V.M., 2009. Inflammasomes: guardians of cytosolic sanctity. Immunol. Rev. 227, 95–105.

Lassig, C., Matheisl, S., Sparrer, K.M., de Oliveira Mann, C.C., Moldt, M., Patel, J.R., Goldeck, M., Hartmann, G., Garcia-Sastre, A., Hornung, V., Conzelmann, K.K., Beckmann, R., Hopfner, K.P., 2015. ATP hydrolysis by the viral RNA sensor RIG-I prevents unintentional recognition of self-RNA. elife 4, e10859.

Latz, E., Xiao, T.S., Stutz, A., 2013. Activation and regulation of the inflammasomes. Nat. Rev. Immunol. 13. https://doi.org/10.1038/nri3452.

Li, X.N., Song, J., Zhang, L., Lemaire, S.A., Hou, X., Zhang, C., Coselli, J.S., Chen, L., Wang, X.L., Zhang, Y., Shen, Y.H., 2009. Activation of the AMPK-FOXO3 pathway reduces fatty acid-induced increase in intracellular reactive oxygen species by upregulating thioredoxin. Diabetes 58, 2246–2257.

Li, X., Shu, C., Yi, G., Chaton, C.T., Shelton, C.L., Diao, J., Zuo, X., Kao, C.C., Herr, A.B., Li, P., 2013. Cyclic GMP-AMP synthase is activated by double-stranded DNA-induced oligomerization. Immunity 39, 1019–1031.

Liao, J.K., 2013. Linking endothelial dysfunction with endothelial cell activation. J. Clin. Invest. 123, 540–541.

Liao, J.C., Lam, R., Brazda, V., Duan, S., Ravichandran, M., Ma, J., Xiao, T., Tempel, W., Zuo, X., Wang, Y.X., Chirgadze, N.Y., Arrowsmith, C.H., 2011. Interferon-inducible protein 16: insight into the interaction with tumor suppressor p53. Structure 19, 418–429.

Liu, X., Yao, M., Li, N., Wang, C., Zheng, Y., Cao, X., 2008. CaMKII promotes TLR-triggered proinflammatory cytokine and type I interferon production by directly binding and activating TAK1 and IRF3 in macrophages. Blood 112, 4961–4970.

Liu, X., Zhang, Z., Ruan, J., Pan, Y., Magupalli, V.G., Wu, H., Lieberman, J., 2016a. Inflammasome—activated gasdermin D causes pyroptosis by forming membrane pores. Nature 535, 153–158.

Liu, Y., Goulet, M.L., Sze, A., Hadj, S.B., Belgnaoui, S.M., Lababidi, R.R., Zheng, C., Fritz, J.H., Olagnier, D., Lin, R., 2016b. RIG-I-mediated STING upregulation restricts herpes simplex virus 1 infection. J. Virol. 90, 9406–9419.

Lolicato, M., Bucchi, A., Arrigoni, C., Zucca, S., Nardini, M., Schroeder, I., Simmons, K., Aquila, M., Difrancesco, D., Bolognesi, M., Schwede, F., Kashin, D., Fishwick, C.W., Johnson, A.P., Thiel, G., Moroni, A., 2014. Cyclic dinucleotides bind the C-linker of HCN4 to control channel cAMP responsiveness. Nat. Chem. Biol. 10, 457–462.

Loo, Y.M., Gale Jr., M., 2011. Immune signaling by RIG-I-like receptors. Immunity 34, 680–692.

Loo, Y.M., Fornek, J., Crochet, N., Bajwa, G., Perwitasari, O., Martinez-Sobrido, L., Akira, S., Gill, M.A., Garcia-Sastre, A., Katze, M.G., Gale Jr., M., 2008. Distinct RIG-I and MDA5 signaling by RNA viruses in innate immunity. J. Virol. 82, 335–345.

Lozano-Ruiz, B., Bachiller, V., Garcia-Martinez, I., Zapater, P., Gomez-Hurtado, I., Moratalla, A., Gimenez, P., Bellot, P., Frances, R., Such, J., Gonzalez-Navajas, J.M., 2015. Absent in melanoma 2 triggers a heightened inflammasome response in ascitic fluid macrophages of patients with cirrhosis. J. Hepatol. 62, 64–71.

Ludlow, L.E., Johnstone, R.W., Clarke, C.J., 2005. The HIN-200 family: more than interferon-inducible genes? Exp. Cell Res. 308, 1–17.

Luthra, P., Sun, D., Silverman, R.H., He, B., 2011. Activation of IFN-β expression by a viral mRNA through RNase L and MDA5. Proc. Natl. Acad. Sci. U. S. A. 108, 2118–2123.

Magnusson, M., Zare, F., Tarkowski, A., 2006. Requirement of type I interferon signaling for arthritis triggered by double-stranded RNA. Arthritis Rheum. 54, 148–157.

Maharaj, N.P., Wies, E., Stoll, A., Gack, M.U., 2012. Conventional protein kinase C-alpha (PKC-alpha) and PKC-beta negatively regulate RIG-I antiviral signal transduction. J. Virol. 86, 1358–1371.
Man, S.M., Karki, R., Malireddi, R.K., Neale, G., Vogel, P., Yamamoto, M., Lamkanfi, M., Kanneganti, T.D., 2015a. The transcription factor IRF1 and guanylate-binding proteins target activation of the AIM2 inflammasome by *Francisella* infection. Nat. Immunol. 16, 467–475.
Man, S.M., Zhu, Q., Zhu, L., Liu, Z., Karki, R., Malik, A., Sharma, D., Li, L., Malireddi, R.K., Gurung, P., Neale, G., Olsen, S.R., Carter, R.A., McGoldrick, D.J., Wu, G., Finkelstein, D., Vogel, P., Gilbertson, R.J., Kanneganti, T.D., 2015b. Critical role for the DNA sensor AIM2 in stem cell proliferation and Cancer. Cell 162, 45–58.
Man, S.M., Karki, R., Kanneganti, T.-D., 2016a. AIM2 inflammasome in infection, cancer and autoimmunity: role in DNA sensing, inflammation and innate immunity. Eur. J. Immunol. 46, 269–280.
Man, S.M., Karki, R., Sasai, M., Place, D.E., Kesavardhana, S., Temirov, J., Frase, S., Zhu, Q., Malireddi, R.K.S., Kuriakose, T., Peters, J.L., Neale, G., Brown, S.A., Yamamoto, M., Kanneganti, T.D., 2016b. IRGB10 liberates bacterial ligands for sensing by the AIM2 and caspase-11-NLRP3 inflammasomes. Cell 167, 382-396.e17.
Martin, H.M., Campbell, B.J., Hart, C.A., Mpofu, C., Nayar, M., Singh, R., Englyst, H., Williams, H.F., Rhodes, J.M., 2004. Enhanced *Escherichia coli* adherence and invasion in crohn's disease and colon cancer. Gastroenterology 127, 80–93.
Martinez-Medina, M., Denizot, J., Dreux, N., Robin, F., Billard, E., Bonnet, R., Darfeuille-Michaud, A., Barnich, N., 2014. Western diet induces dysbiosis with increased *E coli* in CEABAC10 mice, alters host barrier function favouring AIEC colonisation. Gut 63, 116–124.
McCartney, S. A., Thackray, L. B., Gitlin, L., Gilfillan, S., Virgin, H. W. & Colonna, M. 2008. MDA-5 recognition of a murine norovirus. PLoS Pathog., 4**b** e1000108.
Medzhitov, R., 2009. Approaching the asymptote: 20 years later. Immunity 30, 766–775.
Menetret, J.F., Hegde, R.S., Heinrich, S.U., Chandramouli, P., Ludtke, S.J., Rapoport, T.A., Akey, C.W., 2005. Architecture of the ribosome-channel complex derived from native membranes. J. Mol. Biol. 348, 445–457.
Meunier, E., Wallet, P., Dreier, R.F., Costanzo, S., Anton, L., Ruhl, S., Dussurgey, S., Dick, M.S., Kistner, A., Rigard, M., Degrandi, D., Pfeffer, K., Yamamoto, M., Henry, T., Broz, P., 2015. Guanylate-binding proteins promote activation of the AIM2 inflammasome during infection with *Francisella novicida*. Nat. Immunol. 16, 476–484.
Molodecky, N.A., Soon, I.S., Rabi, D.M., Ghali, W.A., Ferris, M., Chernoff, G., Benchimol, E.I., Panaccione, R., Ghosh, S., Barkema, H.W., Kaplan, G.G., 2012. Increasing incidence and prevalence of the inflammatory bowel diseases with time, based on systematic review. Gastroenterology 142, 46-54 e42; quiz e30.
Munoz, L.E., Lauber, K., Schiller, M., Manfredi, A.A., Herrmann, M., 2010. The role of defective clearance of apoptotic cells in systemic autoimmunity. Nat. Rev. Rheumatol. 6, 280–289.
Muruve, D.A., Petrilli, V., Zaiss, A.K., White, L.R., Clark, S.A., Ross, P.J., Parks, R.J., Tschopp, J., 2008. The inflammasome recognizes cytosolic microbial and host DNA and triggers an innate immune response. Nature 452, 103–107.
Nagata, S., 2010. Apoptosis and autoimmune diseases. Ann. N. Y. Acad. Sci. 1209, 10–16.
Nejentsev, S., Walker, N., Riches, D., Egholm, M., Todd, J.A., 2009. Rare variants of IFIH1, a gene implicated in antiviral responses, protect against type 1 diabetes. Science 324, 387–389.
Nestle, F.O., Kaplan, D.H., Barker, J., 2009. Psoriasis. N. Engl. J. Med. 361, 496–509.

Ohman, L., Franzen, L., Rudolph, U., Birnbaumer, L., Hornquist, E.H., 2002. Regression of peyer's patches in G alpha i2 deficient mice prior to colitis is associated with reduced expression of Bcl-2 and increased apoptosis. Gut 51, 392–397.
Onoguchi, K., Yoneyama, M., Fujita, T., 2011. Retinoic acid-inducible gene-I-like receptors. J. Interf. Cytokine Res. 31, 27–31.
Ori, D., Murase, M., Kawai, T., 2017. Cytosolic nucleic acid sensors and innate immune regulation. Int. Rev. Immunol. 36, 74–88.
Ortiz-Fernandez, L., Garcia-LOZANO, J.R., Montes-Cano, M.A., Conde-Jaldon, M., Ortego-Centeno, N., Garcia-Hernandez, F.J., Espinosa, G., Grana-Gil, G., Sanchez-Burson, J., Blanco, R., Barnosi-Marin, A.C., Solans, R., Fanlo, P., Rodriguez-Carballeira, M., Camps, T., Castaneda, S., Nunez-Roldan, A., Martin, J., Gonzalez-Escribano, M.F., 2015. Variants of the IFI16 gene affecting the levels of expression of mRNA are associated with susceptibility to Behcet disease. J. Rheumatol. 42, 695–701.
Osborn, O., Olefsky, J.M., 2012. The cellular and signaling networks linking the immune system and metabolism in disease. Nat. Med. 18, 363–374.
Ouchi, N., Parker, J.L., Lugus, J.J., Walsh, K., 2011. Adipokines in inflammation and metabolic disease. Nat. Rev. Immunol. 11, 85–97.
Pages, F., Kirilovsky, A., Mlecnik, B., Asslaber, M., Tosolini, M., Bindea, G., Lagorce, C., Wind, P., Marliot, F., Bruneval, P., Zatloukal, K., Trajanoski, Z., Berger, A., Fridman, W.H., Galon, J., 2009. In situ cytotoxic and memory T cells predict outcome in patients with early-stage colorectal cancer. J. Clin. Oncol. 27, 5944–5951.
Panchanathan, R., Shen, H., Duan, X., Rathinam, V.A., Erickson, L.D., Fitzgerald, K.A., Choubey, D., 2011. Aim2 deficiency in mice suppresses the expression of the inhibitory Fcgamma receptor (FcgammaRIIB) through the induction of the IFN-inducible p202, a lupus susceptibility protein. J. Immunol. 186, 6762–6770.
Park, E., NA, H.S., Song, Y.R., Shin, S.Y., Kim, Y.M., Chung, J., 2014. Activation of NLRP3 and AIM2 inflammasomes by *Porphyromonas gingivalis* infection. Infect. Immun. 82, 112–123.
Parvatiyar, K., Zhang, Z., Teles, R.M., Ouyang, S., Jiang, Y., Iyer, S.S., Zaver, S.A., Schenk, M., Zeng, S., Zhong, W., Liu, Z.J., Modlin, R.L., Liu, Y.J., Cheng, G., 2012. The helicase DDX41 recognizes the bacterial secondary messengers cyclic di-GMP and cyclic di-AMP to activate a type I interferon immune response. Nat. Immunol. 13, 1155–1161.
Perry, R.J., Samuel, V.T., Petersen, K.F., Shulman, G.I., 2014. The role of hepatic lipids in hepatic insulin resistance and type 2 diabetes. Nature 510, 84–91.
Petrasek, J., Iracheta-Vellve, A., Csak, T., Satishchandran, A., Kodys, K., Kurt-Jones, E.A., Fitzgerald, K.A., Szabo, G., 2013. STING-IRF3 pathway links endoplasmic reticulum stress with hepatocyte apoptosis in early alcoholic liver disease. Proc. Natl. Acad. Sci. U. S. A. 110, 16544–16549.
Plumet, S., Herschke, F., Bourhis, J.M., Valentin, H., Longhi, S., Gerlier, D., 2007. Cytosolic 5′-triphosphate ended viral leader transcript of measles virus as activator of the RIG I-mediated interferon response. PLoS One 2, e279.
Polprasert, C., Schulze, I., Sekeres, M.A., Makishima, H., Przychodzen, B., Hosono, N., Singh, J., Padgett, R.A., Gu, X., Phillips, J.G., Clemente, M., Parker, Y., Lindner, D., Dienes, B., Jankowsky, E., Saunthararajah, Y., Du, Y., Oakley, K., Nguyen, N., Mukherjee, S., Pabst, C., Godley, L.A., Churpek, J.E., Pollyea, D.A., Krug, U., Berdel, W.E., Klein, H.U., Dugas, M., Shiraishi, Y., Chiba, K., Tanaka, H., Miyano, S., Yoshida, K., Ogawa, S., Muller-Tidow, C., Maciejewski, J.P., 2015. Inherited and somatic defects in DDX41 in myeloid neoplasms. Cancer Cell 27, 658–670.
Qiao, B., Wu, J., Chu, Y.W., Wang, Y., Wang, D.P., Wu, H.S., Xiong, S.D., 2005. Induction of systemic lupus erythematosus-like syndrome in syngeneic mice by immunization with activated lymphocyte-derived DNA. Rheumatology (Oxford) 44, 1108–1114.

Qiao, J.T., Cui, C., Qing, L., Wang, L.S., He, T.Y., Yan, F., Liu, F.Q., Shen, Y.H., Hou, X.G., Chen, L., 2018. Activation of the STING-IRF3 pathway promotes hepatocyte inflammation, apoptosis and induces metabolic disorders in nonalcoholic fatty liver disease. Metabolism 81, 13–24.

Rakoff-Nahoum, S., Hao, L., Medzhitov, R., 2006. Role of toll-like receptors in spontaneous commensal-dependent colitis. Immunity 25, 319–329.

Rathinam, V.A., Jiang, Z., Waggoner, S.N., Sharma, S., Cole, L.E., Waggoner, L., Vanaja, S.K., Monks, B.G., Ganesan, S., Latz, E., Hornung, V., Vogel, S.N., Szomolanyi-Tsuda, E., Fitzgerald, K.A., 2010. The AIM2 inflammasome is essential for host defense against cytosolic bacteria and DNA viruses. Nat. Immunol. 11, 395–402.

Ratsimandresy, R.A., Indramohan, M., Dorfleutner, A., Stehlik, C., 2017. The AIM2 inflammasome is a central regulator of intestinal homeostasis through the IL-18/IL-22/STAT3 pathway. Cell. Mol. Immunol. 14, 127–142.

Rebsamen, M., Heinz, L.X., Meylan, E., Michallet, M.-C., Schroder, K., Hofmann, K., Vazquez, J., Benedict, C.A., Tschopp, J., 2009. DAI/ZBP1 recruits RIP1 and RIP3 through RIP homotypic interaction motifs to activate NF-κB. EMBO Rep. 10, 916–922.

Reinholz, M., Kawakami, Y., Salzer, S., Kreuter, A., Dombrowski, Y., Koglin, S., Kresse, S., Ruzicka, T., Schauber, J., 2013. HPV16 activates the AIM2 inflammasome in keratinocytes. Arch. Dermatol. Res. 305, 723–732.

Renne, J., Schafer, V., Werfel, T., Wittmann, M., 2010. Interleukin-1 from epithelial cells fosters T cell-dependent skin inflammation. Br. J. Dermatol. 162, 1198–1205.

Roberto, C.A., Swinburn, B., Hawkes, C., Huang, T.T., Costa, S.A., Ashe, M., Zwicker, L., Cawley, J.H., Brownell, K.D., 2015. Patchy progress on obesity prevention: emerging examples, entrenched barriers, and new thinking. Lancet 385, 2400–2409.

Roberts, T.L., Idris, A., Dunn, J.A., Kelly, G.M., Burnton, C.M., Hodgson, S., Hardy, L.L., Garceau, V., Sweet, M.J., Ross, I.L., Hume, D.A., Stacey, K.J., 2009. HIN-200 proteins regulate caspase activation in response to foreign cytoplasmic DNA. Science 323, 1057–1060.

Romeo, G.R., Lee, J., Shoelson, S.E., 2012. Metabolic syndrome, insulin resistance, and roles of inflammation—mechanisms and therapeutic targets. Arterioscler. Thromb. Vasc. Biol. 32, 1771–1776.

Roth-Cross, J.K., Bender, S.J., Weiss, S.R., 2008. Murine coronavirus mouse hepatitis virus is recognized by MDA5 and induces type I interferon in brain macrophages/microglia. J. Virol. 82, 9829–9838.

Rudolph, U., Finegold, M.J., Rich, S.S., Harriman, G.R., Srinivasan, Y., Brabet, P., Boulay, G., Bradley, A., Birnbaumer, L., 1995. Ulcerative colitis and adenocarcinoma of the colon in G alpha i2-deficient mice. Nat. Genet. 10, 143–150.

Rutsch, F., MacDougall, M., Lu, C., Buers, I., Mamaeva, O., Nitschke, Y., Rice, G.I., Erlandsen, H., Kehl, H.G., Thiele, H., Nurnberg, P., Hohne, W., Crow, Y.J., Feigenbaum, A., Hennekam, R.C., 2015. A specific IFIH1 gain-of-function mutation causes singleton-merten syndrome. Am. J. Hum. Genet. 96, 275–282.

Saiga, H., Kitada, S., Shimada, Y., Kamiyama, N., Okuyama, M., Makino, M., Yamamoto, M., Takeda, K., 2012. Critical role of AIM2 in *Mycobacterium tuberculosis* infection. Int. Immunol. 24, 637–644.

Saito, T., Hirai, R., Loo, Y.M., Owen, D., Johnson, C.L., Sinha, S.C., Akira, S., Fujita, T., Gale Jr., M., 2007. Regulation of innate antiviral defenses through a shared repressor domain in RIG-I and LGP2. Proc. Natl. Acad. Sci. U. S. A. 104, 582–587.

Saitoh, T., Fujita, N., Hayashi, T., Takahara, K., Satoh, T., Lee, H., Matsunaga, K., Kageyama, S., Omori, H., Noda, T., Yamamoto, N., Kawai, T., Ishii, K., Takeuchi, O., Yoshimori, T., Akira, S., 2009. Atg9a controls dsDNA-driven dynamic translocation of STING and the innate immune response. Proc. Natl. Acad. Sci. U. S. A. 106, 20842–20846.

Salcedo, R., Cataisson, C., Hasan, U., Yuspa, S.H., Trinchieri, G., 2013. MyD88 and its divergent toll in carcinogenesis. Trends Immunol. 34, 379–389.

Samanta, M., Iwakiri, D., Kanda, T., Imaizumi, T., Takada, K., 2006. EB virus-encoded RNAs are recognized by RIG-I and activate signaling to induce type I IFN. EMBO J. 25, 4207–4214.

Sato, S., Hoshino, K., Satoh, T., Fujita, T., Kawakami, Y., Fujita, T., Kuwana, M., 2009. RNA helicase encoded by melanoma differentiation-associated gene 5 is a major autoantigen in patients with clinically amyopathic dermatomyositis: association with rapidly progressive interstitial lung disease. Arthritis Rheum. 60, 2193–2200.

Sauer, J.D., Sotelo-Troha, K., von Moltke, J., Monroe, K.M., Rae, C.S., Brubaker, S.W., Hyodo, M., Hayakawa, Y., Woodward, J.J., Portnoy, D.A., Vance, R.E., 2011. The N-ethyl-N-nitrosourea-induced goldenticket mouse mutant reveals an essential function of STING in the in vivo interferon response to *Listeria monocytogenes* and cyclic dinucleotides. Infect. Immun. 79, 688–694.

Schlee, M., 2013. Master sensors of pathogenic RNA—RIG-I like receptors. Immunobiology 218, 1322–1335.

Schlee, M., Hartmann, G., 2016. Discriminating self from non-self in nucleic acid sensing. Nat. Rev. Immunol. 16, 566–580.

Schwartz, T., Behlke, J., Lowenhaupt, K., Heinemann, U., Rich, A., 2001. Structure of the DLM-1-Z-DNA complex reveals a conserved family of Z-DNA-binding proteins. Nat. Struct. Biol. 8, 761–765.

Seelig, H.P., Ehrfeld, H., Renz, M., 1994. Interferon-gamma-inducible protein p16. A new target of antinuclear antibodies in patients with systemic lupus erythematosus. Arthritis Rheum. 37, 1672–1683.

Sharma, S., Tenoever, B.R., Grandvaux, N., Zhou, G.P., Lin, R., Hiscott, J., 2003. Triggering the interferon antiviral response through an IKK-related pathway. Science 300, 1148–1151.

Shepherd, J., Little, M.C., Nicklin, M.J., 2004. Psoriasis-like cutaneous inflammation in mice lacking interleukin-1 receptor antagonist. J. Invest. Dermatol. 122, 665–669.

Shi, J., Zhao, Y., Wang, K., Shi, X., Wang, Y., Huang, H., Zhuang, Y., Cai, T., Wang, F., Shao, F., 2015. Cleavage of GSDMD by inflammatory caspases determines pyroptotic cell death. Nature 526, 660–665.

Singh, V.V., Kerur, N., Bottero, V., Dutta, S., Chakraborty, S., Ansari, M.A., Paudel, N., Chikoti, L., Chandran, B., 2013. Kaposi's sarcoma-associated herpesvirus latency in endothelial and B cells activates gamma interferon-inducible protein 16-mediated inflammasomes. J. Virol. 87, 4417–4431.

Solis, M., Goubau, D., Hiscott, J., 2007. RIG-I has guts: identification of a role for RIG-I in colitis development. Cell Res. 17, 974–975.

Spahn, T.W., Herbst, H., Rennert, P.D., Lugering, N., Maaser, C., Kraft, M., Fontana, A., Weiner, H.L., Domschke, W., Kucharzik, T., 2002. Induction of colitis in mice deficient of peyer's patches and mesenteric lymph nodes is associated with increased disease severity and formation of colonic lymphoid patches. Am. J. Pathol. 161, 2273–2282.

Sridharan, H., Ragan, K.B., Guo, H., Gilley, R.P., Landsteiner, V.J., Kaiser, W.J., Upton, J.W., 2017. Murine cytomegalovirus IE3-dependent transcription is required for DAI/ZBP1-mediated necroptosis. EMBO Rep. 18, 1429–1441.

Steinberg, H.O., Paradisi, G., Hook, G., Crowder, K., Cronin, J., Baron, A.D., 2000. Free fatty acid elevation impairs insulin-mediated vasodilation and nitric oxide production. Diabetes 49, 1231–1238.

Stratmann, S.A., Morrone, S.R., van Oijen, A.M., Sohn, J., 2015. The innate immune sensor IFI16 recognizes foreign DNA in the nucleus by scanning along the duplex. elife 4, e11721.

Sumpter Jr., R., Loo, Y.M., Foy, E., Li, K., Yoneyama, M., Fujita, T., Lemon, S.M., Gale Jr., M., 2005. Regulating intracellular antiviral defense and permissiveness to hepatitis C virus RNA replication through a cellular RNA helicase, RIG-I. J. Virol. 79, 2689–2699.
Sun, X., Yin, J., Starovasnik, M.A., Fairbrother, W.J., Dixit, V.M., 2002. Identification of a novel homotypic interaction motif required for the phosphorylation of receptor-interacting protein (RIP) by RIP3. J. Biol. Chem. 277, 9505–9511.
Sun, W., Li, Y., Chen, L., Chen, H., You, F., Zhou, X., Zhou, Y., Zhai, Z., Chen, D., Jiang, Z., 2009. ERIS, an endoplasmic reticulum IFN stimulator, activates innate immune signaling through dimerization. Proc. Natl. Acad. Sci. U. S. A. 106, 8653–8658.
Sun, Z., Ren, H., Liu, Y., Teeling, J.L., Gu, J., 2011. Phosphorylation of RIG-I by casein kinase II inhibits its antiviral response. J. Virol. 85, 1036–1047.
Sun, L., Wu, J., Du, F., Chen, X., Chen, Z.J., 2013. Cyclic GMP-AMP synthase is a cytosolic DNA sensor that activates the type I interferon pathway. Science 339, 786–791.
Takaoka, A., Wang, Z., Choi, M.K., Yanai, H., Negishi, H., Ban, T., Lu, Y., Miyagishi, M., Kodama, T., Honda, K., Ohba, Y., Taniguchi, T., 2007. DAI (DLM-1/ZBP1) is a cytosolic DNA sensor and an activator of innate immune response. Nature 448, 501–505.
Tamayo, R., Pratt, J.T., Camilli, A., 2007. Roles of cyclic diguanylate in the regulation of bacterial pathogenesis. Annu. Rev. Microbiol. 61, 131–148.
Tanaka, Y., Chen, Z.J., 2012. STING specifies IRF3 phosphorylation by TBK1 in the cytosolic DNA signaling pathway. Sci. Signal. 5, ra20.
Tedeschi, S.K., Bermas, B., Costenbader, K.H., 2013. Sexual disparities in the incidence and course of SLE and RA. Clin. Immunol. 149, 211–218.
Thapa, R.J., Ingram, J.P., Ragan, K.B., Nogusa, S., Boyd, D.F., Benitez, A.A., Sridharan, H., Kosoff, R., Shubina, M., Landsteiner, V.J., Andrake, M., Vogel, P., Sigal, L.J., Tenoever, B.R., Thomas, P.G., Upton, J.W., Balachandran, S., 2016. DAI senses influenza a virus genomic RNA and activates RIPK3-dependent cell death. Cell Host Microbe 20, 674–681.
Udden, S.M.N., Peng, L., Gan, J.L., Shelton, J.M., Malter, J.S., Hooper, L.V., Zaki, M.H., 2017. NOD2 suppresses colorectal tumorigenesis via downregulation of the TLR pathways. Cell Rep. 19, 2756–2770.
Unterholzner, L., Keating, S.E., Baran, M., Horan, K.A., Jensen, S.B., Sharma, S., Sirois, C.M., Jin, T., Latz, E., Xiao, T.S., Fitzgerald, K.A., Paludan, S.R., Bowie, A.G., 2010. IFI16 is an innate immune sensor for intracellular DNA. Nat. Immunol. 11, 997–1004.
Upton, J.W., Kaiser, W.J., Mocarski, E.S., 2012. DAI/ZBP1/DLM-1 complexes with RIP3 to mediate virus-induced programmed necrosis that is targeted by murine cytomegalovirus vIRA. Cell Host Microbe 11, 290–297.
Vanhove, W., Peeters, P.M., Staelens, D., Schraenen, A., van der Goten, J., Cleynen, I., de Schepper, S., van Lommel, L., Reynaert, N.L., Schuit, F., van Assche, G., Ferrante, M., de Hertogh, G., Wouters, E.F., Rutgeerts, P., Vermeire, S., Nys, K., Arijs, I., 2015. Strong upregulation of AIM2 and IFI16 inflammasomes in the mucosa of patients with active inflammatory bowel disease. Inflamm. Bowel Dis. 21, 2673–2682.
Venereau, E., Ceriotti, C., Bianchi, M.E., 2015. DAMPs from cell death to new life. Front. Immunol. 6, 422.
Villaret, A., Galitzky, J., Decaunes, P., Esteve, D., Marques, M.A., Sengenes, C., Chiotasso, P., Tchkonia, T., Lafontan, M., Kirkland, J.L., Bouloumie, A., 2010. Adipose tissue endothelial cells from obese human subjects: differences among depots in angiogenic, metabolic, and inflammatory gene expression and cellular senescence. Diabetes 59, 2755–2763.
Walport, M.J., 2000. Lupus, DNase and defective disposal of cellular debris. Nat. Genet. 25, 135–136.

Wang, X.L., Zhang, L., Youker, K., Zhang, M.X., Wang, J., Lemaire, S.A., Coselli, J.S., Shen, Y.H., 2006. Free fatty acids inhibit insulin signaling-stimulated endothelial nitric oxide synthase activation through upregulating PTEN or inhibiting Akt kinase. Diabetes 55, 2301–2310.

Wang, Y., Zhang, H.X., Sun, Y.P., Liu, Z.X., Liu, X.S., Wang, L., Lu, S.Y., Kong, H., Liu, Q.L., Li, X.H., Lu, Z.Y., Chen, S.J., Chen, Z., Bao, S.S., DAI, W., Wang, Z.G., 2007. Rig-I$^{-/-}$ mice develop colitis associated with downregulation of G alpha i2. Cell Res. 17, 858–868.

Wang, F., Gao, X., Barrett, J.W., Shao, Q., Bartee, E., Mohamed, M.R., Rahman, M., Werden, S., Irvine, T., Cao, J., Dekaban, G.A., McFadden, G., 2008. RIG-I mediates the co-induction of tumor necrosis factor and type I interferon elicited by myxoma virus in primary human macrophages. PLoS Pathog. 4, e1000099.

Wassermann, R., Gulen, M.F., Sala, C., Perin, S.G., Lou, Y., Rybniker, J., Schmid-Burgk, J.L., Schmidt, T., Hornung, V., Cole, S.T., Ablasser, A., 2015. *Mycobacterium tuberculosis* differentially activates cGAS- and inflammasome-dependent intracellular immune responses through ESX-1. Cell Host Microbe 17, 799–810.

Wies, E., Wang, M.K., Maharaj, N.P., Chen, K., Zhou, S., Finberg, R.W., Gack, M.U., 2013. Dephosphorylation of the RNA sensors RIG-I and MDA5 by the phosphatase PP1 is essential for innate immune signaling. Immunity 38, 437–449.

Wilson, J.E., Petrucelli, A.S., Chen, L., Koblansky, A.A., Truax, A.D., Oyama, Y., Rogers, A.B., Brickey, W.J., Wang, Y., Schneider, M., Muhlbauer, M., Chou, W.C., Barker, B.R., Jobin, C., Allbritton, N.L., Ramsden, D.A., Davis, B.K., Ting, J.P., 2015. Inflammasome-independent role of AIM2 in suppressing colon tumorigenesis via DNA-PK and Akt. Nat. Med. 21, 906–913.

Woo, S.R., Fuertes, M.B., Corrales, L., Spranger, S., Furdyna, M.J., Leung, M.Y., Duggan, R., Wang, Y., Barber, G.N., Fitzgerald, K.A., Alegre, M.L., Gajewski, T.F., 2014. STING-dependent cytosolic DNA sensing mediates innate immune recognition of immunogenic tumors. Immunity 41, 830–842.

Woodward, J.J., Iavarone, A.T., Portnoy, D.A., 2010. c-di-AMP secreted by intracellular *Listeria monocytogenes* activates a host type I interferon response. Science 328, 1703–1705.

Wu, J., Sun, L., Chen, X., Du, F., Shi, H., Chen, C., Chen, Z.J., 2013. Cyclic GMP-AMP is an endogenous second messenger in innate immune signaling by cytosolic DNA. Science 339, 826–830.

Wu, X., Yang, J., NA, T., Zhang, K., Davidoff, A.M., Yuan, B.Z., Wang, Y., 2017. RIG-I and IL-6 are negative-feedback regulators of STING induced by double-stranded DNA. PLoS One 12, e0182961.

Yang, Y., Zhou, X., Kouadir, M., Shi, F., Ding, T., Liu, C., Liu, J., Wang, M., Yang, L., Yin, X., Zhao, D., 2013. The AIM2 inflammasome is involved in macrophage activation during infection with virulent Mycobacterium bovis strain. J. Infect. Dis. 208, 1849–1858.

Yang, C.A., Huang, S.T., Chiang, B.L., 2015. Sex-dependent differential activation of NLRP3 and AIM2 inflammasomes in SLE macrophages. Rheumatology (Oxford) 54, 324–331.

Yki-Jarvinen, H., 2014. Non-alcoholic fatty liver disease as a cause and a consequence of metabolic syndrome. Lancet Diabetes Endocrinol. 2, 901–910.

Yoneyama, M., Kikuchi, M., Natsukawa, T., Shinobu, N., Imaizumi, T., Miyagishi, M., Taira, K., Akira, S., Fujita, T., 2004. The RNA helicase RIG-I has an essential function in double-stranded RNA-induced innate antiviral responses. Nat. Immunol. 5, 730–737.

Yoneyama, M., Kikuchi, M., Matsumoto, K., Imaizumi, T., Miyagishi, M., Taira, K., Foy, E., Loo, Y.M., Gale Jr., M., Akira, S., Yonehara, S., Kato, A., Fujita, T., 2005. Shared and unique functions of the DExD/H-box helicases RIG-I, MDA5, and LGP2 in antiviral innate immunity. J. Immunol. 175, 2851–2858.

Yoshida, H., Okabe, Y., Kawane, K., Fukuyama, H., Nagata, S., 2005. Lethal anemia caused by interferon-beta produced in mouse embryos carrying undigested DNA. Nat. Immunol. 6, 49–56.

Zaki, M.H., Boyd, K.L., Vogel, P., Kastan, M.B., Lamkanfi, M., Kanneganti, T.D., 2010. The NLRP3 inflammasome protects against loss of epithelial integrity and mortality during experimental colitis. Immunity 32, 379–391.

Zaki, M.H., Vogel, P., Malireddi, R.K., Body-Malapel, M., Anand, P.K., Bertin, J., Green, D.R., Lamkanfi, M., Kanneganti, T.D., 2011. The NOD-like receptor NLRP12 attenuates colon inflammation and tumorigenesis. Cancer Cell 20, 649–660.

Zevini, A., Olagnier, D., Hiscott, J., 2017. Crosstalk between cytoplasmic RIG-I and STING sensing pathways. Trends Immunol. 38, 194–205.

Zhang, W., Xu, W., Xiong, S., 2010. Blockade of Notch1 signaling alleviates murine lupus via blunting macrophage activation and M2b polarization. J. Immunol. 184, 6465–6478.

Zhang, Z., Yuan, B., Bao, M., Lu, N., Kim, T., Liu, Y.-J., 2011. The helicase DDX41 senses intracellular DNA mediated by the adaptor STING in dendritic cells. Nat. Immunol. 12, 959–965.

Zhang, W., Zhou, Q., Xu, W., Cai, Y., Yin, Z., Gao, X., Xiong, S., 2013a. DNA-dependent activator of interferon-regulatory factors (DAI) promotes lupus nephritis by activating the calcium pathway. J. Biol. Chem. 288, 13534–13550.

Zhang, X., Shi, H., Wu, J., Zhang, X., Sun, L., Chen, C., Chen, Z.J., 2013b. Cyclic GMP-AMP containing mixed phosphodiester linkages is an endogenous high-affinity ligand for STING. Mol. Cell 51, 226–235.

Zhang, X., Wu, J., Du, F., Xu, H., Sun, L., Chen, Z., Brautigam, C.A., Zhang, X., Chen, Z.J., 2014. The cytosolic DNA sensor cGAS forms an oligomeric complex with DNA and undergoes switch-like conformational changes in the activation loop. Cell Rep. 6, 421–430.

Zhang, S.H., Zhao, Y., Xie, Q.B., Jiang, Y., Wu, Y.K., Yan, B., 2018. Aberrant activation of type I interferon system may contribute to the pathogenesis of anti-MDA5 dermatomyositis. Br. J. Dermatol. https://doi.org/10.1111/bjd.16917

Zhong, B., Yang, Y., Li, S., Wang, Y.Y., Li, Y., Diao, F., Lei, C., He, X., Zhang, L., Tien, P., Shu, H.B., 2008. The adaptor protein MITA links virus-sensing receptors to IRF3 transcription factor activation. Immunity 29, 538–550.

Zhou, S., Cerny, A.M., Zacharia, A., Fitzgerald, K.A., Kurt-Jones, E.A., Finberg, R.W., 2010. Induction and inhibition of type I interferon responses by distinct components of lymphocytic choriomeningitis virus. J. Virol. 84, 9452–9462.

Zhu, Q., Man, S.M., Gurung, P., Liu, Z., Vogel, P., Lamkanfi, M., Kanneganti, T.D., 2014. Cutting edge: STING mediates protection against colorectal tumorigenesis by governing the magnitude of intestinal inflammation. J. Immunol. 193, 4779–4782.

Zouk, H., Marchand, L., Li, Q., Polychronakos, C., 2014. Functional characterization of the Thr946Ala SNP at the type 1 diabetes IFIH1 locus. Autoimmunity 47, 40–45.

FSC
www.fsc.org
MIX
Papier | Fördert
gute Waldnutzung
FSC® C083411

Kolibri 360 GmbH
Ferdinand-Jühlke-Straße 7
99095 Erfurt, Deutschland
produktsicherheit@kolibri360.de